D1207584

# INTRODUCTION TO MECHATRONICS AND MEASUREMENT SYSTEMS

Michael B. Histand
and
David G. Alciatore

*Department of Mechanical Engineering*
*Colorado State University*

Boston Burr Ridge, IL Dubuque, IA Madison, WI
New York San Francisco St. Louis
Bangkok Bogotá Caracas Lisbon London Madrid Mexico City
Milan New Delhi Seoul Singapore Sydney Taipei Toronto

*WCB/McGraw-Hill*

*A Division of The* **McGraw-Hill** *Companies*

INTRODUCTION TO MECHATRONICS AND MEASURMENT SYSTEMS

Copyright © 1999 by the McGraw-Hill Companies, Inc. All rights reserved. Printed in the United States of America. Except as permitted under the United States Copyright Act of 1976, no part of this publication may be reproduced or distributed in any form or by an means, or stored in a data base or retrieval system, without the prior written permission of the publisher.

This book is printed on acid-free paper.

2 3 4 5 6 7 8 9 0 DOC / DOC 9 3 2 1 0 9

ISBN 0-07-029089-X

Vice president and editorial director: *Kevin T. Kane*
Publisher: *Tom Casson*
Senior sponsoring editor: *Debra Riegert*
Marketing manager: *John T. Wannemacher*
Project manager: *Jim Labeots*
Production supervisor: *Scott M. Hamilton*
Freelance design coordinator: *Laurie J. Entringer*
Cover design: *Z-Graphics*
Photo research coordinator: *Sharon Miller*
Supplement coordinator: *Linda Huenecke*
Compositor: *Lachina Publishing Services*
Typeface: *10/12 Times Roman*
Printer: *R.R. Donnelley & Sons Company*

**Library of Congress Cataloging-in-publication Data**

Histand, Michael B.
Introduction to mechatronics and measurement systems / Michael B. Histand, David G. Alciatore.
p. cm.
Includes index.
ISBN 0-07-029089-X
1. Mechatronics. 2. Measuring instruments. I. Alciatore, David G. II. Title.
TJ163.12.H57 1998
621--dc21 98-6438

http://www.mhhe.com

## McGraw-Hill Series in Mechanical Engineering

CONSULTING EDITORS
***Jack P. Holman, Southern Methodist University***
***John R. Lloyd, Michigan State University***

Anderson: *Computational Fluid Dynamics: The Basics with Applications*
Anderson: *Modern Compressible Flow: With Historical Perspective*
Arora: *Introduction to Optimum Design*
Borman and Ragland: *Combustion Engineering*
Burton: *Introduction to Dynamic Systems Analysis*
Culp: *Principles of Energy Conversion*
Dieter: *Engineering Design: A Materials & Processing Approach*
Doebelin: *Engineering Experimentation: Planning, Execution, Reporting*
Driels: *Linear Control Systems Engineering*
Edwards and McKee: *Fundamentals of Mechanical Component Design*
Gebhart: *Heat Conduction and Mass Diffusion*
Gibson: *Principles of Composite Material Mechanics*
Hamrock: *Fundamentals of Fluid Film Lubrication*
Heywood: *Internal Combustion Engine Fundamentals*
Hinze: *Turbulence*
Histand and Alciatore: *Introduction to Mechatronics and Measurement Systems*
Holman: *Experimental Methods for Engineers*
Howell and Buckius: *Fundamentals of Engineering Thermodynamics*
Jaluria: *Design and Optimization of Thermal Systems*
Juvinall: *Engineering Considerations of Stress, Strain, and Strength*
Kays and Crawford: *Convective Heat and Mass Transfer*
Kelly: *Fundamentals of Mechanical Vibrations*
Kimbrell: *Kinematics Analysis and Synthesis*
Kreider and Rabl: *Heating and Cooling of Buildings*
Martin: *Kinematics and Dynamics of Machines*
Mattingly: *Elements of Gas Turbine Propulsion*
Modest: *Radiative Heat Transfer*
Norton: *Design of Machinery*
Oosthuizen and Carscallen: *Compressible Fluid Flow*
Oosthuizen and Naylor: *Introduction to Convective Heat Transfer Analysis*
Phelan: *Fundamentals of Mechanical Design*
Reddy: *An Introduction to Finite Element Method*
Rosenberg and Karnopp: *Introduction to Physical Systems Dynamics*
Schlichting: *Boundary-Layer Theory*
Shames: *Mechanics of Fluids*

Shigley: *Kinematic Analysis of Mechanisms*
Shigley and Mischke: *Mechanical Engineering Design*
Shigley and Uicker: *Theory of Machines and Mechanisms*
Stiffler: *Design with Microprocessors for Mechanical Engineers*
Stoecker and Jones: *Refrigeration and Air Conditioning*
Turns: *An Introduction to Combustion: Concepts and Applications*
Ullman: *The Mechanical Design Process*
Wark: *Advanced Thermodynamics for Engineers*
White: *Viscous Fluid Flow*
Zeid: *CAD/CAM Theory and Practice*

# PREFACE

## Approach

The formal boundaries of traditional engineering disciplines have become fuzzy following the advent of integrated circuits and computers. Nowhere is this more evident than in mechanical and electrical engineering, where products today include an assembly of interdependent electrical and mechanical components. The field of mechatronics has broadened the scope of the traditional field of electromechanics. Mechatronics is defined as the field of study involving the analysis, design, synthesis, and selection of systems which combine electronic and mechanical components with modern controls and microprocessors.

This book is designed to serve as a text for (1) a modern instrumentation and measurements course, (2) a hybrid electrical and mechanical engineering course replacing traditional circuits and instrumentation courses, (3) a mechatronics course, or (4) the first course in a mechatronics sequence. The second option, the hybrid course, provides an opportunity to reduce the number of credit hours in a typical mechanical engineering curriculum. Options three and four could involve the development of new interdisciplinary courses.

Currently, most curricula do not include a mechatronics course but include some of the elements in other, more traditional courses. The purpose of a course in mechatronics is to provide a focused interdisciplinary experience for undergraduates that encompasses important elements from traditional courses as well as contemporary developments in electronics and computer control. These elements include measurement theory, electronic circuits, computer interfacing, sensors, actuators, and the design, analysis, and synthesis of mechatronic systems. This approach is valuable to students since virtually every newly designed engineering product is a mechatronic system.

## Content

Chapter 1 introduces mechatronics and measurement systems. Chapter 2 provides a review of basic electrical relations, circuit elements, and circuit analysis. Chapter 3 deals with semiconductor electronics. Chapter 4 presents approaches to analyzing and characterizing the response of mechatronic and measurement systems. Chapter 5 covers the basics of analog signal processing and the design and analysis of operational amplifier circuits. Chapter 6 presents the basics of digital logic and the use of integrated circuits. Chapter 6 also provides an introduction to the microprocessor and microcontroller. Chapter 7 deals with issues involved with coupling microprocessors and computers to measurement systems in order to automate data acquisition and analysis. Chapter 8 provides an overview of the many sensors common in mechatronic systems. Chapter 9 introduces a number of devices used for actuating mechatronic systems. Finally, Chapter 10 provides an overview of mechatronic system control architectures and presents some case studies. The Appendices review the fundamentals of unit systems, statistics, error analysis, and mechanics of materials to support and supplement measurement systems topics in the book. Two mechatronics topics this book does not cover directly are control theory and microprocessor programming. These topics are important and should be included in a curriculum that emphasizes mechatronics, but we decided it would be impractical to adequately cover these topics in this book.

## Learning Tools

Class discussion items are included throughout the book to serve as thought-provoking exercises for the students as instructor-led cooperative learning activities in the classroom. They can also be used as out-of-class homework assignments to supplement the questions at the end of each chapter. Analysis and design examples are also provided throughout the book to improve a student's ability to apply the material. To enhance student learning, carefully designed laboratory exercises coordinated with the lectures should accompany a course using this text. The combination of class discussion items, design examples, and laboratory exercises exposes a student to a real-world practical approach and provides a useful framework for future design work.

## Supplements

More information, including a recommended course outline, a typical laboratory syllabus, MathCAD files for examples from the book, and other supplemental material, is available on the Internet at:

http://www.engr.colostate.edu/~dga/mechatronics.html

An instructor guide is also available from the publisher. It includes complete solutions for all end-of-chapter problems.

## ACKNOWLEDGMENTS

To ensure accuracy of this text, it has been class tested at Colorado State University for four years and at the University of Wyoming for two years. We'd like to thank all of the students at both institutions and Dave Walrath at the University of Wyoming, who have given us valuable feedback throughout this process. In addition, we'd like to thank our many manuscript reviewers for their valuable input. The reviewers included the following people in the field of mechatronics:

| | |
|---|---|
| Ramendra P. Roy | *Arizona State University* |
| Charles Ume | *Georgia Institute of Technology* |
| Jawarharlal Mariappan | *GMI Engineering and Management Institute* |
| Melvin R. Corley | *Louisiana Tech University* |
| Donald G. Morin | *Rose-Hulman Institute of Technology* |
| J. Edward Carryer | *Stanford University* |
| Ahmad Smaili | *Tennessee Technological University* |
| Louis Everett | *Texas A&M University* |
| Gregory P. Starr | *University of New Mexico* |
| David E. Walrath | *University of Wyoming* |

# ACKNOWLEDGMENTS

[illegible]

# CONTENTS

# LIST OF FIGURES

# LIST OF TABLES

# LIST OF CLASS DISCUSSION ITEMS

# LIST OF EXAMPLES

# LIST OF DESIGN EXAMPLES

# 1

# INTRODUCTION TO MECHATRONICS AND MEASUREMENT SYSTEMS

**OBJECTIVES.** After you read, discuss, study, and apply ideas in this chapter, you will be able to:

- Define mechatronics and appreciate its relevance to contemporary engineering design
- Identify a mechatronic system and its primary elements
- Define the elements of a general measurement system

## 1.1 INTRODUCTION TO MECHATRONICS

Mechanical engineering, as a widespread professional practice, experienced a surge of growth during the early 19th century because it provided a necessary foundation for the rapid and successful development of the industrial revolution. At that time, mines needed large pumps never before seen to keep their shafts dry; iron and steel mills required pressures and temperatures beyond levels used commercially until then; transportation systems needed more than real horse power to move goods; structures began to stretch across ever wider abysses and to climb to dizzying heights; manufacturing moved from the shop bench to large factories; and to support these technical feats, people began to specialize and build bodies of knowledge that formed the beginnings of the engineering disciplines.

Now, more than a century later, we are witnessing a new scientific and social revolution known as the information revolution, where engineering specialization ironically seems to be simultaneously focusing and diversifying. This contemporary revolution was spawned by the engineering development of miniature semiconductor electronics, which have driven an information and communications explosion that has transformed human life. To practice engineering today, we must understand new ways to process information and be able to utilize semiconductor electronics within our products, no matter what label we put on ourselves as practitioners. The primary engineering disciplines of the 20th century–mechanical, electrical, civil, and chemical–retained their individual bodies of knowledge, textbooks, and professional journals because the disciplines were viewed as having mutually exclusive intellectual and professional territory. Entering students could assess their individual intellectual talents and choose one of the fields as a profession. That is all changing now as the impact of solid state electronics permeates the traditional fields of engineering both in the way engineers communicate and in what they design. Mechatronics is one of the new and exciting fields on the engineering landscape, subsuming parts of traditional engineering fields and requiring a broader approach to the design of systems that we can formally call mechatronic systems.

Then what precisely is mechatronics? The term ***mechatronics*** is used to denote a rapidly developing, interdisciplinary field of engineering that deals with the design of products whose function relies on the synergistic integration of mechanical, electrical, and electronic components connected by a control architecture. The word *mechatronics* was coined in Japan in the late sixties, spread through Europe, and is growing in use in the United States. The primary disciplines involved in the design of mechatronic systems include mechanics, electronics, controls, and computer science. A mechatronic system designer must assemble analog and digital circuits, microprocessors and computers, mechanical devices, sensors and actuators, and controls so that the final design achieves a desired goal. The subsequent chapters provide an introduction to these components and subsystems and describe aspects of their analysis and design.

More commonly you hear mechatronic systems referred to as *smart devices*. While the term *smart* is elusive in precise definition, in the engineering sense we mean the inclusion of elements such as logic, feedback, and computation which in a complex design may appear to simulate human thinking processes. It is not easy to compartmentalize mechatronic system design within a traditional field of engineering because such design draws from knowledge across many fields. The mechatronic system designer must be a generalist, willing to seek and apply knowledge from a broad range of sources. This may intimidate the student at first, but it offers great benefits for individuality and continued learning during one's career.

Today, there are very few mechanical devices that do not include electrical components and some type of computer monitoring or control. Therefore, the term *mechatronic system* encompasses a myriad of devices and systems. Increasingly, digital circuits and microprocessors are embedded in electromechanical devices, creating much more flexibility and control possibilities in system design. Examples of mechatronic systems include: an aircraft flight control and navigation system, automobile electronic fuel injection and antilock brake systems, automated manufacturing equipment such as robots and numerically controlled (NC) machine tools, smart kitchen and home appliances such as bread machines and clothes washing machines, and even toys.

▼ ***EXAMPLE 1.1. Mechatronic System—Copy Machine.*** An office copy machine is a good example of a contemporary mechatronic system. It includes analog and digital circuits, sensors, actuators, and micro-processors. The copying process works as follows: The user places an original in a loading bin and pushes a button to start the process; the original is transported to the platen glass; a high intensity light source scans the original and transfers the corresponding image as a charge distribution to a metal drum. Next, a blank piece of paper is retrieved from a loading cartridge, and the image is transferred onto the paper with an electrostatic deposition of ink toner powder that is heated to bond to the paper. A sorting mechanism then optionally delivers the copy to an appropriate bin.

Analog circuits control the lamp, heater, and other power circuits in the machine. Digital circuits control the digital displays, indicator lights, buttons, and switches comprising the user interface. Other digital circuits include logic circuits and microprocessors that coordinate all of the functions in the machine. Optical sensors and microswitches detect the presence or absence of paper, its proper positioning, and whether or not doors and latches are in their correct positions. Other sensors include encoders used to track motor rotation. Actuators include servo and stepper motors that load and transport the paper, turn the drum, and index the sorter.

▼▼▼ ***CLASS DISCUSSION ITEM 1.1. Household Mechatronic Systems.*** What typical household items can be characterized as mechatronic systems? What components do they contain that help you identify them as mechatronic systems? If an item contains a microprocessor, describe the functions that are performed by the microprocessor.

To understand how to design and analyze mechatronic systems, a firm grasp of the fundamentals of electrical circuits and measurement systems is a necessity. In the physical world, nearly all devices designed to measure events and control performance are now based on electrical principles. Moreover, with the increasing availability and decreasing cost of integrated circuits (ICs) such as operational amplifiers and digital devices fabricated by large scale integration (LSI) techniques, we are seeing exponential growth in the use of electronic measurement and control systems in all fields of engineering. Although ICs have very complex internal architectures, a non-electrical engineer can nevertheless learn to use them easily to design and construct sophisticated mechatronic systems.

## 1.2 INTRODUCTION TO MEASUREMENT SYSTEMS

A fundamental part of many mechatronic systems is a ***measurement system*** that is composed of the three basic parts illustrated in Figure 1.1. The ***transducer*** is a

**FIGURE 1.1**
Elements of a measurement system.

sensing device that converts a physical input into an output, usually a voltage. The ***signal processor*** performs filtering, amplification, or other signal conditioning on the transducer output. The term ***sensor*** is often used to refer to the transducer or to the combination of transducer and signal processor. Finally, the ***recorder*** is an instrument, a computer, a hardcopy device, or simply a display that maintains the sensor data for online monitoring or subsequent processing.

These three building blocks of measurement systems come in many types with wide variations in cost and performance. It is important for designers and users of measurement systems to develop confidence in their use, to know their important characteristics and limitations, and to be able to select the best elements for the measurement task at hand. In addition to being an integral part of most mechatronic systems, a measurement system is often used as a stand-alone device to acquire data in a laboratory or field environment.

---

▼ ***EXAMPLE 1.2. Measurement System—Digital Thermometer.*** Shown below is an example of a measurement system. The thermocouple is a transducer that converts temperature to a small voltage; the amplifier increases the magnitude of the voltage; the A/D (analog to digital) converter is a device that converts the analog voltage to a digital signal; and the LEDs (light emitting diodes) display the value of the temperature.

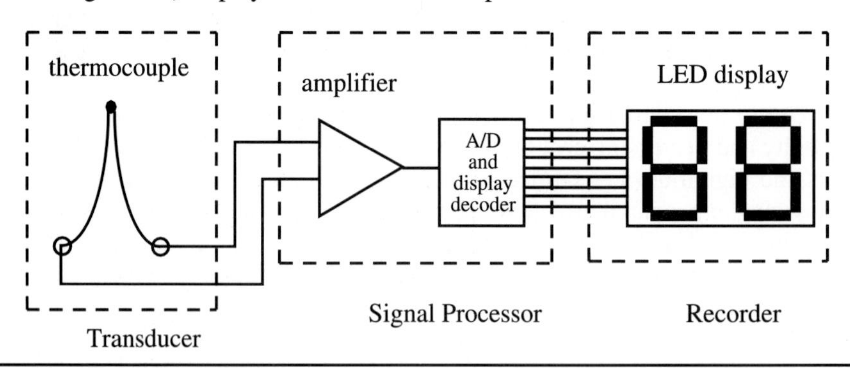


---

Supplemental information important to measurement systems and analysis is provided in Appendix 1. Included are sections on systems of units, numerical precision, and statistics. You should review this material on an as-needed basis.

## BIBLIOGRAPHY


Alciatore, D. and Histand, M., "Mechatronics at Colorado State University," *Journal of Mechatronics*, "Mechatronics Education in the United States" issue, Pergamon Press, May, 1995.

Alciatore, D. and Histand, M., "Mechatronics and Measurement Systems Course at Colorado State University," Proceedings of the Workshop on Mechatronics Education, pp. 7–11, Stanford, CA, July, 1994.

Ashley, S., "Getting a Hold on Mechatronics," *Mechanical Engineering*, pp. 60–63, ASME, New York, May, 1997.

Beckwith, T., Marangoni, R., and Lienhard, J., *Mechanical Measurements,* Addison-Wesley, Reading, MA, 1993.

Craig, K., "Mechatronics System Design at Rensselaer," Proceedings of the Workshop on Mechatronics Education, pp. 24–27, Stanford, CA, July, 1994.

Doeblin, E., *Measurement Systems Applications and Design,* 4th edition, McGraw-Hill, New York, 1990.

Morley, D., "Mechatronics Explained," *Manufacturing Systems*, p. 104, November, 1996.

Shoureshi, R. and Meckl, P., "Teaching MEs to Use Microprocessors," *Mechanical Engineering*, v. 166, n. 4, pp. 71–74, April, 1994.

# 2

# ELECTRIC CIRCUITS AND COMPONENTS

**OBJECTIVES:** After you read, discuss, study, and apply ideas in this chapter, you will be able to:

- Define resistance, capacitance, and inductance
- Define Kirchoff's Voltage and Current Laws and apply them to passive circuits that include resistors, capacitors, inductors, voltage sources, and current sources
- Visualize models for ideal voltage and current sources
- Predict the steady state behavior of circuits with sinusoidal inputs
- Characterize the power dissipated or generated by a circuit
- Predict the effects of mismatched impedances
- Understand how to reduce noise and interference in electrical circuits
- Appreciate the need to pay attention to electrical safety and to properly ground components

## 2.1 INTRODUCTION

Practically all mechatronic and measurement systems contain electrical circuits, and as engineers we need to know how to design and build systems that utilize them. This chapter presents a thorough review of basic electricity and circuit analysis techniques–topics that are fundamental to understanding everything else that follows in this book.

When electrons move they produce an electrical current, and we can do useful things with the energized electrons. The reason they move is that we impose an electrical field that imparts energy by doing work on electrons, which in turn can be delivered by the electrons somewhere else. A measure of the electric field's potential is called ***voltage***; it's like the potential energy in a gravitational field. We can think of voltage as an "across variable," between two points in the field, and the resulting movement of electrons (the flow) is the current, a "through variable," which moves through the field. When we measure current through a circuit we place a meter in the circuit and let the current flow through it. When we measure a voltage we place two conducting probes on the points across which we want to measure the voltage. Voltage is sometimes referred to as ***electromotive force*** or ***emf.***

***Current*** ($I$ or $i$) is defined as the time rate of flow of charge ($q$):

$$I(t) = \frac{dq}{dt} \tag{2.1}$$

The charge is provided by the negatively charged electrons. The SI unit for current is the ***ampere*** (A), and charge is measured in ***coulombs*** (C = A · s). When voltage and current in a circuit are constant (independent of time), their values and the circuit are referred to as ***direct current*** or ***dc***. When the voltage and current vary with time, usually sinusoidally, we refer to their values and the circuit as ***alternating current*** or ***ac***.

An electrical circuit consists of a closed loop, usually made of a conductor interrupted here and there with other components. The conductor itself may be broken by components called switches. Some simple examples of valid circuits are shown in Figure 2.1.

Many of the mechatronic devices we analyze or design will contain electric circuits. The fundamental concepts involved when considering an electrical circuit are illustrated in Figure 2.2a. The voltage source, which provides energy to the circuit, can be a power supply, battery, or generator. The positive side of the source,

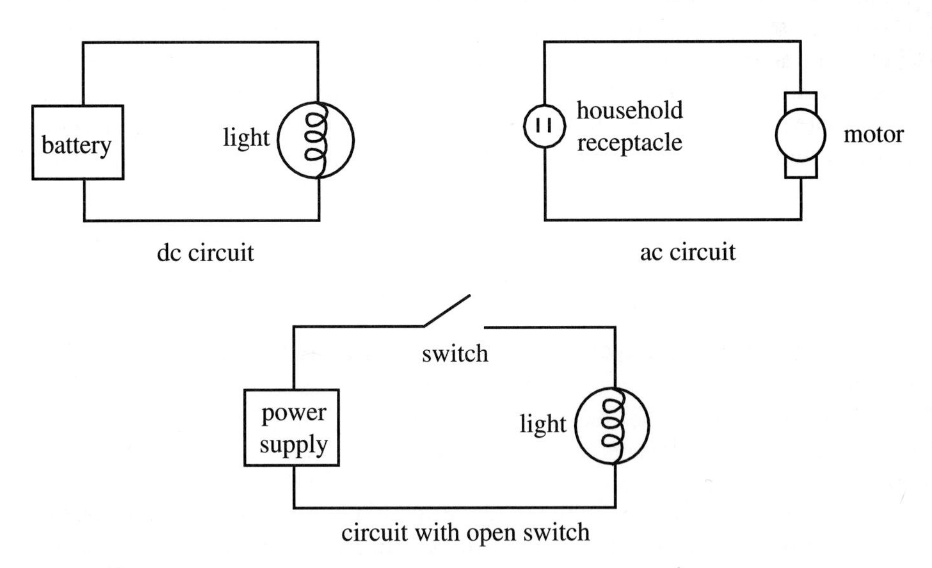


**FIGURE 2.1**
Electrical circuits.

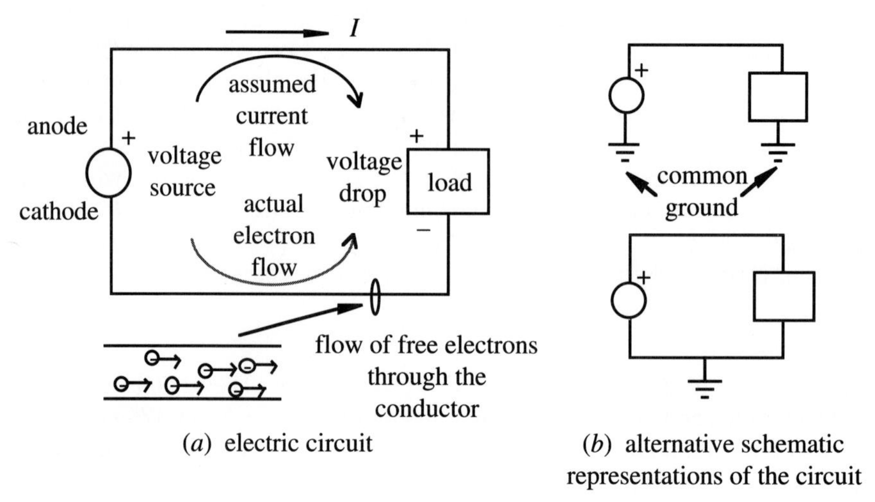


**FIGURE 2.2**
Electric circuit terminology.

where electrons are attracted, is called the ***anode,*** and the negative side, where electrons are released, is called the ***cathode.*** Electrons flow from the cathode to the anode through the circuit, but standard convention assumes that current is the flow of positive charge carriers in the opposite direction. We owe this convention to Benjamin Franklin, who thought current was the result of the motion of positively charged particles. The ***load*** consists of a network of circuit elements that may dissipate or store electrical energy. Figure 2.2b shows two alternative ways to draw a circuit schematic. The ***ground*** indicates a reference point in the circuit where the voltage is assumed to be zero. Even though we don't show a connection between the ground symbols in the top circuit, it is implied that both ground symbols represent a single reference point. This technique can be applied when drawing complicated circuits to reduce the number of lines. The bottom circuit is an equivalent representation indicating the location of the ground reference.

## 2.2 THE BASIC ELECTRICAL ELEMENTS

There are three basic passive electrical elements: the resistor ($R$), capacitor ($C$), and inductor ($L$). Passive elements require no additional power supply, as active devices such as integrated circuits do. All elements are defined by their voltage-current characteristics. There are two types of ideal energy sources: a ***voltage source*** ($V$) and a ***current source*** ($I$). These ideal sources contain no internal resistance, inductance, or capacitance. Figure 2.3 illustrates the schematic symbols for all of these components.

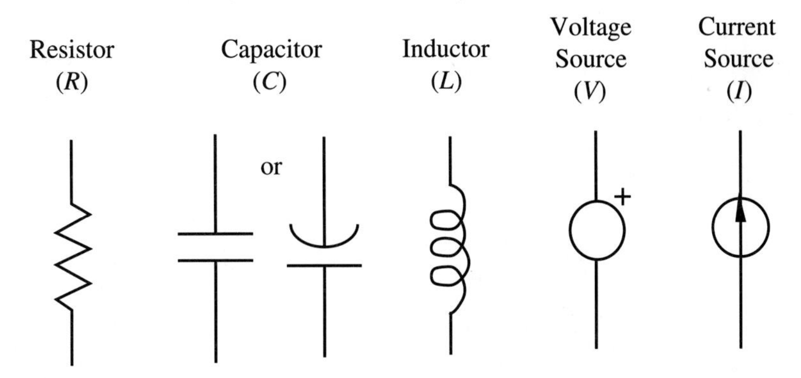


**FIGURE 2.3**
Basic electrical elements.

### 2.2.1 The Resistor

A ***resistor*** is a dissipative element that converts electrical energy into heat. ***Ohm's Law*** defines the voltage-current characteristic of an ideal resistor:

$$V = IR \tag{2.2}$$

The unit of resistance is the ***ohm*** ($\Omega$). Resistance is a material property whose value is the slope of the voltage-current curve (see Figure 2.4). For an ideal resistor, the voltage-current relationship is linear and the resistance is constant. However, real resistors are typically nonlinear at higher currents due to temperature effects. Resistance usually increases with increasing temperatures. Also, a real resistor has a limited power dissipation capability designated in watts, and it may "burn up" after this limit is reached.

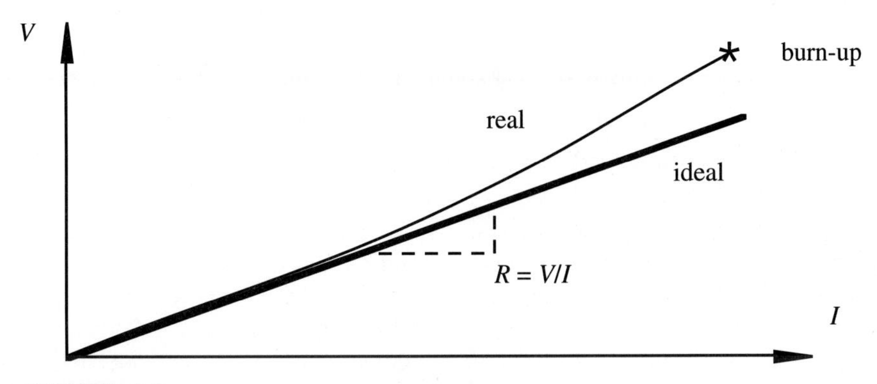


**FIGURE 2.4**
Voltage-current relation for an ideal resistor.

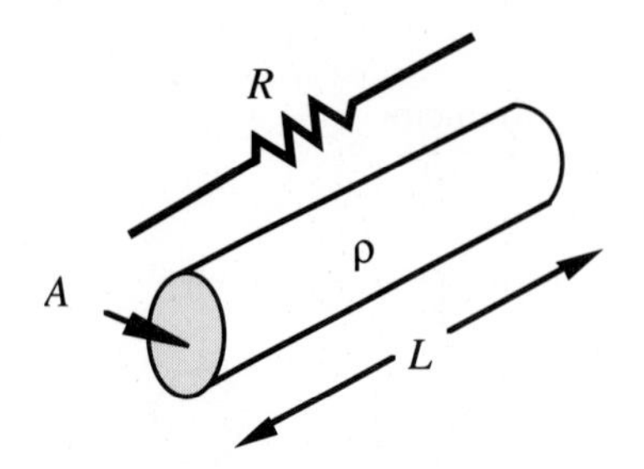


**FIGURE 2.5**
Wire resistance.

If a resistor's material is homogeneous and has constant cross-sectional area, such as a cylindrical wire illustrated in Figure 2.5, then the resistance is given by

$$R = \frac{\rho L}{A} \tag{2.3}$$

where ρ is the ***resistivity***, or specific resistance of the material, $L$ is the wire length, and $A$ is the cross-sectional area. Resistivities for common conductors are given in Table 2.1.

---

▼ ***EXAMPLE 2.1. Resistance of a Wire.*** As an example of the use of Equation 2.3, consider a copper wire 1.0 mm in diameter and 10 m long.

From Table 2.1, the resistivity of copper is

$$\rho = 1.7 \text{ x } 10^{-8}\ \Omega\text{m}$$

Since the wire diameter, area, and length are

$$D = 0.0010 \text{ m}$$
$$A = \pi D^2/4 = 7.8 \text{ x } 10^{-7} \text{ m}^2$$
$$L = 10 \text{ m}$$

the total wire resistance is

$$R = \rho L / A = 0.22\ \Omega$$

---

**TABLE 2.1**
**Resistivities of common conductors**

| Material | Resistivity ($10^{-8}$ Ωm) |
|---|---|
| Aluminum | 2.8 |
| Carbon | 4000 |
| Constantan | 44 |
| Copper | 1.7 |
| Gold | 2.4 |
| Iron | 10 |
| Silver | 1.6 |
| Tungsten | 5.5 |

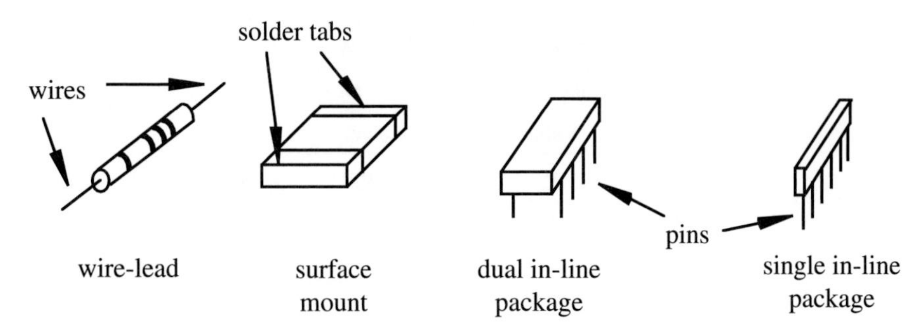


**FIGURE 2.6**
Resistor packaging.

Actual resistors used in building circuits are packaged in various forms including wire-lead components, surface mount components, and the ***dual in-line package (DIP)*** and the ***single in-line package (SIP),*** which contain multiple resistors in a package that conveniently fits into printed circuit boards. These four types are illustrated in Figure 2.6.

The most common wire-lead resistors you will use in ordinary electronic circuity are 1/4 watt 5% tolerance carbon or metal-film resistors. Resistors with higher power ratings are also available.

A wire-lead resistor's value and precision are usually coded with four colored bands (*a, b, c, tol*) as illustrated in Figure 2.7. The colors used for bands are listed with their respective values in Table 2.2. A resistor's value and precision are expressed as

$$R = ab \times 10^{c} \pm \text{tolerance}\,(\%) \tag{2.4}$$

where the *a* band represents the tens digit, the *b* band represents the ones digit, the *c* band represents the power of 10, and the *tol* band represents the tolerance or uncertainty as a percentage of the coded resistance value. Resistor values of this type range in value between 1 Ω and 24 MΩ. The set of ***standard values*** for the first two digits are: 10, 11, 12, 13, 14, 15, 16, 18, 20, 22, 24, 27, 30, 33, 36, 39, 43, 47, 51, 56, 62, 68, 75, 82, and 91.

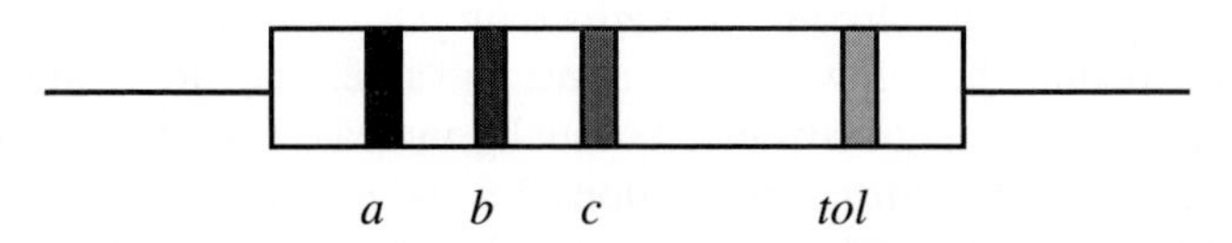


**FIGURE 2.7**
Wire-lead resistor color bands.

**TABLE 2.2**
**Resistor color band codes**

| *a, b,* and *c* Bands | | *tol* Band | |
|---|---|---|---|
| **Color** | **Value** | **Color** | **Value** |
| Black | 0 | Gold | ±5% |
| Brown | 1 | Silver | ±10% |
| Red | 2 | Nothing | ±20% |
| Orange | 3 | | |
| Yellow | 4 | | |
| Green | 5 | | |
| Blue | 6 | | |
| Violet | 7 | | |
| Gray | 8 | | |
| White | 9 | | |

Precision metal-film resistors have 1% or smaller uncertainties and are available in a wider range of values than the lower tolerance resistors. They usually have a numerical four-digit code printed directly on the body of the resistor. The first three digits denote the value of the resistor, and the last digit indicates the power of 10 by which to multiply.

---

▼ ***EXAMPLE 2.2. Resistance Color Codes.*** A wire-lead resistor has the following color bands:

$$a = \text{green},\ b = \text{brown},\ c = \text{red, and } tol = \text{gold}$$

From Equation 2.4 and Table 2.2, the range of possible resistance values is

$$R = 51 \times 10^2\ \Omega \pm 5\% = 5100 \pm (0.05 \times 5100)\ \Omega$$

or

$$4800\ \Omega < R < 5300\ \Omega$$

---

Variable resistors are available that provide a range of resistance values controlled by a mechanical screw, knob, or linear slide. The most common type is called a ***potentiometer*** or *pot*. The various schematic symbols for a potentiometer are shown in Figure 2.8. A potentiometer that is included in a circuit to adjust (trim) the resistance in the circuit is called a ***trim pot***. A trim pot is shown with a little symbol to denote the screw used to adjust its value. The direction to rotate the potentiometer for increasing resistance is usually indicated on the component. Potentiometers are discussed further in Sections 4.8 and 8.2.2.

***Conductance*** ($G$), defined as the reciprocal of resistance ($1/R$), is sometimes used as an alternative to resistance to characterize a dissipative circuit element. It is

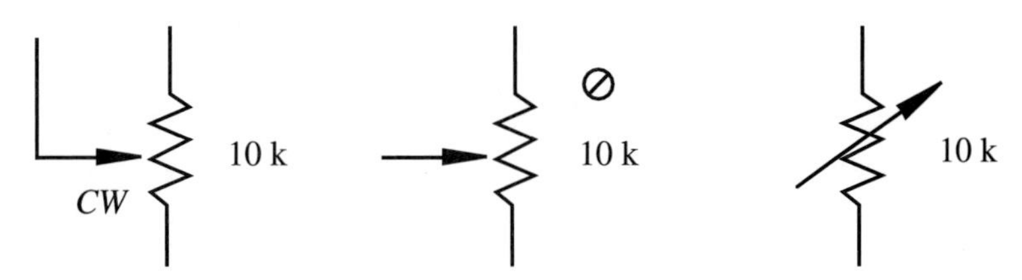


**FIGURE 2.8**
Potentiometer schematic symbols.

a measure of how easily an element conducts current as opposed to how much it resists it. The unit of conductance is the ***siemen*** ($S = 1/\Omega = mho$).

### 2.2.2 The Capacitor

A ***capacitor*** is a passive element that stores energy in the form of an electric field. This field is the result of a separation of electric charge. The simplest capacitor consists of a pair of parallel conducting plates separated by a dielectric material as illustrated in Figure 2.9. The ***dielectric material*** is an insulator that increases the capacitance as a result of permanent or induced electric dipoles in the material. Strictly, dc current does not flow through a capacitor; rather, charges are displaced from one side of the capacitor through the conducting circuit to the other side, establishing the electric field. The displacement of charge is called a ***displacement current*** since current appears to flow momentarily through the device. The capacitor's voltage-current relationship is defined as

$$V(t) = \frac{1}{C}\int_0^t I(\tau)d\tau = \frac{Q(t)}{C} \tag{2.5}$$

where $Q(t)$ is the amount of accumulated charge measured in coulombs and $C$ is the ***capacitance*** measured in ***farads*** ($F$ = coulombs/volts). By differentiating this equation, we can relate the displacement current to the rate of change of voltage:

$$I(t) = C\frac{dV}{dt} \tag{2.6}$$

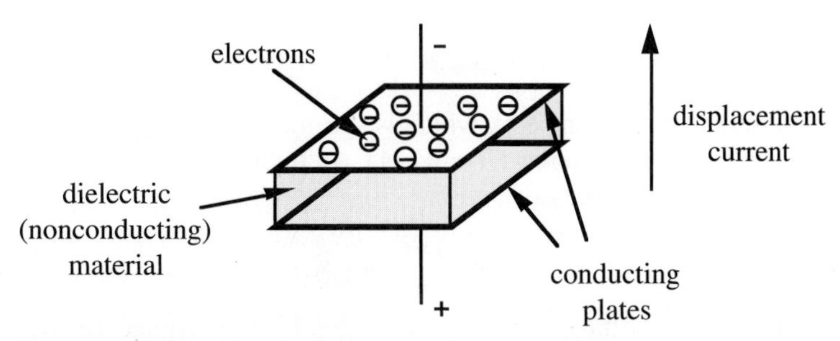


**FIGURE 2.9**
Parallel plate capacitor.

Two capacitors in parallel add (see Question 2.4):

$$C_{eq} = C_1 + C_2 \tag{2.7}$$

and two capacitors in series combine as

$$C_{eq} = \frac{C_1 C_2}{C_1 + C_2} \tag{2.8}$$

Capacitance is a property of the dielectric material and the plate geometry and separation. Values for typical capacitors range from 1 pF to 1000 μF. Since the voltage across a capacitor is the integral of the displacement current (see Equation 2.5), the voltage cannot change instantaneously. This characteristic can be controlled for timing purposes in electrical circuits such as a simple RC circuit.

The primary types of commercial capacitors are electrolytic capacitors, tantalum capacitors, ceramic disk capacitors, and mylar capacitors. A polarized capacitor has a positive and negative end; electrolytic capacitors are polarized. Capacitors come in many sizes and shapes. Often the capacitance is printed directly on the component, typically in μF or pF, but sometimes a three-digit code is used. The first two digits are the value and the third the power of 10 multiplied times picofarads.

### 2.2.3 The Inductor

An ***inductor*** is a passive energy storage element that stores energy in the form of a magnetic field. The simplest form of an inductor is a wire coil, which has a tendency to maintain a magnetic field once established. The inductor's characteristics are a direct result of Faraday's Law of Induction, which states

$$V(t) = \frac{d\lambda}{dt} \tag{2.9}$$

where $\lambda$ is the total ***magnetic flux*** through the coil windings due to the current. Magnetic flux is measured in webers (Wb). The magnetic field lines surrounding an inductor are illustrated in Figure 2.10. The north-to-south direction of the magnetic field lines, shown with arrowheads in the figure, is found using the ***right-hand-rule*** for a coil. The rule states that if you curl the fingers of your right hand in the direction of current flow through the coil, your thumb will point in the direction of the magnetic field from north to south. For an ideal coil, the flux is proportional to the current:

$$\lambda = LI \tag{2.10}$$

where $L$ is the inductance of the coil, which is assumed to be constant. The unit of measure of inductance is the ***henry*** ($H = Wb/A$). An inductor's voltage-current relationship can therefore be expressed as

$$V(t) = L\frac{dI}{dt} \tag{2.11}$$

The magnitude of the voltage across an inductor is proportional to the rate of change of the current through the inductor. If the current through the inductor is in-

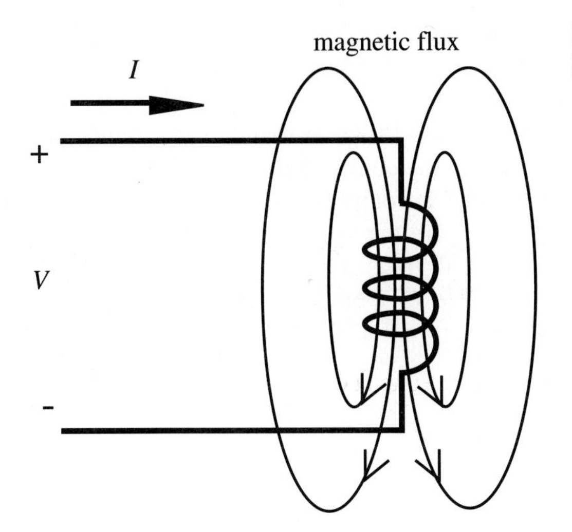


**FIGURE 2.10**
Inductor flux linkage.

creasing ($dI/dt > 0$), the voltage polarity is as shown in Figure 2.10. If the current through the inductor is decreasing ($dI/dt < 0$), the voltage polarity is opposite to that shown. Integrating Equation 2.11 results in an expression for current through an inductor given the voltage:

$$I(t) = \frac{1}{L}\int_0^t V(\tau)d\tau \tag{2.12}$$

where $\tau$ is a dummy variable of integration.

From Equation 2.12 we can infer that the current through an inductor cannot change instantaneously because it is the integral of the voltage. This is important in understanding the function or consequences of inductors in circuits. It takes time to increase or decrease the current flowing through an inductor. An important mechatronic system component, the electric motor, has large inductance, so it is difficult to turn the motor on or off very fast. This is true of electromagnetic relays or solenoids as well.

Two inductors in series add:

$$L_{eq} = L_1 + L_2 \tag{2.13}$$

and two inductors in parallel combine as (see Question 2.5)

$$L_{eq} = \frac{L_1 L_2}{L_1 + L_2} \tag{2.14}$$

Typical inductor components range in value from 1 μH to 100 mH. Inductance is important to consider in motors, relays, solenoids, some power supplies, and high-frequency circuits. Although some manufacturers have coding systems for inductors, there is no standard method. Often, the value is printed on the device directly, typically in μH or mH.

## 2.3 KIRCHOFF'S LAWS

Now we are ready to put circuit elements and sources together in circuits and calculate voltages and currents anywhere in the circuit. Kirchoff's Laws are essential for the analysis of circuits, no matter how complex the circuit elements or how modern their design. In fact, these laws are the basis for even the most complex circuit analysis such as that involved with transistor circuits, operational amplifiers, or integrated circuits with hundreds of elements. ***Kirchoff's Voltage Law (KVL)*** states that the sum of voltages around a closed loop or path is zero (see Figure 2.11):

$$\sum_{i=1}^{N} V_i = 0 \tag{2.15}$$

Note that the loop must be closed, but the conductors themselves need not be closed.

To apply KVL to a circuit, as illustrated in Figure 2.11, you first assume current directions through every element in the circuit. Next, assign the appropriate polarity to the voltage across each passive element assuming that the voltage drops across each element in the direction of the current. (Where assumed current enters a passive element, a plus is shown, and where the assumed current leaves the element, a minus is shown.) The polarity of voltage across a voltage source and the direction of current through a current source must always be maintained as given. Now, starting at any point in the circuit (such as node $A$ in Figure 2.11) and follow-

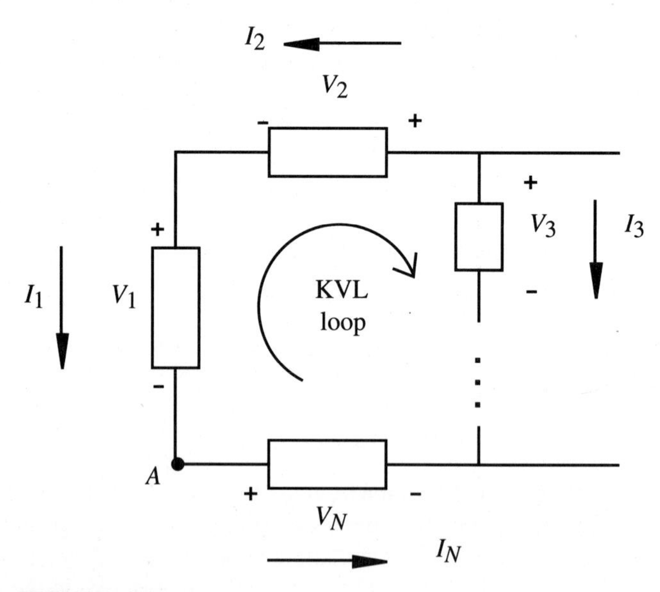


**FIGURE 2.11**
Kirchoff's Voltage Law.

ing either a clockwise or counterclockwise loop direction (clockwise in Figure 2.11), form the sum of the voltages across each element, assigning to each voltage the first algebraic sign encountered at each element in the loop. For Figure 2.11, the result would be

$$-V_1 - V_2 + V_3 + \cdots - V_N = 0 \tag{2.16}$$

---

▼ ***EXAMPLE 2.3. Kirchoff's Voltage Law.*** KVL will be used to find the current $I_R$ in the following circuit.

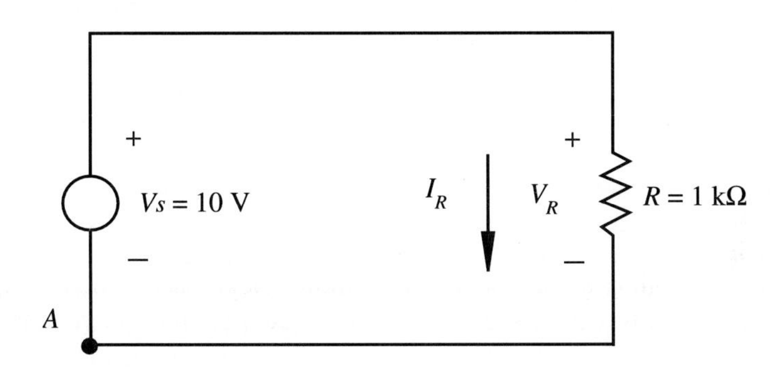


The first step is to assume the current direction for $I_R$. The chosen direction is shown in the figure. Then we use the current direction through the resistor to assign the voltage drop polarity. (If the current were assumed to flow in the opposite direction, the voltage polarity across the resistor would also have to be reversed.) The polarity for the voltage source is fixed regardless of current direction. Starting at point $A$ and progressing clockwise around the loop, we assign the first voltage sign we come to on each element yielding

$$-V_s + V_R = 0$$

Applying Ohm's Law,

$$-V_s + I_R V_R = 0$$

Therefore,

$$I_R = V_s / R = 10/1000 \text{ A} = 10 \text{ mA}$$

---

***Kirchoff's Current Law (KCL)*** states that the sum of the currents flowing into a closed surface or node is zero. Referring to Figure 2.12a,

$$I_1 + I_2 - I_3 = 0 \tag{2.17}$$

More generally, referring to Figure 2.12b,

$$\sum_{i=1}^{N} I_i = 0 \tag{2.18}$$

Note that currents leaving a node or surface are assigned a negative value.

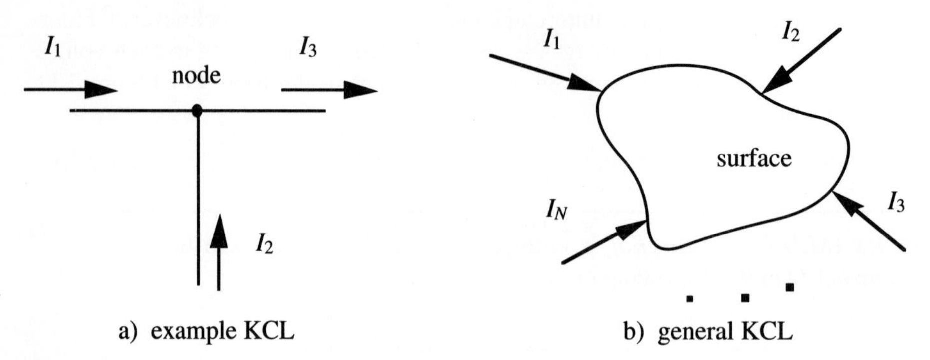


**FIGURE 2.12**
Kirchoff's Current Law.

It is important to note that when analyzing a circuit, you arbitrarily assume current directions and denote the directions with arrows on the schematic. If the calculated result for a current is negative, the current actually flows in the opposite direction. Also, assumed voltage drops must be consistent with the assumed current directions. If a voltage result is negative, its polarity is opposite to that shown.

### 2.3.1 Series Resistance Circuit

Applying KVL to a simple series resistor circuit illustrated in Figure 2.13 yields some useful results. Assuming a current direction $I$, starting at node $A$, and following a clockwise direction yields

$$-V_s + V_{R_1} + V_{R_2} = 0 \tag{2.19}$$

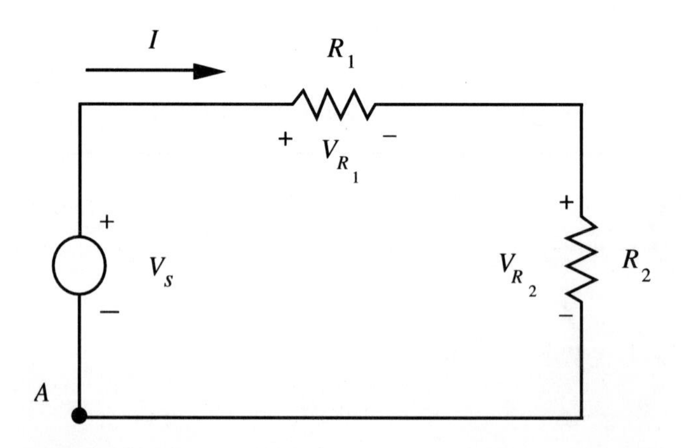


**FIGURE 2.13**
Series resistance circuit.

From Ohm's Law,

$$V_{R_1} = IR_1 \tag{2.20}$$

and

$$V_{R_2} = IR_2 \tag{2.21}$$

Substituting these two equations into Equation 2.19 gives

$$-V_s + IR_1 + IR_2 = 0 \tag{2.22}$$

and solving for $I$ yields

$$I = \frac{V_s}{(R_1 + R_2)} \tag{2.23}$$

Note that if we had a single resistor of value $R_1 + R_2$, we would have the same result. Therefore, resistors in series add, and the equivalent resistance of a series resistance circuit is

$$R_{eq} = R_1 + R_2 \tag{2.24}$$

In general, $N$ resistors connected in series can be replaced by a single equivalent resistance given by

$$R_{eq} = \sum_{i=1}^{N} R_i \tag{2.25}$$

A circuit containing two resistors in series is referred to as a ***voltage divider*** since the source voltage $V_s$ divides between each resistor. Expressions for the resistor voltages can be obtained by substituting Equation 2.23 into Equations 2.20 and 2.21 giving

$$V_{R_1} = \frac{R_1}{R_1 + R_2} V_s, \quad V_{R_2} = \frac{R_2}{R_1 + R_2} V_s \tag{2.26}$$

In general, for $N$ resistors connected in series with a total applied voltage of $V_s$, the voltage $V_{R_i}$ across any resistor $R_i$ is

$$V_{R_i} = \frac{R_i}{R_{eq}} V_s = \frac{R_i}{\sum_{i=1}^{N} R_j} V_s \tag{2.27}$$

Voltage dividers are useful since they allow us to create different reference voltages in a circuit even if the circuit is energized only by a single output supply. However, care must be exercised that attached loads do not drain significant current and affect the voltage references produced with the dividers (see Class Discussion Item 2.1).

▼▼▼ ***Class Discussion Item 2.1. Improper Application of a Voltage Divider.*** Your car has a 12 V battery that powers some circuits in the car at lower voltage levels. Is it appropriate to use a simple voltage divider to create a lower voltage level for a circuit that requires significant current?

### 2.3.2 Parallel Resistance Circuit

Applying KCL to a simple parallel resistor circuit illustrated in Figure 2.14 also yields some useful results. Since each resistor experiences the same voltage $V_s$ from KVL, Ohm's Law gives

$$I_1 = \frac{V_s}{R_1} \tag{2.28}$$

and

$$I_2 = \frac{V_s}{R_2} \tag{2.29}$$

Applying KCL at node $A$ gives

$$I - I_1 - I_2 = 0 \tag{2.30}$$

Substituting the currents from Equations 2.28 and 2.29 yields

$$I = \frac{V_s}{R_1} + \frac{V_s}{R_2} = V_s\left(\frac{1}{R_1} + \frac{1}{R_2}\right) \tag{2.31}$$

Replacing the resistance values $R_1$ and $R_2$ with their conductance equivalents $1/G_1$ and $1/G_2$ gives

$$I = V_s(G_1 + G_2) \tag{2.32}$$

A single resistor with a conductance of value $(G_1 + G_2)$ would have given the same result; therefore, conductances in parallel add. We can write Equation 2.32 as

$$I = V_s G_{eq} = \frac{V_s}{R_{eq}} \tag{2.33}$$

where $G_{eq}$ is the equivalent conductance and $R_{eq}$ is the equivalent resistance. By comparing the right-hand side of this equation to Equation 2.31, we get

$$\frac{1}{R_{eq}} = \frac{1}{R_1} + \frac{1}{R_2} \tag{2.34}$$

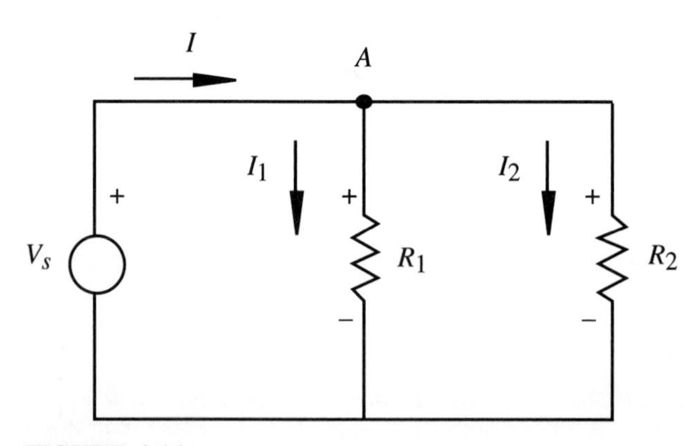


**FIGURE 2.14**
Parallel resistance circuit.

or
$$R_{eq} = \frac{R_1 R_2}{R_1 + R_2} \tag{2.35}$$

In general, $N$ resistors connected in parallel can be replaced by a single equivalent resistance given by

$$\frac{1}{R_{eq}} = \sum_{i=1}^{N} \frac{1}{R_i} \tag{2.36}$$

or
$$R_{eq} = 1 \Big/ \sum_{i=1}^{N} \frac{1}{R_i} \tag{2.37}$$

A circuit containing two resistors connected in parallel is called a ***current divider*** since the source current $I$ divides between each resistor. Expressions for the divided currents can be obtained by solving Equation 2.31 for $V_s$ and substituting into Equation 2.28 and 2.29 giving

$$I_1 = \frac{R_2}{R_1 + R_2} I, \quad I_2 = \frac{R_1}{R_1 + R_2} I \tag{2.38}$$

---

▼ ***EXAMPLE 2.4. Circuit Analysis.*** As an example of how the tools presented in the previous sections apply to a nontrivial circuit, consider the network below where the goal is to find $I_{out}$ and $V_{out}$. At any node in the circuit such as the one labeled by $V_{out}$, the voltage is defined with respect to the ground reference denoted by the ground symbol ⏚. Voltage differences between any two points can be obtained by taking the difference between the ground-referenced values at the points.

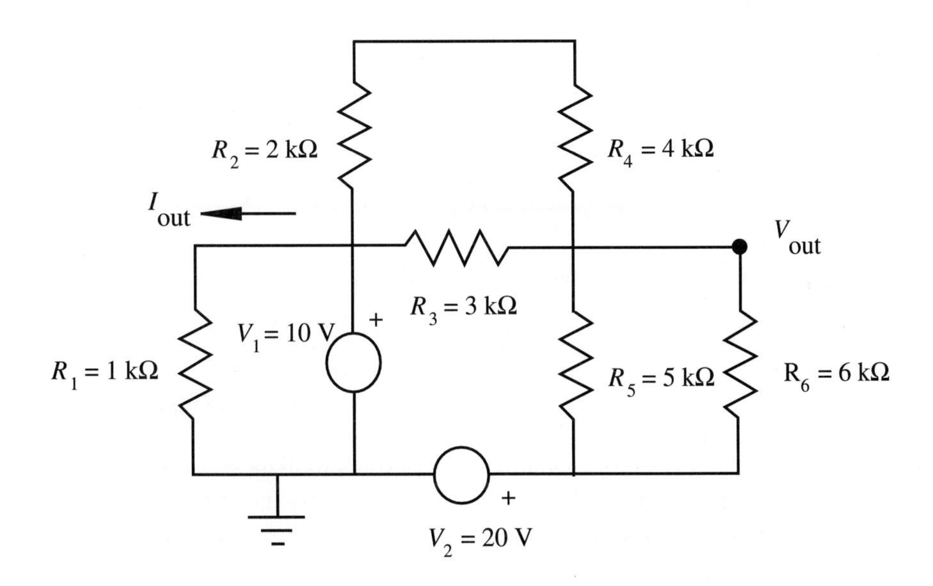

The first step is to combine resistor clusters between and around the sources ($V_1$ and $V_2$) and the branches of interest (those dealing with $I_{out}$ and $V_{out}$) using the series and parallel resistance formulas (Equations 2.24 and 2.35). The resultant resistances for the equivalent circuit shown below are

$$R_{234} = \frac{(R_2 + R_4)R_3}{(R_2 + R_4) + R_3} = 2.00 \text{ k}\Omega$$

$$R_{56} = \frac{R_5 R_6}{R_5 + R_6} = 2.73 \text{ k}\Omega$$

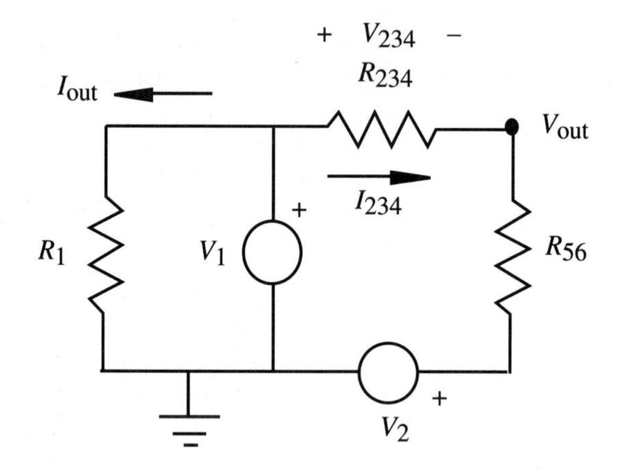


Applying KVL to the left loop gives

$$V_1 = I_{out} R_1$$

so
$$I_{out} = V_1 / R_1 = 10 \text{ V} / 1 \text{ k}\Omega = 10 \text{ mA}$$

Applying KVL to the right loop tells us that the total voltage across $R_{234}$ and $R_{56}$ in the assumed direction of $I_{234}$ is $(V_1 - V_2)$. Voltage division (Equation 2.26) can then be used to determine the voltage drop across $R_{234}$ in the assumed direction of $I_{234}$:

$$V_{234} = \frac{R_{234}}{R_{234} + R_{56}}(V_1 - V_2) = -4.23 \text{ V}$$

Since $V_1$ is referenced to ground, the desired output voltage is

$$V_{out} = V_1 - V_{234} = 14.2 \text{ V}$$

Note that since $V_{234}$ was found to be negative, the actual flow of current through $R_{234}$ would be in the opposite direction from that assumed in this solution.

There are a myriad of methods to solve this problem, and the one presented here is just an example solution, not necessarily the best method.

## 2.4
## VOLTAGE AND CURRENT SOURCES AND METERS

When we analyze electrical networks on paper, we usually assume that sources and meters are ideal. However, actual physical devices are not ideal, and it is sometimes necessary to account for the limitations when circuits contain these devices. The following ideal behavior is usually assumed:

- An ***ideal voltage source*** has zero output resistance and can supply infinite current.
- An ***ideal current source*** has infinite output resistance and can supply infinite voltage.
- An ***ideal voltmeter*** has infinite input resistance and draws no current.
- An ***ideal ammeter*** has zero input resistance and no voltage drop across it.

Unfortunately, real sources and meters have terminal characteristics that are somewhat different from the ideal cases. However, the terminal characteristics of the real sources and meters can be modeled using ideal sources and meters with their associated input and output resistances.

As shown in Figure 2.15, a "real" voltage source can be modeled as an ideal voltage source in series with a resistance called the ***output impedance*** of the device. When a load is attached to the source and current flows, the output voltage $V_{out}$ will be different from the ideal source voltage $V_s$ due to voltage division. The output impedance of most commercially available voltage sources (e.g., a power supply) is very small, usually a fraction of an ohm. For most applications, this impedance is small enough to be neglected. However, the output impedance can be important when driving a circuit with small resistance since the impedance adds to

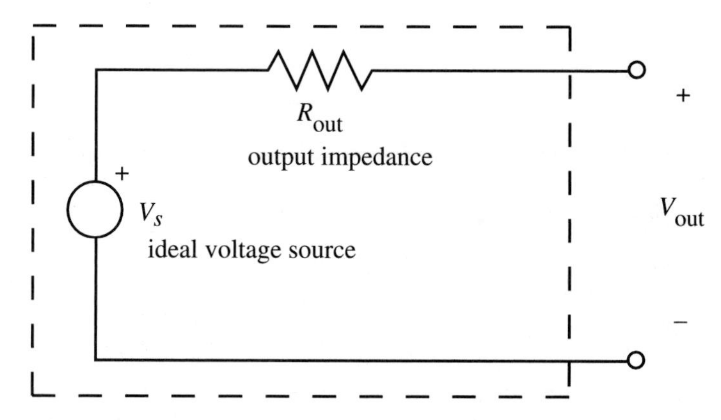


**FIGURE 2.15**
Real voltage source with output impedance.

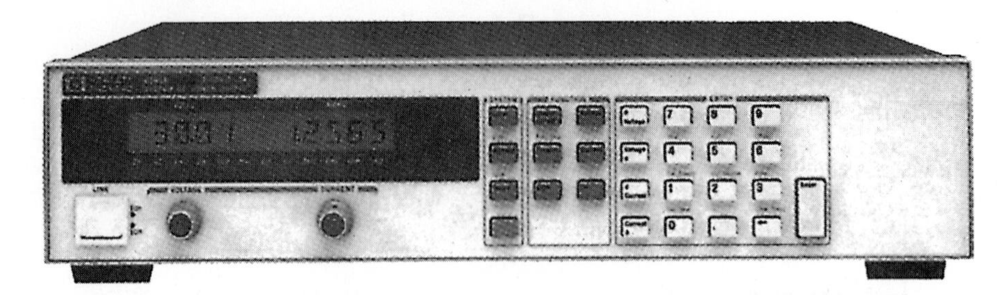

*The front panel of the HP 6641A is typical for HP system power supplies. Most functions and features can be set from the front panel or programmed through the HP-IB interface. A similar front panel is used with the HP 6500 Series power supplies, but these do not have the HP-IB interface.*

**FIGURE 2.16**
Example of a commercially available power supply. *(Courtesy of Hewlett Packard, Santa Clara, CA)*

the resistance of the circuit. Figure 2.16 shows an example of a commercially available power supply that contains digitally controlled voltage sources.

As shown in Figure 2.17, a "real" current source can be modeled as an ideal current source in parallel with an output impedance. When a load is attached to the source, the source current $I_s$ divides between the output impedance and the load. The output impedance of most commercially available current sources is very large, minimizing the current division effect. However, this impedance can be important when driving a circuit with a large resistance.

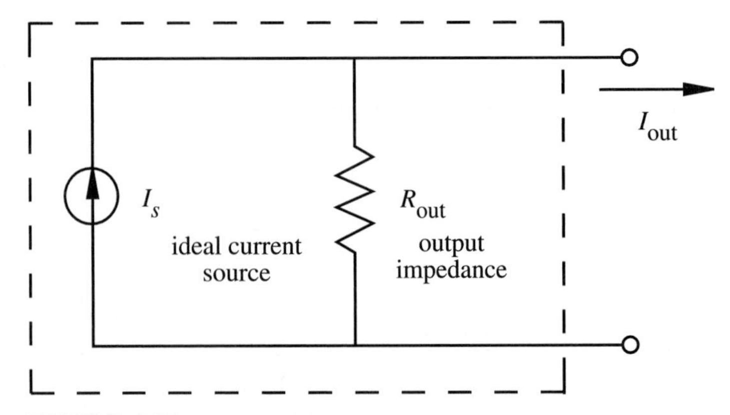


**FIGURE 2.17**
Real current source with output impedance.

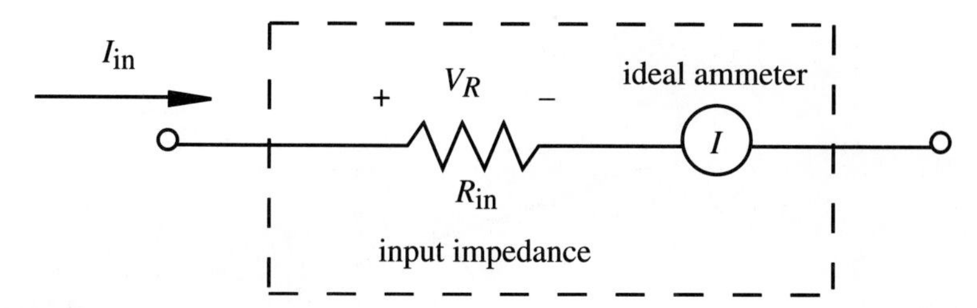


**FIGURE 2.18**
Real ammeter with input impedance.

As shown in Figure 2.18, a "real" ammeter can be modeled as an ideal ammeter in series with a resistance called the ***input impedance*** of the device. The input impedance of most commercially available ammeters is very small, minimizing the voltage drop $V_R$ added in the circuit. However, this resistance can be important when making a current measurement through a circuit branch with small resistance since the output impedance adds to the resistance of the branch.

As shown in Figure 2.19, a "real" voltmeter can be modeled as an ideal voltmeter in parallel with an input impedance. The input impedance of most commercially available voltmeters (e.g., an oscilloscope or multimeter) is very large, usually on the order of 1 to 10 MΩ. However, this resistance must be considered when making a voltage measurement across a circuit branch with large resistance since the parallel combination of the meter input impedance and the circuit branch would result in significant error in the measured value. Figure 2.20 shows examples of commercially available ***digital multimeters (DMMs)*** that contain, among other things, ammeters and voltmeters. Figure 2.21 shows an example of a commercially available oscilloscope that contains a voltmeter capable of digitizing, displaying, and recording dynamic measurements.

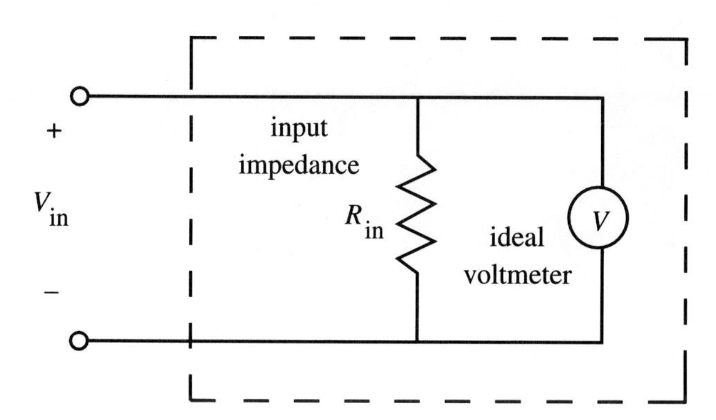


**FIGURE 2.19**
Real voltmeter with input impedance.

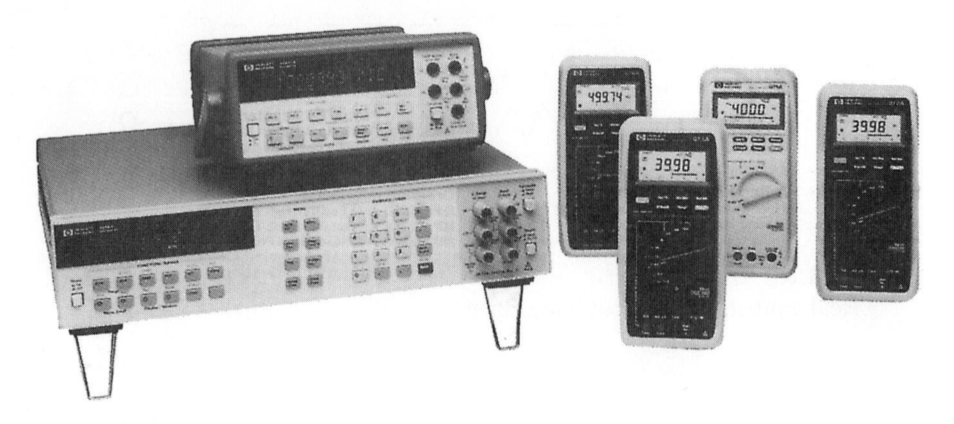

*HP offers a full line of digital multimeters*

**FIGURE 2.20**
Examples of commercially available digital multimeters. *(Courtesy of Hewlett Packard, Santa Clara, CA)*

- *20-GHz bandwidth*
- *> 100-fs timing resolution*
- *5-ps time interval accuracy*
- *Fast data acquisition and throughput*
- *Modular system design*
- *2.5-GHz edge trigger*

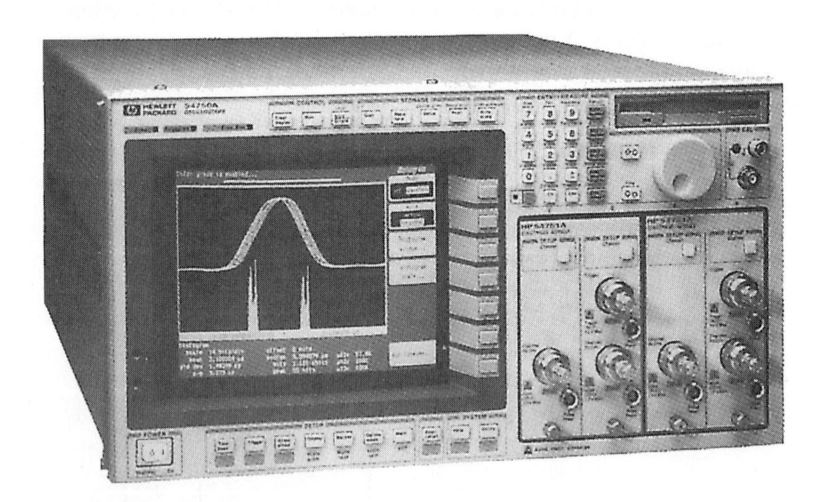

*HP 54750A*

**FIGURE 2.21**
Example of a commercially available oscilloscope. *(Courtesy of Hewlett Packard, Santa Clara, CA)*

▼ ***EXAMPLE 2.5. Input and Output Impedance.*** This example illustrates the effects of source and meter output and input impedance on making measurements in a circuit. Consider the circuit below with voltage source $V_s$ and voltage meter $V_m$.

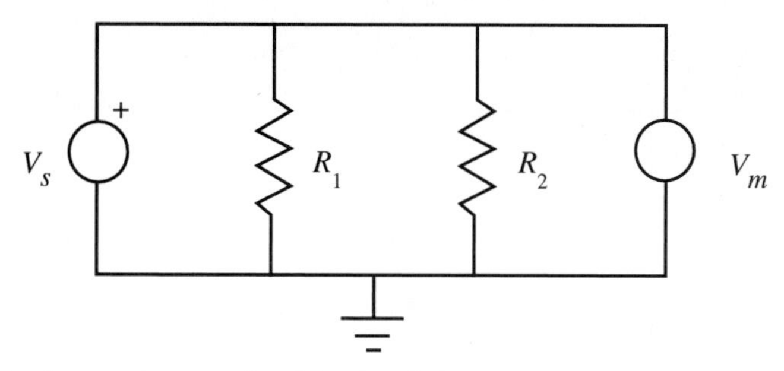


The equivalent resistance for this circuit is

$$R_{\text{eq}} = \frac{R_1 R_2}{R_1 + R_2}$$

If the source and meter were both ideal, the measured voltage, $V_m$, would be equal to $V_s$ and the equivalent circuit would look like:

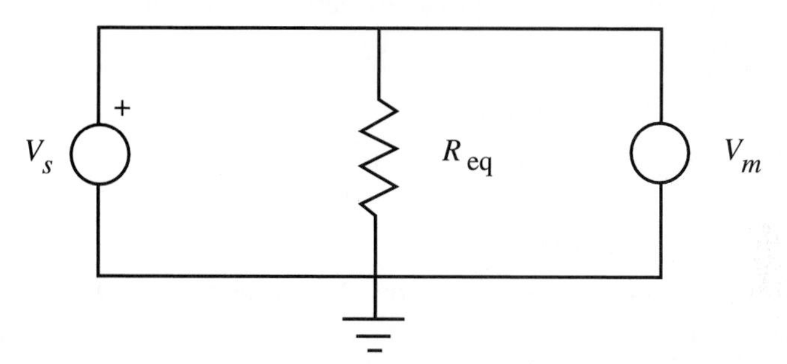


However, if the source has output impedance $Z_{\text{out}}$ and the meter has input impedance $Z_{\text{in}}$, the "real" circuit actually looks like:

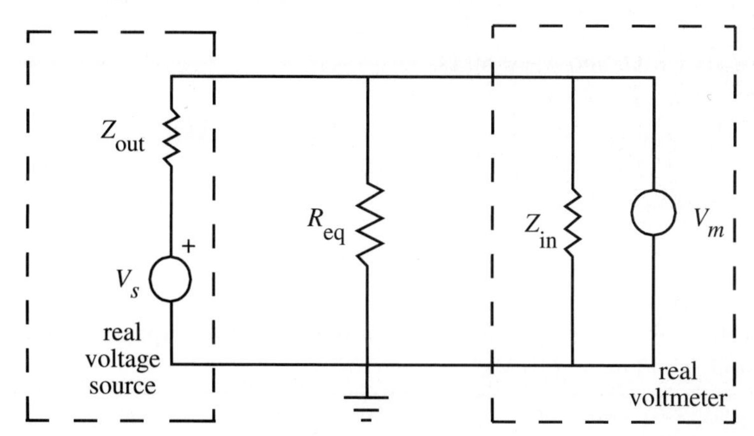

The parallel combination of $R_{eq}$ and $Z_{in}$ yields circuit (*a*) below. $Z_{out}$ and the parallel combination of $R_{eq}$ and $Z_{in}$ are now effectively in series since no current flows into the ideal meter $V_m$. Thus, the total equivalent resistance shown in circuit (*b*) is

$$R'_{eq} = \frac{R_{eq}Z_{in}}{R_{eq}+Z_{in}} + Z_{out}$$

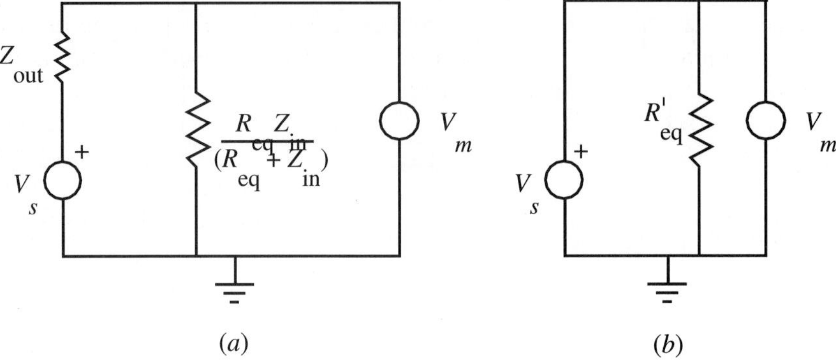


(*a*) (*b*)

Note that $R'_{eq}$ defined in the previous equation approaches $R_{eq}$ as $Z_{in}$ approaches infinity and as $Z_{out}$ approaches zero. From voltage division in circuit (*a*) above, the voltage measured by the actual meter would be

$$V_m = \frac{\dfrac{R_{eq}Z_{in}}{(R_{eq}+Z_{in})}}{\dfrac{R_{eq}Z_{in}}{(R_{eq}+Z_{in})} + Z_{out}} V_s = \frac{R'_{eq} - Z_{out}}{R'_{eq}} V_s$$

The measured voltage, $V_m$, equals $V_s$ for $Z_{in} = \infty$ and $Z_{out} = 0$, but with a real source and real meter, the measured voltage could differ appreciably from the expected ideal result. For example, if $R_1 = R_2 = 1\ \text{k}\Omega$,

$$R_{eq} = \frac{1 \cdot 1}{1+1}\ \text{k}\Omega = 0.5\ \text{k}\Omega$$

and if $Z_{in} = 1\ \text{M}\Omega$, and $Z_{out} = 50\ \Omega$,

$$R'_{eq} = \frac{0.5 \cdot 1000}{0.5 + 1000} + 0.05\ \text{k}\Omega = 0.550\ \text{k}\Omega$$

Therefore, if $V_s = 10$ V,

$$V_m = \left(\frac{0.550 - 0.05}{0.550}\right) 10\ \text{V} = 9.09\ \text{V}$$

This differs substantially from the result that would be expected with an ideal source and meter (10 V).

## 2.5 THEVENIN AND NORTON EQUIVALENT CIRCUITS

Often, to simplify the analysis of more complex circuits, we wish to replace voltage sources and resistor networks with an equivalent voltage source and series resistor. This is called a ***Thevenin equivalent*** of the circuit. Thevenin's Theorem states that given a pair of terminals in a linear network, the network may be replaced by an ideal voltage source $V_{OC}$ in series with a resistance $R_{TH}$. $V_{OC}$ is equal to the open circuit voltage across the terminals, and $R_{TH}$ is the equivalent resistance across the terminals when independent voltage sources are shorted and independent current sources are replaced with open circuits.

The circuit shown in Figure 2.22 will be used as an example. The part of the circuit in the dashed box will be replaced by its Thevenin equivalent. The open circuit voltage $V_{OC}$ is found by disconnecting the rest of the circuit and determining the voltage across the terminals of the remaining open circuit. For this example, the voltage divider rule gives

$$V_{OC} = \frac{R_2}{R_1 + R_2} V_s \tag{2.39}$$

To find $R_{TH}$ the supply $V_s$ is shorted (i.e., $V_s = 0$), grounding the left end of $R_1$. If there were current sources in the circuit, they would be replaced with open circuits. Since $R_1$ and $R_2$ are in parallel relative to the open terminals, the equivalent resistance is

$$R_{TH} = \frac{R_2 R_1}{R_2 + R_1} \tag{2.40}$$

The Thevenin equivalent circuit is shown in Figure 2.23.

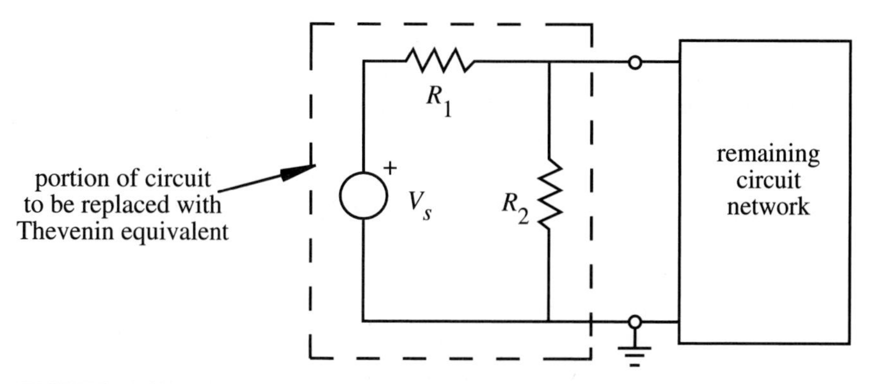


**FIGURE 2.22**
Example illustrating Thevenin's Theorem.

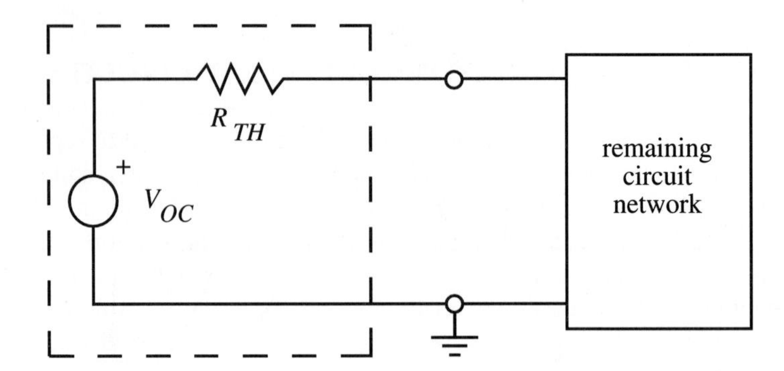


**FIGURE 2.23**
Thevenin equivalent circuit.

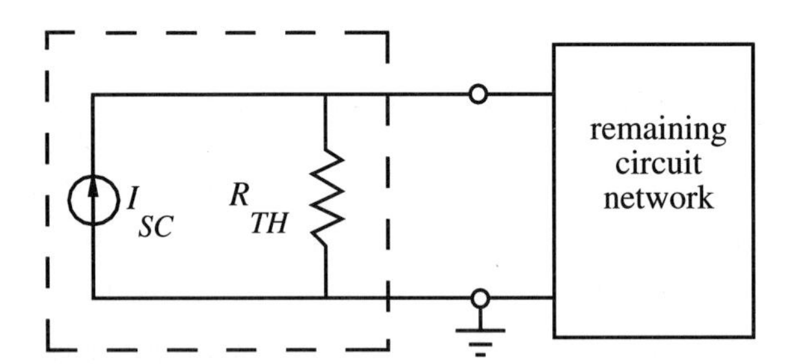


**FIGURE 2.24**
Norton equivalent circuit.

Another equivalent circuit representation is the ***Norton equivalent,*** shown in Figure 2.24. Here the linear network is replaced by an ideal current source $I_{SC}$ and the Thevenin resistance $R_{TH}$ in parallel with this source. $I_{SC}$ is found by calculating the current that would flow through the terminals if they were shorted together having removed the remaining load circuit. It can be shown that the current $I_{SC}$ flowing through $R_{TH}$ produces the Thevenin voltage $V_{OC}$ discussed above.

The Thevenin and Norton equivalents are independent of the remaining circuit network representing a load. This is useful because it is possible to make changes in the load without reanalyzing the Thevenin or Norton equivalent.

## 2.6 ALTERNATING CURRENT CIRCUIT ANALYSIS

When linear circuits are excited by alternating current (ac) signals of a given frequency, the current through and voltage across every element in the circuit will be ac signals of the same frequency. A sinusoidal ac voltage $V(t)$ is illustrated in Figure 2.25 and can be expressed mathematically as

$$V(t) = V_m \sin(\omega t + \phi) \tag{2.41}$$

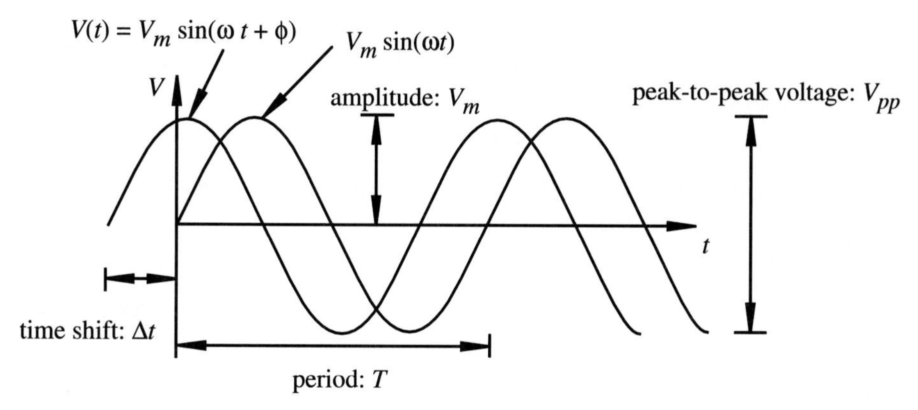


**FIGURE 2.25**
Sinusoidal waveform.

where $V_m$ is the signal ***amplitude***, ω is the ***radian frequency*** measured in radians/sec, and ϕ is the ***phase angle*** relative to the reference sinusoid $V_m \sin(\omega t)$ measured in radians. The phase angle is related to the ***time shift*** ($\Delta t$) between the signal and the reference:

$$\phi = \omega \Delta t \tag{2.42}$$

A positive phase angle ϕ implies a ***leading*** waveform (i.e., it occurs earlier on the time axis) and a negative angle implies a ***lagging*** waveform (i.e., it occurs later on the time axis). The ***period*** $T$ of the waveform is the time required for a full cycle. The frequency of the signal, measured in hertz (Hz = 1/sec), is related to the period and radian frequency as

$$f = \frac{1}{T} = \frac{\omega}{2\pi} \tag{2.43}$$

---

▼ ***EXAMPLE 2.6. ac Signal Parameters.*** As an example of how the ac signal parameters are discerned in a signal equation, consider the following ac voltage:

$$V(t) = 5.00 \sin(t + 1)\ \text{V}$$

The signal amplitude is

$$V_m = 5.00\ \text{V}$$

The signal radian frequency is

$$\omega = 1.00\ \text{rad/sec}$$

ω is the coefficient of the time variable $t$ in the sinusoid. Likewise, the frequency in hertz is

$$f = \frac{\omega}{2\pi} = \frac{1}{2\pi}\text{ Hz} = 0.159\text{ Hz}$$

and the phase angle is

$$\phi = 1\text{ rad} = 57.3°$$

The arguments of the sinsusoids are always assumed to be specified in radians for computational purposes.

---

Alternating current (ac) power is used in many applications where direct current (dc) power is impractical or infeasible. Principal reasons for using ac power include:

- Ac power is more efficient to transmit over long distances because it is easily transformed to a high voltage, low current form minimizing power losses (see Section 2.7) during transmission. In residential areas it is easily transformed back to required levels. Note that the voltage drop in the transmission line is small compared to the voltage level at the source.
- Ac power is easy to generate with rotating machinery (e.g., an electric generator).
- Ac power is easy to use to drive rotating machinery (e.g., an electric motor).
- Ac power provides a fixed frequency signal (60 Hz in the United States, 50 Hz in Europe) that can be used for timing purposes and synchronization (e.g., an electric clock based on counting periods).

▼▼▼ *CLASS DISCUSSION ITEM 2.2.* ***Reasons for ac.*** Justify and fully explain the reasons ac power is used in virtually all commercial and public utility systems. Refer to the reasons listed above.

The steady state analysis of ac circuits is simplified by the use of ***phasor*** analysis, which uses complex numbers to represent sinusoidal signals. **Euler's Formula** forms the basis for this analysis:

$$e^{j(\omega t + \phi)} = \cos(\omega t + \phi) + j\sin(\omega t + \phi) \tag{2.44}$$

where $j = \sqrt{-1}$. This implies that sinusoidal signals can be expressed as real and imaginary components of ***complex exponentials***. Because of the relative ease of manipulating exponential expressions vs. trigonometric expressions, this form of analysis is convenient for making and interpreting calculations.

Once all transients have dissipated in an ac circuit, the voltage across and current through each element will oscillate with the same frequency $\omega$ as the input. The amplitude of the voltage and current for each element will be constant but may differ in phase from the input. This fact lets us treat circuit variables $V$ and $I$ as complex exponentials with magnitudes $V_m$ and $I_m$ and phases $\phi$. A phasor (e.g., voltage $V$) is a vector representation of the complex exponential:

$$V = V_m e^{j(\omega t + \phi)} = V_m\langle\phi\rangle = V_m[\cos(\omega t + \phi) + j\sin(\omega t + \phi)] \tag{2.45}$$

where $V_m e^{j(\omega t + \phi)}$ is the complex exponential form, $V_m\langle\phi\rangle$ is the ***polar form***, and $V_m[\cos(\omega t + \phi) + j\sin(\omega t + \phi)]$ is the complex ***rectangular form*** of the phasor.

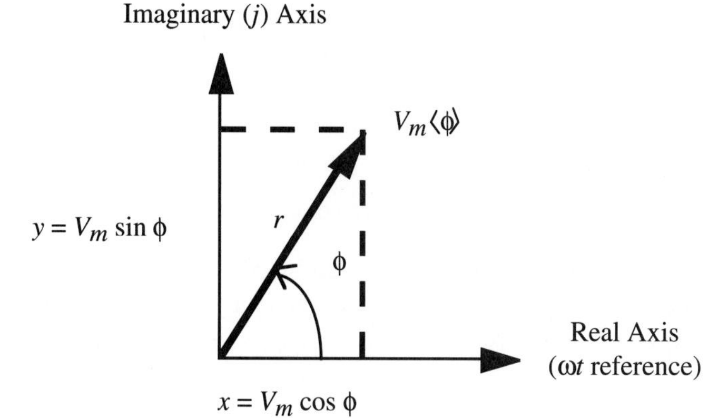


**FIGURE 2.26**
Phasor representation of a sinusoidal signal.

A graphical interpretation of these quantities is shown on the ***complex plane*** in Figure 2.26. Note that the phase angle $\phi$ is measured from the $\omega t$ reference.

Useful mathematical relations for manipulating complex phasors include

$$r = \sqrt{x^2 + y^2} \tag{2.46}$$

$$\phi = \tan^{-1}\left(\frac{y}{x}\right) \tag{2.47}$$

$$r_1\langle\phi_1\rangle \cdot r_2\langle\phi_2\rangle = r_1 \cdot r_2\langle\phi_1 + \phi_2\rangle \tag{2.48}$$

$$r_1\langle\phi_1\rangle / r_2\langle\phi_2\rangle = r_1/r_2\langle\phi_1 - \phi_2\rangle \tag{2.49}$$

where $r$ is the phasor magnitude, $\phi$ is the phasor angle, $x$ is the real component, and $y$ is the imaginary component. Note that the quadrant determined by the arguments $(x, y)$ of the arctangent function must be carefully considered when converting from rectangular to polar form. For example, if $x = y = -1$, $\phi = -225°$, not $45°$ that you would get if you carelessly entered a single argument $\tan^{-1}$ function on a calculator or in a computer program.

Ohm's Law can be extended to the ac circuit analysis of resistor, capacitor, and inductor elements as

$$V = ZI \tag{2.50}$$

where $Z$ is called the ***impedance*** of the element. This is a complex number, and you can imagine $Z$ as a complex resistance. This means its magnitude may change as a function of frequency. Impedances can be derived from the fundamental constitutive equations for the elements using complex exponentials. The unit of impedance is the ohm ($\Omega$).

For the resistor, since $V = IR$,

$$Z_R = R \tag{2.51}$$

For the inductor, since $V = L\frac{dI}{dt}$, if $I = I_m e^{j(\omega t + \phi)}$, then

$$V = Lj\omega I_m e^{j(\omega t + \phi)} = (Lj\omega)I \tag{2.52}$$

Therefore, the impedance of an inductor is given by

$$Z_L = j\omega L = \omega L\langle 90°\rangle \tag{2.53}$$

which implies that the voltage will lead the current by 90°. Note that since a dc signal can be considered an ac signal with zero frequency ($\omega = 0$), the impedance of an inductor in a dc circuit is zero. Therefore, it acts as a short in a dc circuit. At very high ac frequencies ($\omega = \infty$), the inductor has infinite impedance so it behaves as an open circuit.

For the capacitor, since $I = C\frac{dV}{dt}$, if $V = V_m e^{j(\omega t + \phi)}$, then

$$I = Cj\omega V_m e^{j(\omega t + \phi)} = (Cj\omega)V \tag{2.54}$$

giving

$$V = \left(\frac{1}{Cj\omega}\right)I \tag{2.55}$$

Therefore, the impedance of a capacitor is given by

$$Z_C = \frac{1}{j\omega C} = \frac{-j}{\omega C} = \frac{1}{\omega C}\langle -90°\rangle \tag{2.56}$$

which implies the voltage will lag the current by 90°. The impedance of a capacitor in a dc circuit ($\omega = 0$) is infinite, so it acts as an open circuit. At very high ac frequencies ($\omega = \infty$), the capacitor has zero impedance, so it acts as a short circuit.

Every result presented in previous sections for analyzing simple dc circuits including Ohm's Law, series and parallel resistance combinations, voltage division, and current division applies to the ac impedances presented above!

---

▼ ***EXAMPLE 2.7. ac Circuit Analysis.*** The following is an illustrative example of ac circuit analysis. The goal is to find the steady state current $I$ through the capacitor in the circuit below:

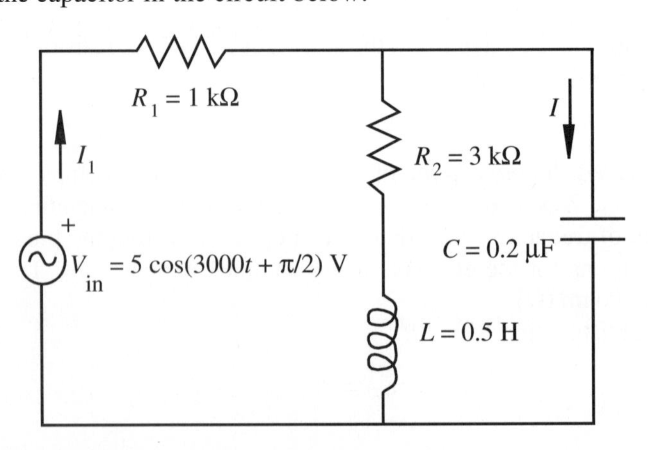

Since the input voltage source is

$$V_{\text{in}} = 5\cos\left(3000\,t + \frac{\pi}{2}\right)\text{ V}$$

each element in the circuit will respond at the radian frequency:

$$\omega = 3000 \text{ rad/sec}$$

The phasor and complex form of the voltage source is

$$V_{\text{in}} = 5\langle 90°\rangle\ V = (0+5j)\ V$$

The complex and phasor form of the capacitor impedance is

$$Z_C = -j/\omega C = -1666.67j\ \Omega = 1666.67\langle -90°\rangle\ \Omega$$

The complex and phasor form of the inductor impedance is

$$Z_L = j\omega L = 1500j\ \Omega = 1500\langle 90°\rangle\ \Omega$$

Combining all of the impedances in series and parallel yields the equivalent impedance of the entire circuit:

$$Z_{\text{eq}} = R_1 + \frac{(R_2+Z_L)Z_C}{(R_2+Z_L)+Z_C}$$

The complex and phasor form of $(R_2 + Z_L)$ is

$$(R_2+Z_L) = (3000+1500j)\ \Omega = 3354.1\langle 26.57°\rangle\ \Omega$$

so
$$Z_{\text{eq}} = 1000 + \frac{3354.1\langle 26.57°\rangle \cdot 1666.67\langle -90°\rangle}{3000-166.67j}\ \Omega$$

Using the relations in Equations 2.46 through 2.49 we get

$$Z_{\text{eq}} = 1000 + \frac{5{,}590{,}180\ \langle -63.43°\rangle}{3004.63\langle -3.18°\rangle}\Omega = 1000 + 1860.52\langle -60.25°\rangle\Omega$$

The rectangular form of the right-hand side is

$$Z_{\text{eq}} = 1000 + 923.22 - 1615.30j\Omega = 1923.22 - 1615.30j\,)\Omega$$

Therefore, the phasor form is

$$Z_{\text{eq}} = 2511.57\langle -40.03°\rangle\ \Omega$$

We can now find $I_1$ from Ohm's Law:

$$I_1 = \frac{V_{\text{in}}}{Z_{\text{eq}}} = \frac{5\langle 90°\rangle}{2511.57\langle -40.03°\rangle} = 1.991\langle 130.03°\rangle\ \text{mA}$$

Current division is used to find $I$

$$I = \frac{(R_2+Z_L)}{(R_2+Z_L)+Z_C}I_1 = \frac{3354.1\langle 26.57°\rangle}{3004.63\langle -3.18°\rangle}1.991\langle 130.03°\rangle\ \text{mA}$$

which gives

$$I = 2.22\langle 159.8°\rangle \text{ mA}$$

so the capacitor current leads the input voltage by 159.8° or 2.789 rad, and the resulting current is

$$I(t) = 2.22 \cos(3000t + 2.789) \text{ mA}$$

Note that if the input voltage were $V_{in} = 5 \sin(3000t + \pi/2)$ V instead, the resulting current would be $I(t) = 2.22 \sin(3000t + 2.789)$ mA.

---

In circuits with multiple sources, it is important to express them all in either their sine or cosine form so that the phase relationships are consistent. The following trigonometric identities are useful in accomplishing this:

$$\sin(\omega t + \phi) = \cos(\omega t + \phi - \pi/2) \tag{2.57}$$

$$\cos(\omega t + \phi) = \sin(\omega t + \phi + \pi/2) \tag{2.58}$$

## 2.7 POWER IN ELECTRICAL CIRCUITS

All circuit elements dissipate, store, or deliver power through the physical interaction between charges and electromagnetic fields. An expression for power can be derived by first looking at the infinitesimal work ($dW$) done when an infinitesimal charge ($dq$) moves through an electric field resulting in a change in potential represented by a voltage $V$. This infinitesimal work is given by

$$dW = Vdq \tag{2.59}$$

Since ***power*** is the rate of work done,

$$P = \frac{dW}{dt} = V\frac{dq}{dt} = VI \tag{2.60}$$

Therefore, the power consumed or generated by an element is simply the product of the voltage across and the current through the element. If the current flows in the direction of decreasing voltage as shown in Figure 2.27, $P$ is negative, implying that the element is dissipating or storing energy. If the current flows in the direction of increasing voltage, $P$ is positive, implying that the element is generating or releasing energy. The instantaneous power in a resistive circuit can be expressed as

$$P = VI = I^2R = V^2/R \tag{2.61}$$

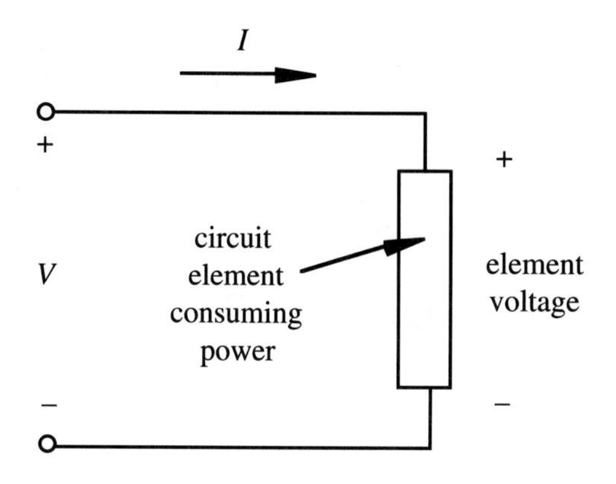


**FIGURE 2.27**
Power in a circuit element.

For ac circuits, since $V = V_m \sin(\omega t + \phi_V)$ and $I = I_m \sin(\omega t + \phi_I)$, the power changes continuously over a period of the ac waveform. Instantaneous power is not a useful quantity by itself, but if we look at the average power delivered over a period, we get a good measure of the circuit's or component's overall power characteristics. It can be shown (see Question 2.19) that the average power over a period is

$$P_{avg} = \frac{V_m I_m}{2} \cos(\theta) \tag{2.62}$$

where $\theta$ is the difference between the voltage and current phase angles ($\phi_V - \phi_I$), which is the phase angle of the complex impedance $Z = V/I$.

If we use the ***rms*** or ***root-mean-square*** values of the voltage and current defined by

$$I_{rms} = \sqrt{\frac{1}{T}\int_0^T I^2 dt} = \frac{I_m}{\sqrt{2}} \quad \text{and} \quad V_{rms} = \sqrt{\frac{1}{T}\int_0^T V^2 dt} = \frac{V_m}{\sqrt{2}} \tag{2.63}$$

the average ac power consumed by a resistor can be expressed in the same form as with dc circuits (see Question 2.20):

$$P_{avg} = V_{rms} I_{rms} = R I_{rms}^2 = V_{rms}^2 / R \tag{2.64}$$

For ac networks with inductance and capacitance in addition to resistance, the average power consumed by the network can be expressed as

$$P_{avg} = I_{rms} V_{rms} \cos\theta = I_{rms}^2 |Z| \cos\theta = (V_{rms}^2 / |Z|) \cos\theta \tag{2.65}$$

where $|Z|$ is the magnitude of the complex impedance. Cos$\theta$ is called the ***power factor*** since the average power dissipated by the network is dependent on this term.

▼▼▼ ***CLASS DISCUSSION ITEM 2.3. ac Line Waveform.*** Draw a figure that represents one cycle of the ac voltage signal present at a typical household wall receptacle. What is the amplitude, frequency, period, and rms value for the voltage? Also, what is a typical rms current capacity for a household circuit?

▼▼▼ ***CLASS DISCUSSION ITEM 2.4. International ac.*** In European countries, the household ac signal is 220 $V_{rms}$ at 50 Hz. What effect does this have on electrical devices such as an electric razor or an electric clock purchased in the United States but used in these countries?

## 2.8 TRANSFORMER

A ***transformer*** is a device used to change the relative amplitudes of voltage and current in an ac circuit. As illustrated in Figure 2.28, it consists of primary and secondary windings whose magnetic fluxes are linked by a ferromagnetic core.

Using Faraday's Law of Induction, and neglecting magnetic losses, the voltage per turn of wire is the same for both the primary and secondary windings since the windings experience the same alternating magnetic flux. Therefore, the primary and secondary voltages ($V_P$ and $V_S$) are related by

$$\frac{V_P}{N_P} = \frac{V_S}{N_S} = -\frac{d\phi}{dt} \tag{2.66}$$

where $N_P$ is the number of turns in the primary winding, $N_S$ is the number of turns in the secondary winding, and $\phi$ is the magnetic flux linked between the two coils. Thus, the secondary voltage is related to the primary voltage by

$$V_S = \frac{N_S}{N_P} V_P \tag{2.67}$$

where $N_S/N_P$ is the turns ratio of the transformer. If $N_S > N_P$, the transformer is called a ***step-up transformer*** since the voltage increases. If $N_S < N_P$, it is called a ***step-down transformer*** since the voltage decreases. If $N_S = N_P$, it is called an ***isolation transformer,*** and the output voltage is the same as the input voltage. Transformers electrically isolate the output circuit from the input circuit.

If we neglect losses in the transformer due to winding resistance and magnetic losses, the power in the primary and secondary circuits is equal:

$$I_P V_P = I_S V_S \tag{2.68}$$

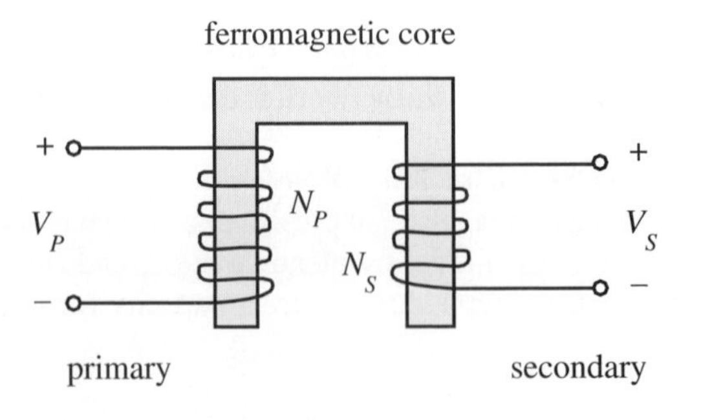


**FIGURE 2.28** Transformer.

Substituting Equation 2.67 results in the following relation between the secondary and primary currents:

$$I_S = \frac{N_P}{N_S} I_P \tag{2.69}$$

Thus a step-up transformer results in lower current in the secondary and a step-down transformer results in higher current. An isolation transformer has equal ac currents in both the primary and secondary. Note that any dc component of current in a transformer primary will not appear in the secondary. Only ac currents are transformed.

▼▼▼ *CLASS DISCUSSION ITEM 2.5.* ***dc Transformer.*** Can a transformer be used to increase voltage in a dc circuit? Why or why not?

## 2.9 IMPEDANCE MATCHING

Often we must be careful when connecting different devices and circuits together. For example, when using certain function generators to drive a circuit, proper ***signal termination***, or loading, may be required as illustrated in Figure 2.29. Placing the 50 Ω termination resistance in parallel with a higher impedance network helps match the receiving network input impedance to the generator output impedance. This is called ***impedance matching***. If we do not match impedances, a high impedance network will reflect frequency components of the driving circuit (e.g., the function generator), especially the high-frequency components. A good analogy to this effect is a thin string attached to a thicker string.

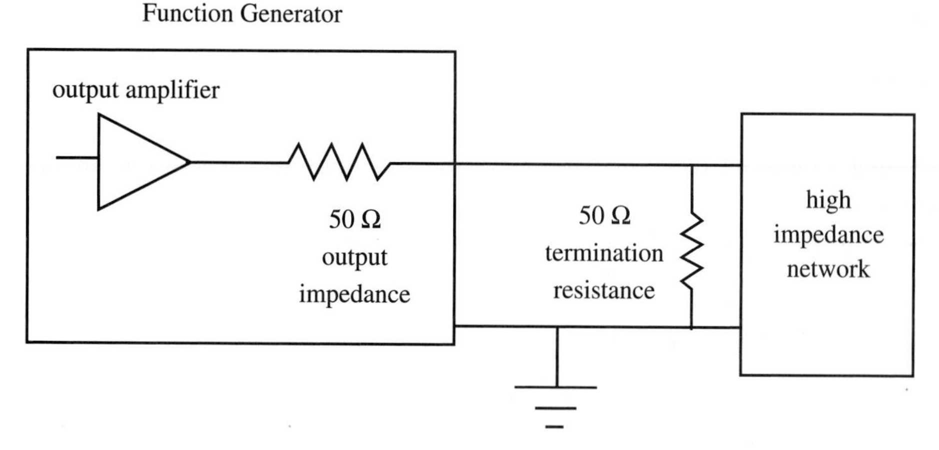


**FIGURE 2.29**
Signal termination.

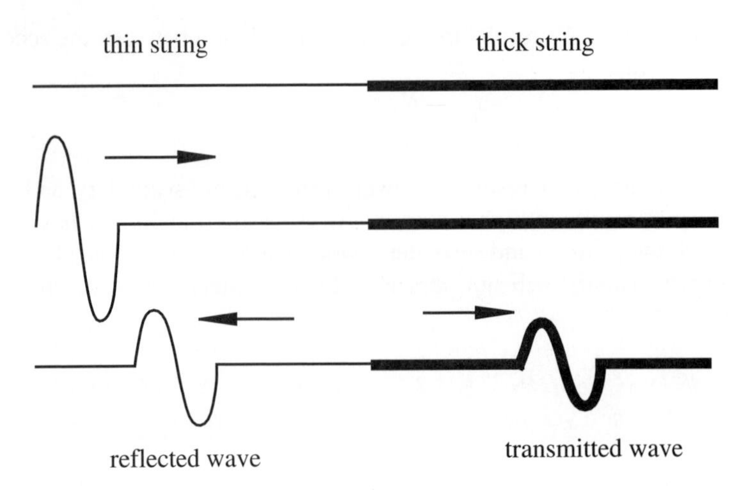


**FIGURE 2.30**
Impedance matching—string analogy.

As illustrated in Figure 2.30, if you propagate transverse vibrations along the thin string, there will be partial transmission to the thick string and partial reflection back to the source. This is a result of the mismatch of the properties at the interface between the two strings.

In addition to the signal termination concerns, another place where impedance matching is important is in applications where it is desired to transmit maximum power to a load from a source. This concept is easily illustrated with the simple resistive circuit shown in Figure 2.31 with source voltage $V_s$, source output impedance $R_i$, and load resistance $R_L$. The voltage across the load is given by voltage division:

$$V_L = \frac{R_L}{R_L + R_i} V_s \tag{2.70}$$

Therefore, the power transmitted to the load is

$$P_L = \frac{V_L^2}{R_L} = \frac{R_L}{(R_L + R_i)^2} V_s^2 \tag{2.71}$$

To find the load resistance that maximizes this power, we set the derivative of the power equal to zero and solve for the load resistance:

$$\frac{dP_L}{dR_L} = V_s^2 \frac{(R_L + R_i)^2 - 2R_L(R_L + R_i)}{(R_L + R_i)^4} = 0 \tag{2.72}$$

The derivative is zero only when the numerator is zero, so

$$(R_L + R_i)^2 = 2R_L(R_L + R_i) \tag{2.73}$$

Solving for $R_L$ gives

$$R_L = R_i \tag{2.74}$$

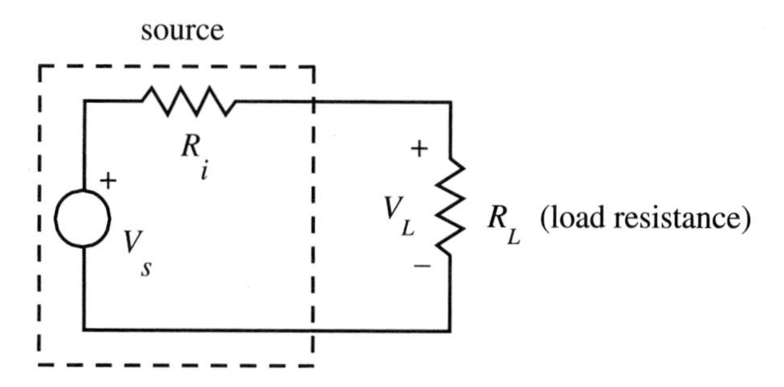


**FIGURE 2.31**
Impedance matching.

The second derivative of power can be checked to verify that this solution results in a maximum and not a minimum. The result of this analysis is as follows: To maximize power transmission to a load, the load's impedance should match the source's impedance.

▼▼▼ ***CLASS DISCUSSION ITEM 2.6. Audio Stereo Amplifier Impedances.*** Why are audio stereo amplifier output impedances important specifications when selecting speakers?

▼▼▼ ***CLASS DISCUSSION ITEM 2.7. Common Usage of Electrical Components.*** Cite specific examples in your experience where and how each of the following electrical components is used:

- Battery
- Resistor
- Capacitor
- Inductor
- Voltage divider
- Transformer

## 2.10 GROUNDING AND ELECTRICAL INTERFERENCE

It is important to provide a common ground defining a common voltage reference among all instruments and power sources used in a circuit or system. As illustrated in Figure 2.32, many power supplies have both a positive dc output (+ output) and a negative dc output (– output). These outputs produce both positive and negative voltages referenced to a common ground, usually labeled COM. On other instruments and circuits that may be connected to the power supply, all input and output voltages must be referenced to the same common ground. It is wise to double-check the integrity of each signal ground connection when assembling a group of instruments.

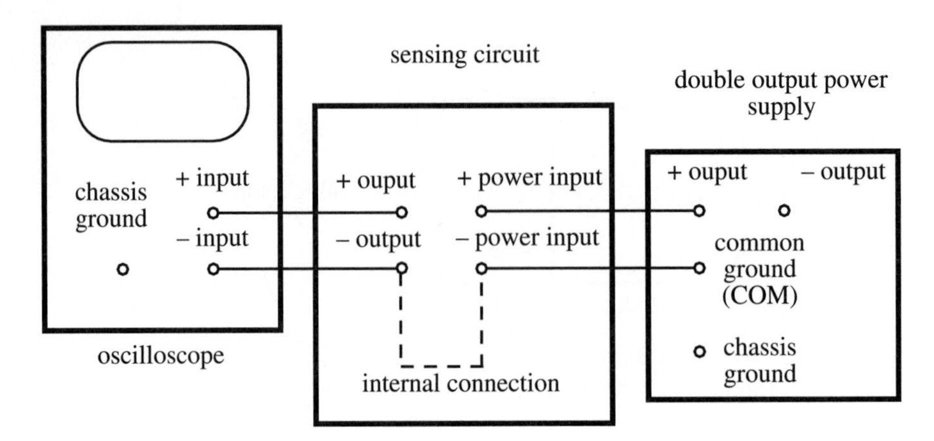


**FIGURE 2.32**
Common ground.

It is important not to confuse the signal ground with the chassis ground. The ***chassis ground*** is internally connected to the ground wire on the power cord and may not be connected to the signal ground (COM). The chassis ground is attached to the metal case enclosing an instrument to provide user safety if there are internal faults in the instruments (see Section 2.10.1).

▼▼▼ ***CLASS DISCUSSION ITEM 2.8. Automotive Circuits.*** Often, electrical components in an automobile such as the alternator or starter motor are grounded to the frame. Explain how this results in an electrical circuit.

Figure 2.33 illustrates another grounding problem where high-frequency noise can be induced in a signal by magnetic induction in the measurement leads. The area circumscribed by the leads will enclose any external magnetic fields from ac magnetic sources in the environment such as electric machinery, 60 Hz power lines, or computer monitors. This will result in an undesirable magnetically induced ac voltage, as a result of Faraday's Law of Induction, given by

$$V_{\text{noise}} = A \cdot \frac{dB}{dt} \tag{2.75}$$

where $A$ is the area enclosed by the leads and $B$ is the external magnetic field. The measured voltage will differ from the actual value according to

$$V_{\text{measured}} = V_{\text{actual}} + V_{\text{noise}} \tag{2.76}$$

Many types of ***electromagnetic interference*** (EMI) can reduce the effectiveness and reliability of a circuit or system. Also, poorly designed connections within a circuit can cause noise and unwanted signals. These effects can be mitigated using a number of standard methods. The first approach is to eliminate or move the

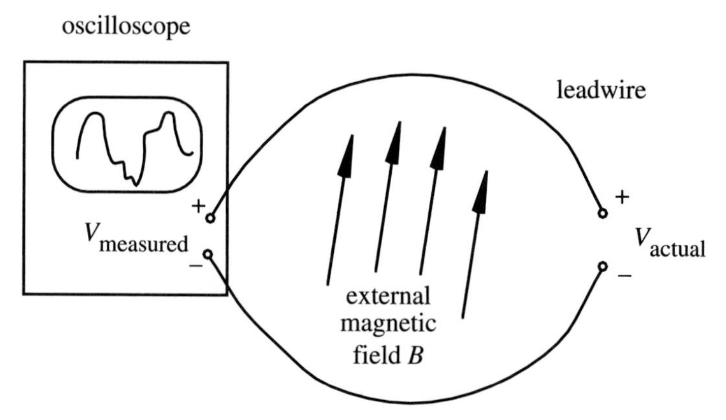


**FIGURE 2.33**
Inductive coupling.

source of the interference, if possible. The source may be a switch, motor, or ac power line in close proximity to the circuit. It may be possible to remove, relocate, shield, or improve grounding of the interference source. However, this is not usually possible, and standard methods to reduce external EMI or internal coupling may be applied. Some standard methods are:

- Eliminate potential differences that are caused by ***multiple point grounding.*** A common ground bus (large conductor, plate, or solder plane) should have a resistance small enough that voltage drops between grounding points are negligibly small. Also, make the multiple point connections close to ensure that each ground point is at approximately the same potential.
- Isolate sensitive signal circuits from high-power circuits using ***optoisolators*** or transformer couplings. Optoisolators are LED–phototransistor pairs that electrically decouple two sides of a circuit by transmitting a signal as light rather than through a solid electrical connection. One advantage is that the sensitive signal circuits are isolated from current spikes existing in the high-power circuit.
- Eliminate inductive coupling caused by ***ground loops.*** When the distance between multiple ground points is large, noise can be inductively coupled to the circuit through the conducting loops created by the multiple ground points.
- Shield sensitive circuits with grounded metal covers to block external electric and magnetic fields.
- Use short leads in connecting all circuits to reduce capacitive and inductive coupling between leads.
- Use bypass capacitors between high supply potentials and ground to provide a short circuit for high-frequency noise on the power supply lines.
- Use coaxial cable or twisted pair cable for high-frequency signal lines to minimize the effects of external magnetic fields.

- If printed circuit boards are being designed, ensure that adequate ***ground planes*** are provided. A ground plane is a large surface conductor that minimizes potential differences between ground points.

### 2.10.1 Electrical Safety

When using and designing electrical systems, safety should always be a concern. In the United States, electrical codes require outlets with three terminals: hot, neutral, and ground. Figure 2.34 illustrates the prongs on a plug that is inserted into an outlet. The wires in the plug cable include a black wire connected to the hot prong, a white wire connected to the neutral prong, and a bare or green wire connected to the ground prong. The two flat prongs (hot and neutral) of a plug complete the active circuit, allowing alternating current to flow from the wall outlet to an electrical device. The round ground prong is connected only to the chassis of the device and not to the power circuit ground in the device. The chassis ground provides an alternative path to earth ground, reducing the danger to a person who may contact the chassis when there is a fault in the power circuit. Without a separation between chassis and power ground, a high voltage can exist on the chassis creating a safety hazard for the user since he or she can complete a path to ground. Removing the ground prong or using a 3-prong-to-2-prong adapter carelessly creates a hazard.

Electricity passing through a person can cause discomfort, injury, and even death. The human body, electrically speaking, is roughly composed of a low-resistance core (on the order of 500 Ω across the abdomen) surrounded by high-resistance skin (on the order of 10 kΩ through the skin when dry). When the skin is wet, its resistance drops dramatically. Currents through the body below 1 milliamp are usually not perceived. Currents as low as 10 milliamps can cause tingling and muscle contractions. Currents through the thorax as low as 100 milliamps can affect normal heart rhythm. Currents above 5 amps can cause tissue burning.

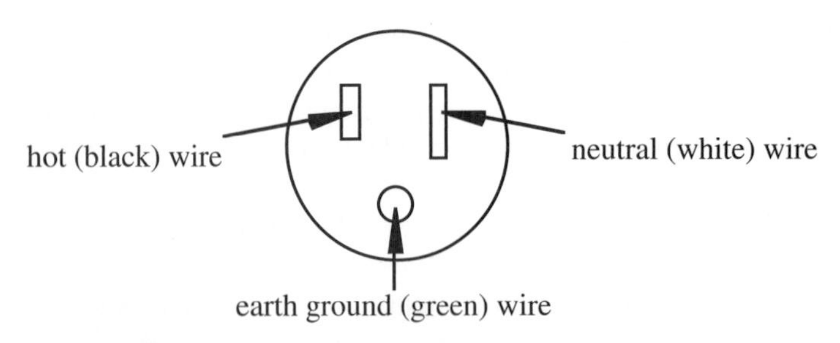


**FIGURE 2.34**
Three-prong ac power plug.

▼▼▼ ***CLASS DISCUSSION ITEM 2.9. Safe Grounding.*** Consider the oscilloscope below whose power cord ground prong has been broken so that the chassis is not connected to ground. If you use this instrument describe the possible danger you face.

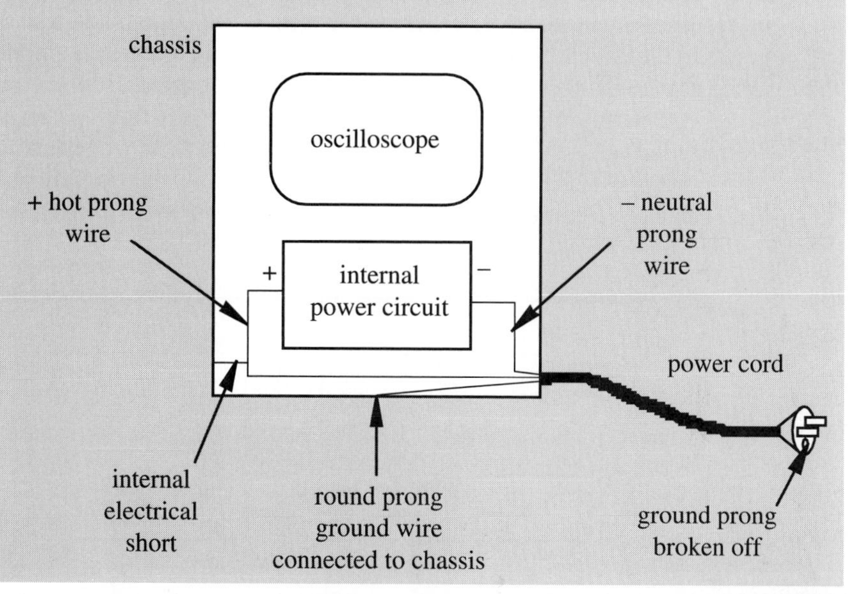


▼▼▼ ***CLASS DISCUSSION ITEM 2.10. Electric Drill Bathtub Experience.*** The electric drill shown below runs on household power and has a metal housing. You use a 3-prong-to-2-prong adapter to plug the drill into the wall socket. You are standing in a wet bathtub drilling a hole in the wall. You are unaware that the black wire's insulation has worn thin and the bare copper black wire is contacting the metal housing of the drill. How have you created a lethal situation for yourself? How could it have been prevented or mitigated?

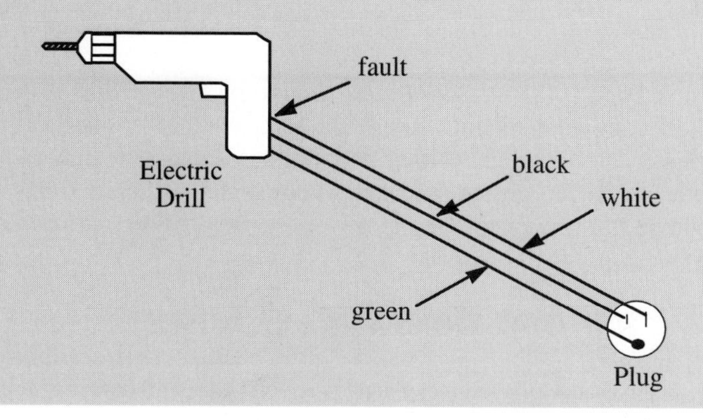

▼▼▼ ***CLASS DISCUSSION ITEM 2.11. Dangerous EKG.*** A cardiac patient is lying in his hospital bed with electrocardiograph (EKG) leads attached to his chest to monitor his cardiac rhythm. An electrical short occurs in the next room, and our patient experiences a cardiac arrest. You and the hospital facilities engineer have determined that there were multiple grounding points in the patient's room (see below) and a fault in electrical equipment in the next room causing current to flow in the ground wire from the piece of the equipment. You are on the scene to determine if there could have been a lethal current through the patient. Consider the fact that ground lines have finite resistance per unit length and that a few microamps through the heart can cause ventricular fibrillation (a fatal malfunction).

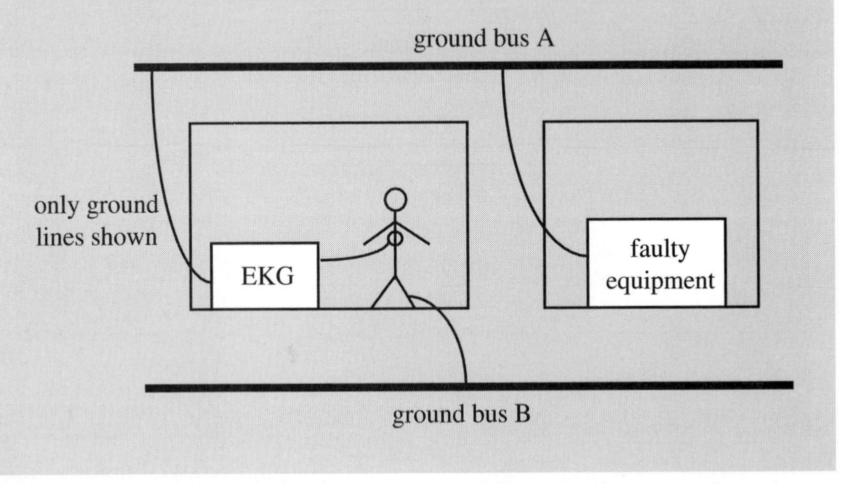


▼▼▼ ***CLASS DISCUSSION ITEM 2.12. High-voltage Measurement Pose.*** When performing a high-voltage test, a creative electrical technician claims that standing on your right foot and using your right hand to hold the probe is the safest posture for making the measurement. What possible logic could support this claim?

▼▼▼ ***CLASS DISCUSSION ITEM 2.13. Lightning Storm Pose.*** A park ranger at Rocky Mountain National Park recommends that if your hair rises when hiking in an open area during a lightning storm, it is imperative to get down low to the ground and keep your feet together. Explain why this might be life-saving advice.

## QUESTIONS AND EXERCISES

**2.1.** What is the resistance of a kilometer-long piece of 14-gage (0.0648 in.) copper wire?

**2.2.** Determine the possible range of resistance values for each of the following:

(*a*) Resistor $R_1$ with color bands: red, brown, yellow.

(*b*) Resistor $R_2$ with color bands: black, violet, orange.

(*c*) The series combination of $R_1$ and $R_2$.

(*d*) The parallel combination of $R_1$ and $R_2$.

For parts *c* and *d*, use Ohm's Law, KVL, and KCL only.

**2.3.** What colors should bands *a, b, c,* and *d* be for circuit B to have the equivalent resistance of circuit A below?

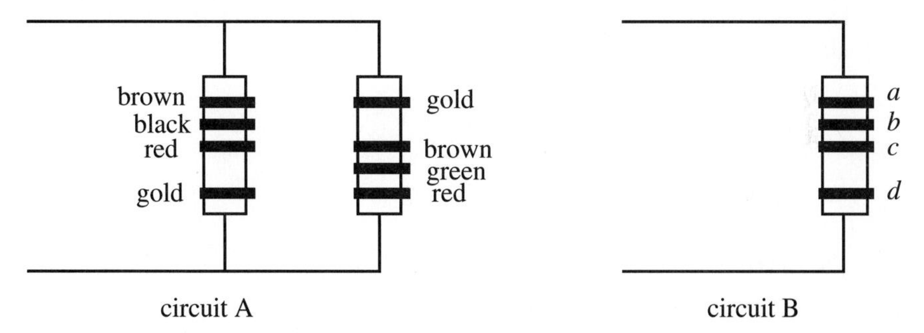


**2.4.** Derive an expression for the equivalent capacitance of two capacitors attached in series.

**2.5.** Derive an expression for the equivalent inductance of two inductors attached in parallel.

**2.6.** Using Ohm's Law, KVL, and KCL, derive an expression for the equivalent resistance of three parallel resistors ($R_1$, $R_2$, and $R_3$).

**2.7.** Derive current division formulas, similar to Equation 2.38, for three resistors in parallel.

**2.8.** Given two resistors $R_1$ and $R_2$, where $R_1$ is much greater than $R_2$, show that the parallel combination is approximately equal to $R_2$.

**2.9.** Find $V_{out}$ in the circuit below:

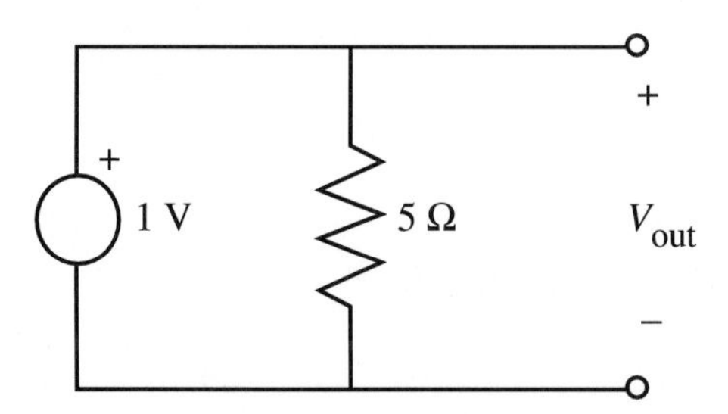

**2.10.** Find $V_{out}$ in the circuit below:

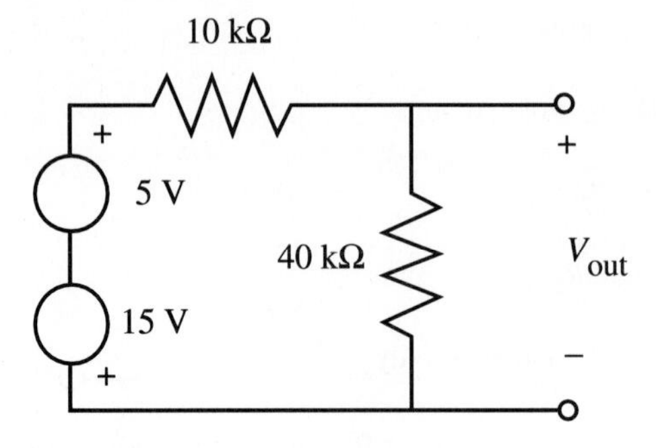


**2.11.** Given the circuit below with $R_1 = 1$ kΩ, $R_2 = 2$ kΩ, $R_3 = 3$ kΩ, $R_4 = 2$ kΩ, $R_5 = 1$ kΩ, and $V_s = 10$ V, determine:

(*a*) The total equivalent resistance seen by $V_s$.

(*b*) The voltage at node *A*.

(*c*) The current through resistor $R_5$.

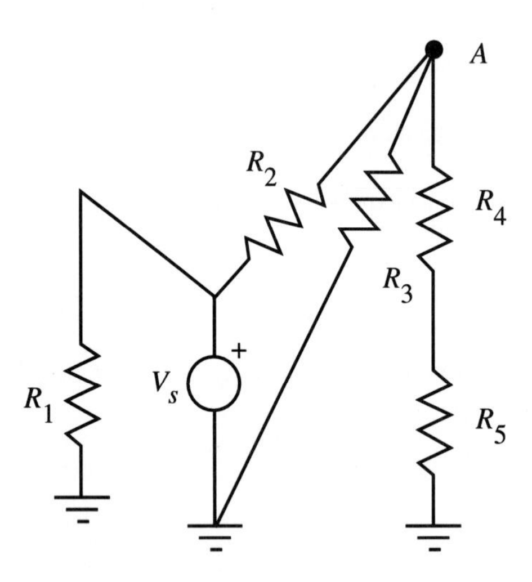


**2.12.** For the circuit below with $R_1 = 1$ kΩ, $R_2 = 2$ kΩ, $R_3 = 2$ kΩ, $R_4 = 1$ kΩ, $V_1 = 10$ V, $V_2 = 5$ V, and $V_3 = 10$ V, find

(*a*) $V_{out}$.

(*b*) The power produced by each voltage source.

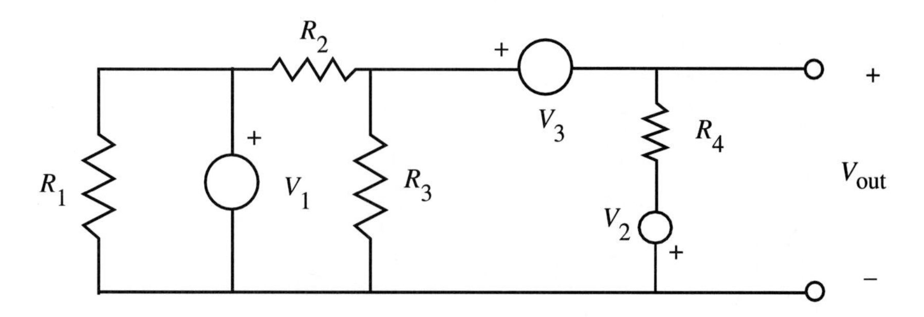


**2.13.** For the circuit below with $V_1 = 1$ V, $i_1 = 1$ A, $R_1 = 10\ \Omega$, and $R_2 = 100\ \Omega$, what is $V_{R_2}$?

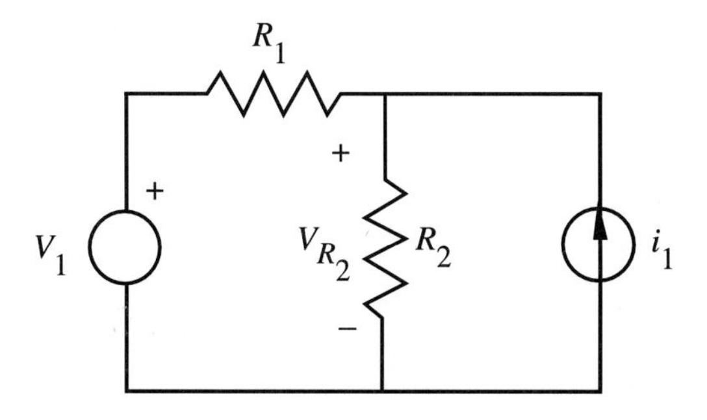


**2.14.** For the circuit below, what is $V_3$ in terms of $V_2$ for

(*a*) $R_3 = 50\ \Omega$, $R_4 = 10\ \text{k}\Omega$, $R_5 = 1.0\ \text{M}\Omega$?

(*b*) $R_3 = 50\ \Omega$, $R_4 = 500\ \text{k}\Omega$, $R_5 = 1.0\ \text{M}\Omega$?

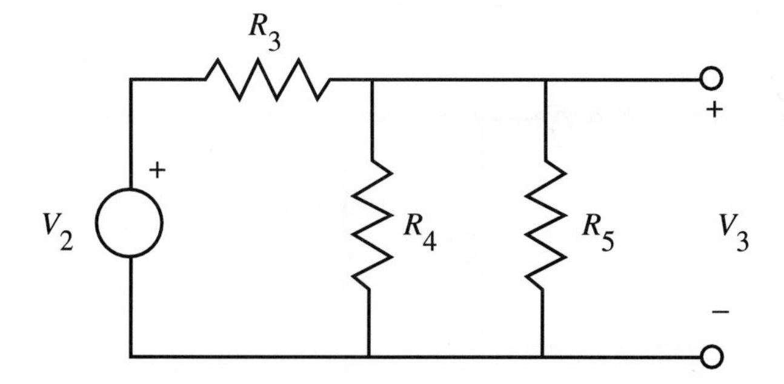


If $R_5$ represents the input impedance associated with a device measuring the voltage across $R_4$, what conclusions can you make about the voltage measurement?

**2.15.** For the circuit below with $R_1 = 1\ \text{k}\Omega$, $R_2 = 9\ \text{k}\Omega$, $R_3 = 10\ \text{k}\Omega$, $R_4 = 1\ \text{k}\Omega$, $R_5 = 1\ \text{k}\Omega$, $V_1 = 5$ V, and $V_2 = 10$ V, find $I$ and the voltage at node $A$.

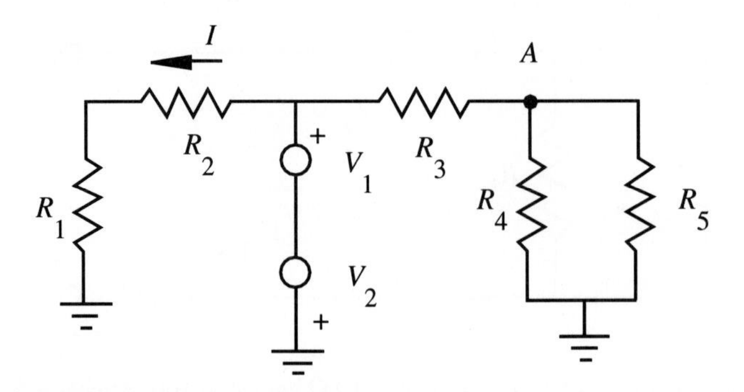


**2.16.** For the circuit below, what are the voltages across $R_1$, $R_2$, and $C$, if $V_s = 10$ V dc, $R_1 = 1\ \text{k}\Omega$, $R_2 = 1\ \text{k}\Omega$, and $C = 0.01\ \mu\text{F}$?

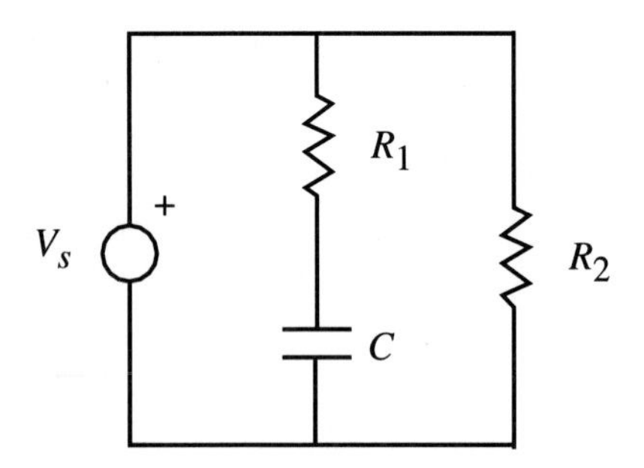


**2.17.** Find the steady state current $I(t)$ in the circuit below, where $R_1 = R_2 = 100\ \text{k}\Omega$, $C = 1\ \mu\text{F}$, and $L = 20$ H for:

(*a*) $V_s = 5$ V dc

(*b*) $V_s = 5\cos(\pi t)$ V

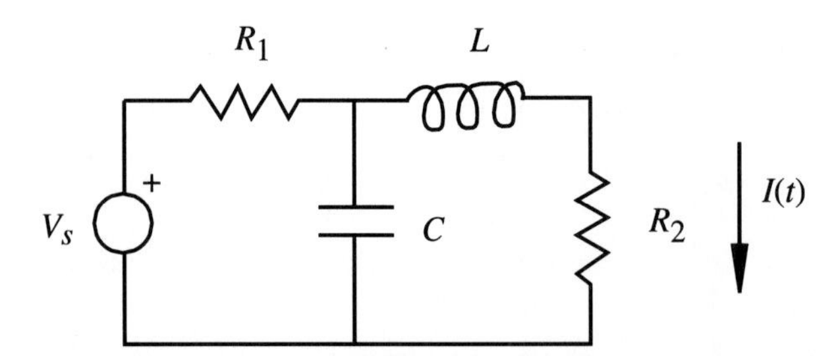


**2.18.** For each waveform expressed below, what is the frequency in Hz and in rad/sec, the peak-to-peak amplitude, and the dc offset?

(*a*) $2.0\sin(\pi t)$

(*b*) $10.0 + \cos(2\pi t)$

(*c*) $3.0 \sin(2\pi t + \pi)$

(*d*) $\sin(\pi) + \cos(\pi)$

**2.19.** Prove Equation 2.62.

**2.20.** Derive Equation 2.63 and show that Equation 2.64 is correct.

**2.21.** Sketch the output waveform for $V_{out}$ in the circuit below on axes as shown below:

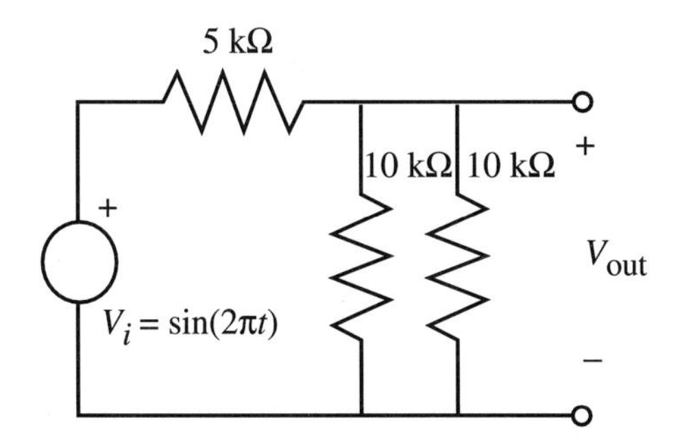


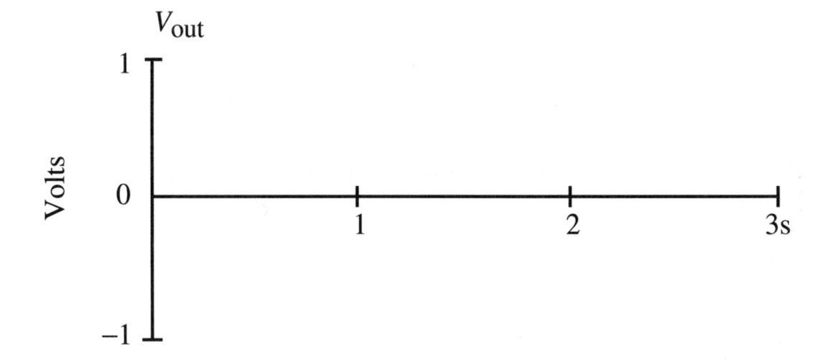


## BIBLIOGRAPHY

Horowitz, P. and Winfield, H., *The Art of Electronics,* 2nd Edition, Cambridge University Press, New York, 1989.

Johnson, D., Hilburn, J., and Johnson, J., *Basic Electric Circuit Analysis,* 2nd Edition, Prentice-Hall, Englewood Cliffs, NJ, 1984.

Lerner, R. and Trigg, G., *Encyclopedia of Physics,* VCH Publishers, New York, 1991.

McWhorter, G. and Evans, A., *Basic Electronics,* Master Publishing, Inc., Richardson, TX, 1994.

Mims, F., *Getting Started in Electronics,* Radio Shack Archer Catalog No. 276-5003A, 1991.

# 3

# SEMICONDUCTOR ELECTRONICS

**OBJECTIVES:** After you read, discuss, study, and apply ideas in this chapter, you will be able to:

- Comprehend the basic physics of semiconductor devices
- Design circuits using diodes, voltage regulators, bipolar transistors, and field effect transistors
- Understand some introductory mechatronic system designs using semiconductors
- Select semiconductor components for your designs

## 3.1 INTRODUCTION

We will examine some extraordinary materials scientists and engineers have transformed into inventions that will affect all aspects of life in the 21st century. To understand the inventions, we need to understand the physical foundations upon which the new designs are based. To begin, we will discuss a class of materials known as semiconductors, which are used extensively in electronic circuits today. They are the basis of components as simple as diodes and as complex as microprocessors. Since they will be included in nearly all mechatronic systems, we will learn the basic differences of some of the primary components, recognize their schematic symbols, and learn how to properly use them in circuit design.

## 3.2 SEMICONDUCTOR PHYSICS AS THE BASIS FOR UNDERSTANDING ELECTRONIC DEVICES

Metals have few valence electrons, which are loosely bound to their atoms. In a metal crystal the electrons become delocalized, that is, they are shared among all the atoms in the crystal. When an electric field is imposed on a metal, the electrons migrate freely as a current through the metal. Because of the ease by which currents can flow in metals they are called ***conductors.*** In contrast, other materials have atoms with valence electrons that are tightly bound and cannot easily be moved by electric fields. These materials are called ***insulators*** and do not normally sustain electric currents. In addition, there is a very useful class of materials, elements, and some compounds that have properties that lie in between conductors and insulators. They are called ***semiconductors.*** Semiconductors such as silicon, germanium, and cadmium sulfide have current-carrying characteristics that are profoundly altered by temperature or the amount of light falling on them. As shown in Figure 3.1, when a voltage is applied across a warm semiconductor, some of the valence electrons easily jump to the conductance band and then move in the electric field to produce a current. This current will be dependent on temperature and, in fact, can be used to sense temperature using a device such as a thermistor.

Furthermore, the properties of semiconductors can be significantly changed by adding small quantities of elements known as ***dopants,*** which can be diffused into the materials or injected with accelerator beams. A tiny piece of silicon, often

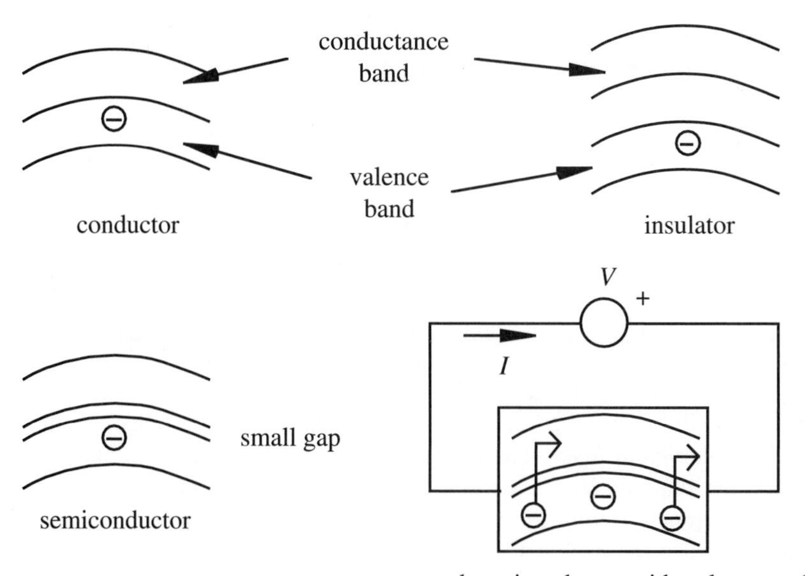


**FIGURE 3.1**
Valence and conduction bands of materials.

called a chip, can have a minute pattern of dopants deposited on and diffused into its surface, resulting in a device that is the basis of all modern electronics.

In a warmed semiconductor, a valence electron may jump to the conduction band, and its absence in the valence band is called a ***hole***. A valence electron from a nearby atom can move to the hole, leaving another hole in its former place. This chain of events can continue, resulting in a current that can be thought of as the movement of holes in one direction or electrons in the other. The net effect is the same, so perhaps old Ben Franklin wasn't completely wrong when he thought currents were the movement of positive charges!

Properties really get interesting when different amounts and types of dopants are added to semiconductors. When the semiconductor is cold there are no electrons in the conduction band, and currents will not flow. But consider what happens if different atoms are embedded in the crystal lattice of a semiconductor such as silicon. Silicon has 4 valence electrons, which form bound symmetrical electron configurations in the crystal. However, if an arsenic or phosphorous atom is added having 5 valence electrons, 1 electron is free when the arsenic settles in the silicon crystal lattice. This extra electron is freer to move around, and this so-called ***donor*** element enhances the electron conductivity of the semiconductor. It is called ***n-type*** silicon due to the extra electrons freed from the crystal lattice as charge carriers. Conversely, if the silicon is doped with an atom of boron or galium that has 3 valence electrons, a ***hole*** is formed due to a missing electron in the lattice when the so-called ***acceptor*** atom boron settles in. This hole can jump from atom to atom effectively producing a positive current. (What really happens is that electrons move to occupy the hole, and this effectively looks like holes moving!) This is called ***p-type*** silicon due to the extra holes, which are effectively positive charge carriers. As we will see shortly, the interaction between n-type and p-type semiconductor materials is the basis for semiconductor electronic devices.

## 3.3 JUNCTION DIODE

Contemporary electronic devices are designed by creating interfaces between differently doped semiconductor material on a microscopic scale. If a p-type region of silicon is created adjacent to an n-type region, a ***pn junction*** with very interesting properties is the result. As illustrated in Figure 3.2, at the pn junction, electrons from the n-type silicon jump to occupy the holes in the p-type silicon, creating what is called a ***depletion layer.*** The effect of the depletion layer is to inhibit current flow because charge carriers are tied up. However, if a voltage source is connected to the pn junction with the anode connected to the p-type silicon and the cathode connected to the n-type silicon, the anode in effect becomes a source of holes and the cathode a source of electrons so that holes and electrons are continuously replenished. Current then flows easily and continuously, and we say the junction is ***forward biased.*** If, on the other hand, the anode is connected to the n-type

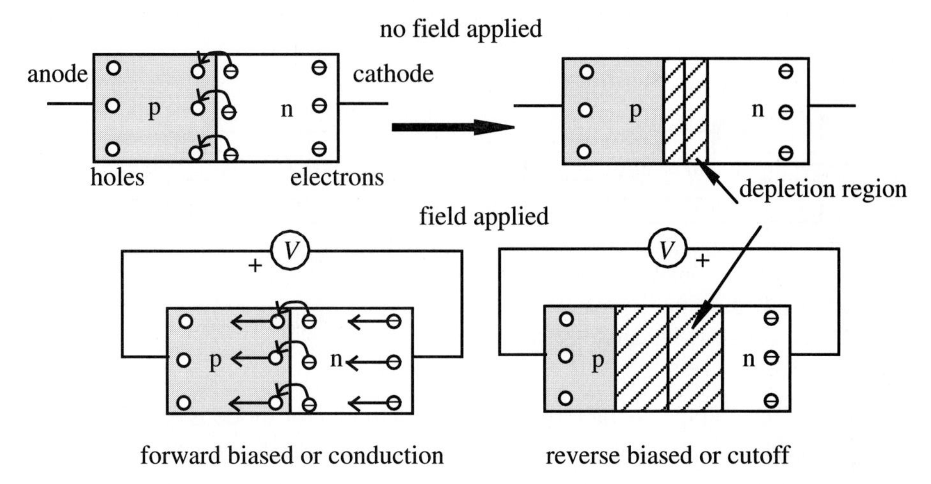


**FIGURE 3.2**
pn Junction characteristics.

silicon and the cathode to the p-type silicon, the depletion layer is enlarged and no current flows, and we say the junction is ***reverse biased.***

The pn junction that passes current in only one direction is known as a silicon ***diode***, sometimes referred to as a ***rectifier.*** The schematic symbol for the silicon diode is shown in Figure 3.3. The diode is analogous to a fluid check valve, which allows fluid to flow only in one direction as illustrated in Figure 3.4. We will soon see that pn junctions also occur in more advanced devices like transistors and integrated circuits where the behavior is more complex.

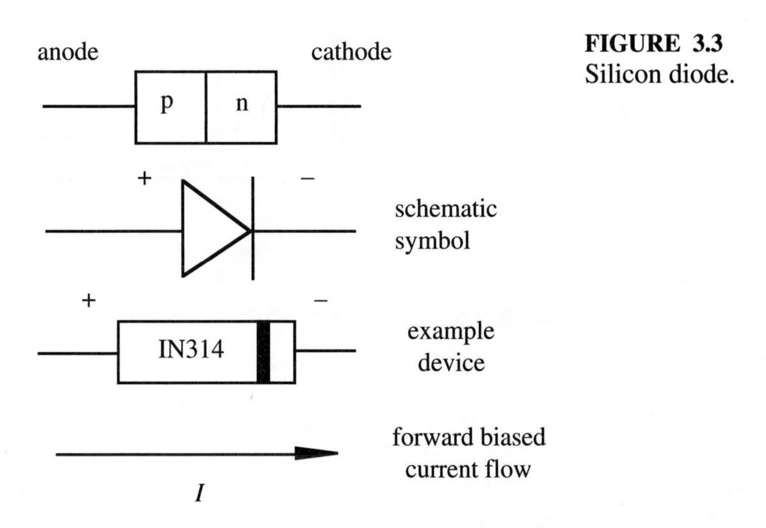


**FIGURE 3.3**
Silicon diode.

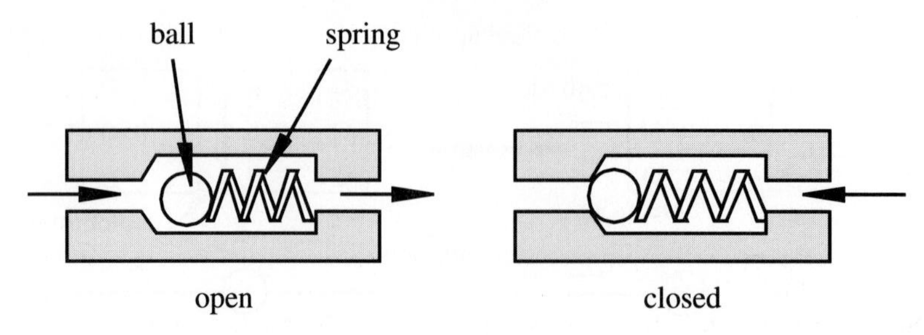


**FIGURE 3.4**
Diode check valve analogy.

The current-voltage characteristic for a semiconductor diode is exponential as shown in Figure 3.5. In reality there is a dramatic nonlinear increase in current with a forward bias of 0.6 to 0.7 V in silicon. Note the different scales used on the positive and negative sides of the voltage axis. In a first analysis, we will approximate the behavior of the semiconductor diode as that of an ***ideal diode***, one whose current-voltage characteristic is shown by the darker lines in Figure 3.5. This implies that the diode is on for any voltage greater than or equal to zero. Later in actual circuit design, a good first approximation for the ***real diode*** is given by the dashed curve as this replicates the real voltage drop of 0.7 V measured across the silicon diode when it is forward biased.

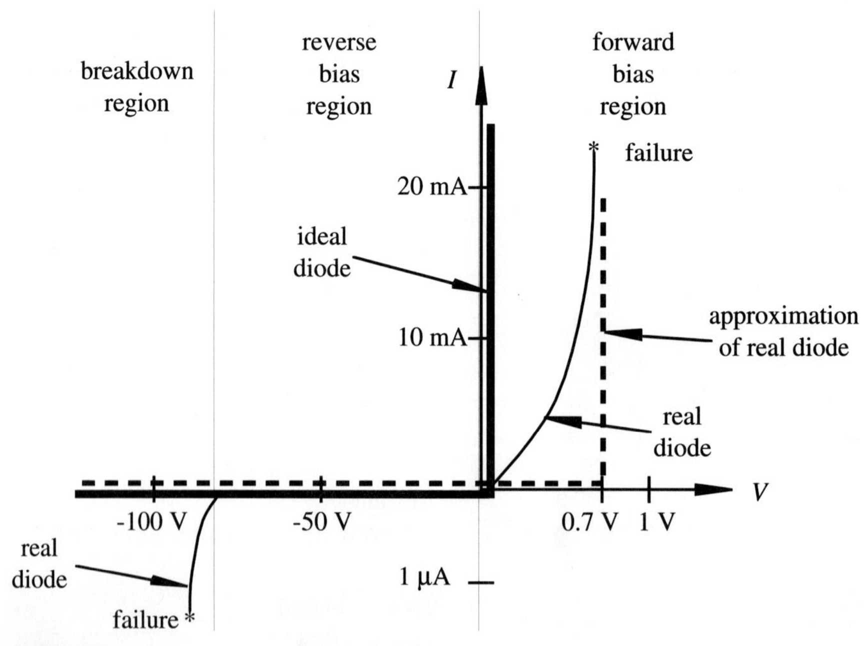


**FIGURE 3.5**
Ideal, approximate, and real diode curves.

A diode is a nonlinear device, and no single equation can simulate its current voltage relationship. An ideal diode has zero resistance when forward biased and infinite resistance when reverse biased. For analytic purposes, it can be replaced by a short circuit if it is forward biased and an open circuit if it is reverse biased. A real diode requires about 0.7 V of forward bias to enable significant current flow. When a real diode is reverse biased, it can withstand a reverse voltage up to a limit known as the ***breakdown voltage***, when the diode will fail as the reverse current increases precipitously.

Diodes are useful in passing only the positive half or the negative half of an ac signal, a process that is called ***rectification.*** This is the reason diodes are sometimes called rectifiers. Example 3.1 below illustrates how to analyze a simple ideal diode circuit called a half-wave rectifier. Rectifier circuits are used in the design of power supplies, when ac power must be transformed into dc power for electronic and computer circuits.

The important specifications that differentiate diodes are the maximum current through the diode and the maximum permissible reverse bias voltage before breakdown occurs. The instantaneous surge current and average current are usually both specified and must exceed maximum calculated values for a circuit by a good margin. Check to see if large reverse bias voltages will occur and compare to the specifications. Rectifier diodes are capable of carrying very large currents and may be designed in a bolt-like stud package to be screwed to heat sinks to dissipate the heat produced in the junction. Diodes switch on and off in the nanosecond range, which is therefore of no consequence unless very high speed circuits are being designed. We will see in the next section that there is a class of diodes designed for use in the reverse bias region for a special application.

---

▼ ***EXAMPLE 3.1. Half-Wave Rectifier Circuit Assuming an Ideal Diode.*** Given the following circuit containing a diode, we will illustrate how to determine the output voltage $V_o$ given a sinusoidal input $V_i$.

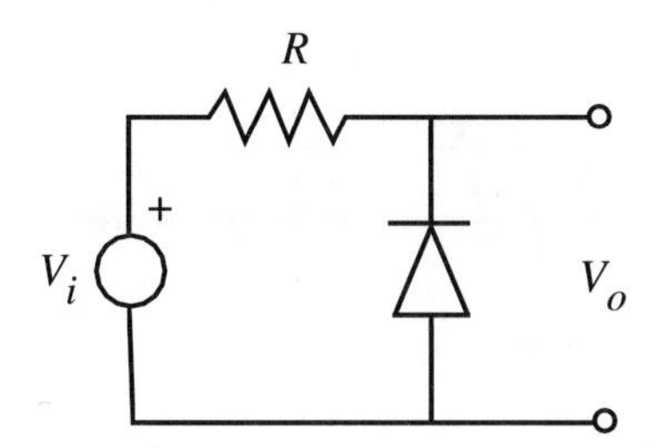


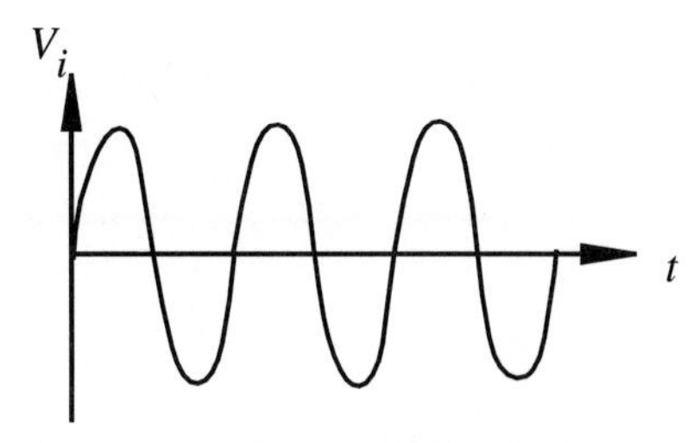


When $V_i$ is positive, the diode is reverse biased and therefore equivalent to an open circuit. No current flows through the resistor and the output $V_o$ equals $V_i$. When $V_i$ is negative, the diode is forward biased, and it is equivalent to a short circuit. Therefore, there is no voltage drop across the diode and $V_o$ is 0 V. The resulting output waveform retains the positive peaks in the sine wave and loses the negative peaks (see the figure on the next page).

Since only the positive half of the wave remains, this circuit is known as a ***half-wave rectifier.*** Question 3.2 at the end of the chapter deals with a full-wave rectifier.

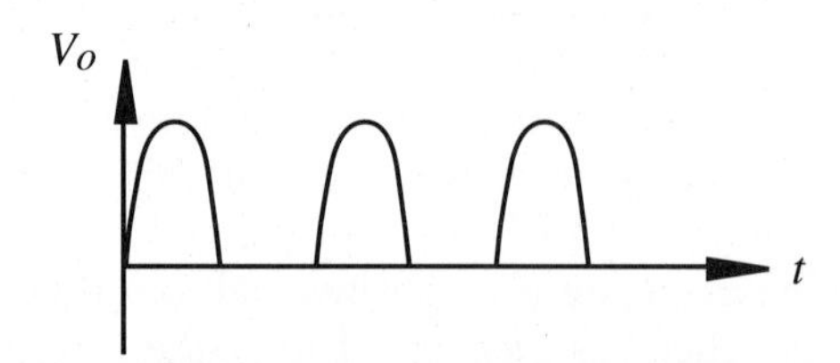


▼▼▼ ***CLASS DISCUSSION ITEM 3.1. Real Silicon Diode in a Half-Wave Rectifier.*** In Example 3.1, we assumed that the diode was ideal. The first approximation to a real diode assumes that 0.7 V is required to forward bias the diode. Using the current voltage relation shown by the dashed curve in Figure 3.5, show how the output of the half-wave rectifier would be changed from that shown in the previous example.

▼▼▼ ***CLASS DISCUSSION ITEM 3.2. Inductive "Kick."*** The circuit shown below illustrates a common application of a diode to reduce current arcs (mini-lightning bolts) between the switch contacts when the switch is opened. Arcs can damage the switch and can create electromagnetic interference (EMI) that can affect surrounding circuits. Why does the switch arc when it is opened? What is the purpose of the diode? Consider the current flow in the inductor and how it can change as a function of time.

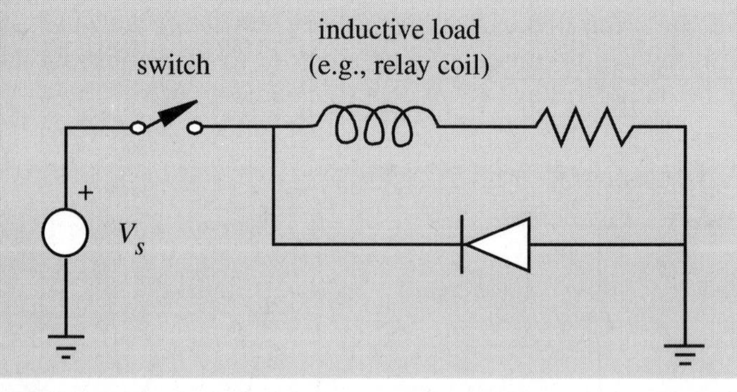


▼▼▼ ***CLASS DISCUSSION ITEM 3.3. Peak Detector.*** The circuit on the next page is known as a ***peak detector.*** When a time-varying signal $V_{in}$ is applied at the input, the output $V_{out}$ retains the maximum positive value of input signal. Under what condition does the capacitor charge? Sketch an arbitrary input signal and the resulting ideal output. What nonideal behavior would you expect from a real circuit where the capacitor is "leaky," i.e., the capacitor's charge is gradually dissipated through a parallel resistance?

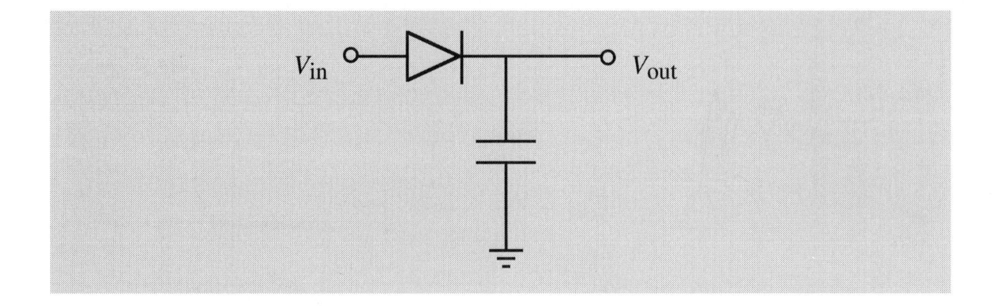


### 3.3.1 Zener Diode

Reflect back on the current-voltage relationship for a diode shown in Figure 3.5. Notice that when a diode is sufficiently reverse biased it will allow a large reverse current to flow and "break down" if the reverse bias voltage is high enough. For most diodes this value is at least 50 V and may extend to kilovolts. There is a special class of diodes that is designed to exploit this characteristic. They are known as ***zener***, avalanche, or voltage-regulator ***diodes***. This family of diodes exhibits steep breakdown curves with well-defined breakdown voltages, and thus they can maintain a nearly constant voltage over a wide range of currents (see Figure 3.6). This characteristic makes them good candidates for building simple voltage regulators, since they can maintain a stable dc voltage in the presence of a variable supply voltage and variable load resistance.

To properly use the zener diode in a circuit, the zener should be reverse biased with a voltage kept in excess of its breakdown or ***zener voltage*** $V_z$. Using a zener diode in series with a resistor as shown in Figure 3.7 results in a simple circuit known as a ***voltage regulator***. The output voltage of the circuit $V_{out}$ is maintained or regulated by the zener diode at the zener voltage $V_z$. Even when the current

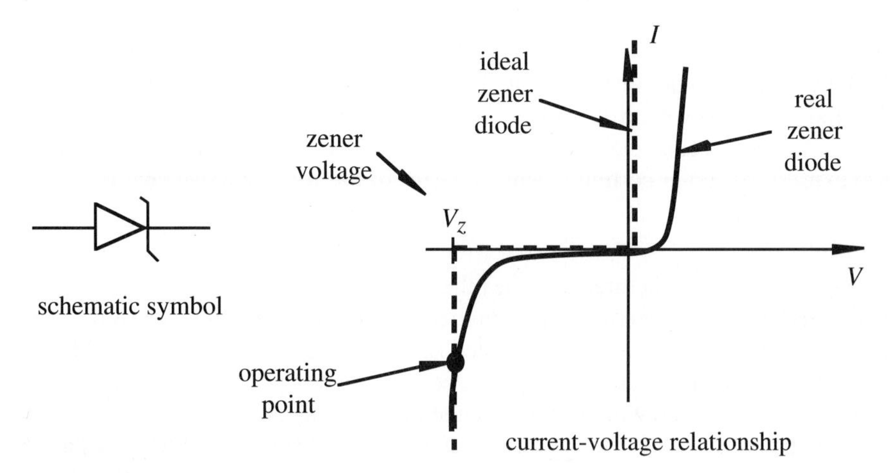


**FIGURE 3.6**
Zener diode symbol and current-voltage relationship.

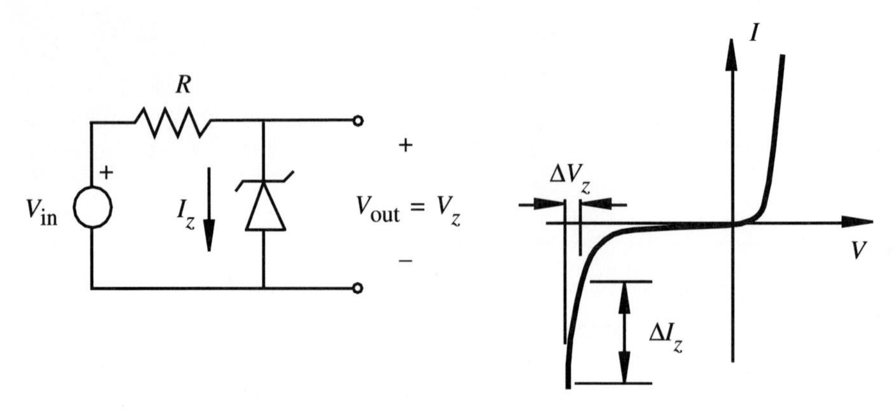


**FIGURE 3.7**
Zener diode voltage regulator.

through the zener diode changes ($\Delta I_z$ in the figure), the output voltage remains relatively constant (i.e., $\Delta V_z$ is small). The narrowness of the voltage range for a given current change is a measure of the voltage regulation of the circuit. If the input voltage and load do not change much, this circuit is effective in obtaining steady and lower dc voltage values from a source that is not well regulated.

Since the load applied to the voltage regulator will change with time in most applications and the voltage source will exhibit fluctuations, careful consideration must be paid to the effect on the regulated voltage $V_z$. For the circuit shown in Figure 3.7, the zener current is related to the circuit voltages according to

$$I_z = \frac{(V_{\text{in}} - V_z)}{R} \tag{3.1}$$

To determine how changes in current are related to changes in voltages, we take the finite differential of Equation 3.1 which yields

$$\Delta I_z = \frac{1}{R}(\Delta V_{\text{in}} - \Delta V_z) \tag{3.2}$$

The zener diode is a nonlinear circuit element, and therefore $\Delta V_z$ is not directly proportional to $\Delta I_z$. However, it is useful to define a dynamic resistance $R_d$ that is the slope of the zener characteristic curve at a particular operating point. This allows us to express the zener current change in terms of the zener voltage change:

$$\Delta I_z = \frac{\Delta V_z}{R_d} \tag{3.3}$$

Normally a manufacturer specifies the nominal zener current $I_{zt}$ and the maximum dynamic impedance at the nominal zener current. In a circuit design using a zener diode the zener current must exceed $I_{zt}$, otherwise the zener may operate near the "knee" of the characteristic curve where regulation is poor (i.e., where there is a large change in voltage with a small change in zener current).

By substituting Equation 3.3 into Equation 3.2 and solving for $\Delta V_z$, we can express changes in the regulator output voltage $\Delta V_{\text{out}}$ in terms of fluctuations in the source voltage $\Delta V_{\text{in}}$:

$$\Delta V_{\text{out}} = \Delta V_z = \frac{R_d}{R_d + R} \Delta V_{\text{in}} \tag{3.4}$$

Therefore, the circuit acts like a voltage divider (for a change in voltage) with the zener diode represented by its dynamic resistance at the operating current of the circuit.

---

▼ ***EXAMPLE 3.2. Zener Regulation Performance.*** We wish to determine the regulation performance of the zener diode circuit shown in Figure 3.7 for a voltage source $V_{\text{in}}$ whose value ranges between 20 and 30 V. For the zener diode we will select a 1N4744A manufactured by National Semiconductor from the family 1N4728A to 1N4752A (having different zener voltage values). It is a 15 V, 1 watt zener diode. We will select a value of $R$ based on the specifications of this diode. To limit the maximum power dissipation to less than 1 watt, the current through the diode must be limited to

$$I_{z_{\max}} = 1 \text{ W}/15 \text{ V} = 66.7 \text{ mA}$$

Therefore, using Equation 3.1, the value for resistance $R$ should be chosen to be at least

$$R_{\min} = (V_{\text{in}_{\max}} - V_z)/I_{z_{\max}} = (30 \text{ V} - 15 \text{ V})/66.7 \text{ mA} = 225\ \Omega$$

The closest acceptable standard resistance value is 240 Ω. From the manufacturer's specifications of this zener diode, its dynamic resistance $R_d$ is 14 Ω at 17 mA. The current $I_z$ in this example is larger than this value, so the operating point of the zener diode is on the well-regulated portion of the characteristic curve. Using the given value for $R_d$ in Equation 3.4, we can approximate the resulting output voltage range:

$$\Delta V_{\text{out}} = \Delta V_z = \frac{R_d}{R_d + R} \Delta V_{\text{in}} = \frac{14}{14 + 240}(30 - 20)\ V = 0.55\ V$$

which is a measure of regulation of this circuit. This can be expressed as a percentage of the output voltage for a relative measure:

$$\frac{\Delta V_{\text{out}}}{V_{\text{out}}} 100\% = \frac{0.55 \text{ V}}{15 \text{ V}} 100\% = 3.7\%$$

---

▼▼▼ ***CLASS DISCUSSION ITEM 3.4. Effects of Load on Voltage Regulator Design.*** Example 3.2 ignored the current that would be drawn by a load. What effect would a load have on the results of the analysis?

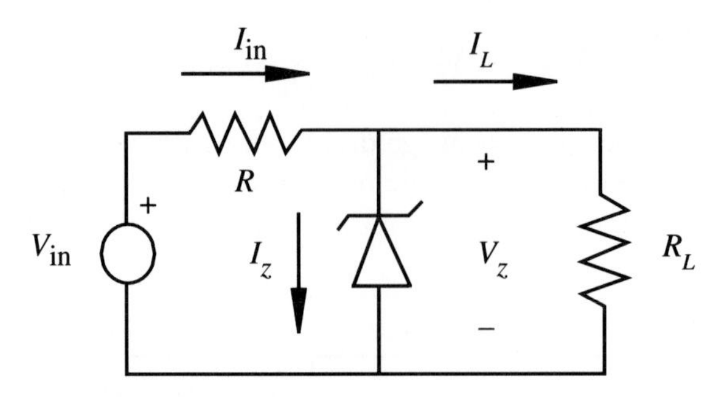


**FIGURE 3.8**
Zener diode voltage regulator circuit.

Figure 3.8 illustrates a simple voltage regulator circuit where $R_L$ is a load resistance and $V_{in}$ is an unregulated source whose value exceeds the zener voltage $V_z$. The purpose of this circuit is to provide a constant dc voltage $V_z$ across the load with a corresponding constant current through the load. Providing a stable regulated voltage to a system containing digital integrated circuits is a common application.

If we assume the zener diode is ideal (its breakdown current-voltage curve is vertical), we can draw some conclusions about the regulator circuit. First, the load voltage will be $V_z$ as long as the zener diode is subject to reverse breakdown. Therefore, the load current $I_L$ is

$$I_L = \frac{V_z}{R_L} \tag{3.5}$$

Second, the load current will be the difference between the unregulated input current $I_{in}$ and the zener diode current $I_z$:

$$I_L = I_{in} - I_z \tag{3.6}$$

As long as $V_z$ is constant and the load doesn't change, $I_L$ will remain constant. This means that the diode current will change to absorb changes from the unregulated source. Third, the unregulated source current $I_{in}$ is given by

$$I_{in} = \frac{(V_{in} - V_z)}{R} \tag{3.7}$$

$R$ is known as a current-limiting resistor since it limits the power dissipated by the zener diode. If $I_z$ gets too large, the zener diode will fail.

---

▼▼ ***DESIGN EXAMPLE 3.1. Zener Diode Voltage Regulator Design.*** Suppose we need to design a regulated 15 V dc source to power a mechatronic system, and we would like to use the voltage regulator circuit shown in Figure 3.8. Furthermore, suppose we only have access to a poorly regulated dc source $V_{in}$ whose nominal value is 24 V.

As the load $R_L$ changes, the zener current $I_z$ will increase for larger $R_L$ and decrease for smaller $R_L$. If we know the maximum possible load resistance (assuming that the output will never be an open circuit), we can size the zener diode with regard to its power dissipation characteristics and select a current-limiting resistor. Combining Equations 3.5 and 3.6 and using the maximum value of the load $R_{L_{max}}$ gives

$$I_{z_{max}} = \left(I_{in} - \frac{V_z}{R_{L_{max}}}\right)$$

This is the largest current the zener will experience. The power dissipated by the zener diode is

$$P_{z_{max}} = I_{z_{max}} V_z = \left(I_{in} - \frac{V_z}{R_{L_{max}}}\right) V_z$$

$I_{in}$ is controlled by the current-limiting resistor $R$. Substituting Equation 3.7 yields

$$P_{z_{max}} = \left(\frac{V_{in} - V_z}{R}\right) V_z - \frac{V_z^2}{R_{L_{max}}}$$

Furthermore, for this design problem, we assume that $R_{L_{max}}$ is 240 Ω and we wish to select a 1 W zener. Therefore,

$$1\text{W} = \frac{24\text{V} - 15\text{V}}{R_{min}}(15\text{V}) - \frac{225\text{V}^2}{240\Omega}$$

We can now solve for the minimum required current-limiting resistance $R$:

$$R_{min} = 69.7\ \Omega$$

The closest acceptable standard resistance value is 75 Ω.

---

In summary, zener diodes are useful in circuits where it is necessary to derive smaller regulated voltages from a single higher voltage source. When designing zener diode circuits, one must select appropriate current-controlling resistors given the power limitations of the diodes. In simple mechatronic designs that may be powered by a 9 V battery and require good 5 V dc supplies for digital devices, a well-designed zener regulator will be a cheap and effective solution if the current requirements are modest.

### 3.3.2 Voltage Regulators

Although the zener diode voltage regulator is cheap and simple to use, it has some drawbacks: The output voltage cannot be set to a precise value, and regulation against source ripple and changes in load is limited. Special semiconductor devices are designed to serve as voltage regulators, some for fixed positive or negative values and others which are easy to adjust to a desired, nonstandard value. One group

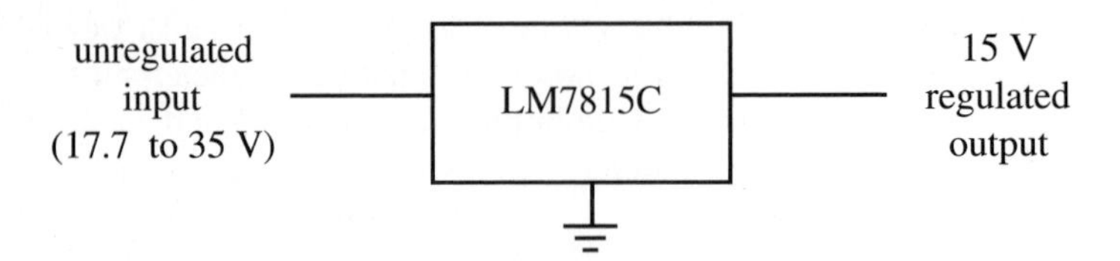


**FIGURE 3.9**
15 V regulated dc supply.

of regulators that is easy to use is the three-terminal regulator designated as the 78XX, where the last two digits (XX) specify a voltage with standard values: 5 (05), 12, or 15 V. Using a regulator such as the LM7815C, a well-regulated 15 V source is easy to create, as shown in Figure 3.9.

We could use this design instead of the zener regulator shown in Design Example 3.1 (see Class Discussion Item 3.5). The 78XX can deliver up to 1 amp of current and is internally protected from overload. Using this device, the designer doesn't have to perform the design calculations shown with the zener diode regulator. The 78XX series of regulators have complementary 79XX series values for the design of +/– voltage supplies.

▼▼▼ ***CLASS DISCUSSION ITEM 3.5. 78xx Series Voltage Regulator.*** In Design Example 3.1 we used a zener diode to provide a desired dc voltage. Now show how a 78XX voltage regulator can do the same job. Specify the regulator and describe its characteristics.

In some cases you may need a regulated voltage source with a value not provided in a manufacturer's standard sequence. Then you may use a three-terminal regulator designed to be adjustable by the addition of external resistors. The LM317L can provide an adjustable output with the addition of two external resistors as shown in Figure 3.10. The output voltage is given by

$$V_{out} = 1.25\left(1 + \frac{R_2}{R_1}\right) \text{ V} \tag{3.8}$$

These adjustable regulators are available in higher current and voltage ratings.

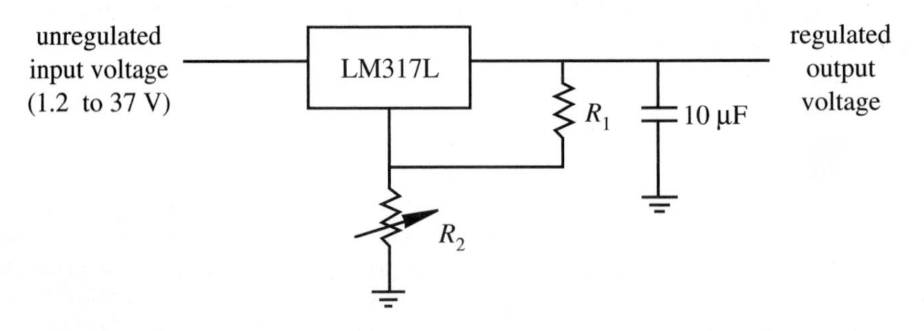


**FIGURE 3.10**
1.2 to 37 V adjustable regulator.

In summary, three-terminal voltage regulators are accurate, reject ripple on the input, reject voltage spikes, have roughly a 0.1% regulation, and are quite stable, making them great for mechatronic system design.

▼▼▼ ***CLASS DISCUSSION ITEM 3.6. Automobile Charging System.*** The typical automobile has a 12 V dc electrical system where a lead-acid battery is charged by a belt-driven ac alternator whose frequency and voltage vary with engine speed. What type of signal conditioning must be performed between the alternator and the battery and how can this be done?

### 3.3.3 Optoelectronic Diodes

You have seen the little red lights that blink and dazzle in computer rooms and big control consoles. They are really ***light-emitting diodes (LEDs)***, diodes that when forward biased emit photons. The typical LED and its schematic symbol are illustrated in Figure 3.11. The positive lead, or anode, is usually the longer of the two leads. The LED is usually encased in a colored plastic material which enhances the wavelength generated by the diode and sometimes helps focus the light into a beam. The intensity of light is related to the amount of current flowing through the device. LEDs are manufactured in red, yellow, and green packages and more recently blue, but the latter are more expensive. It is important to remember that an LED has a voltage drop of 1.5 to 2.5 volts when forward biased, somewhat more than small signal silicon diodes. It only takes a few milliamps of current to see the diode, and it is important to include a series current-limiting resistor in the circuit to prevent excess forward current, which can quickly destroy the diode. Usually a 330 ohm resistor is used in series with the LED when used in digital (5 V) circuit designs.

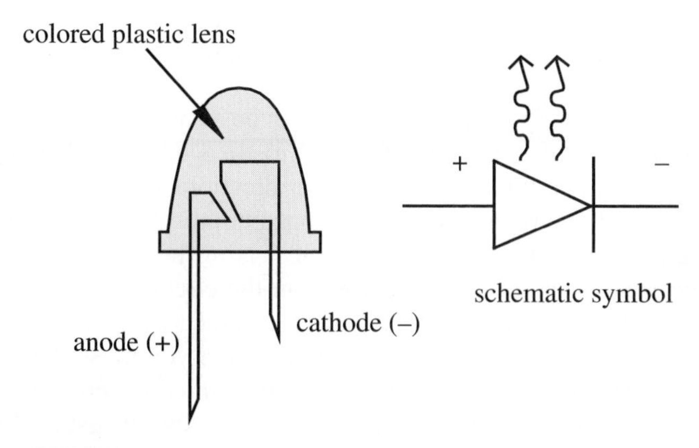


**FIGURE 3.11**
Light-emitting diode (LED).

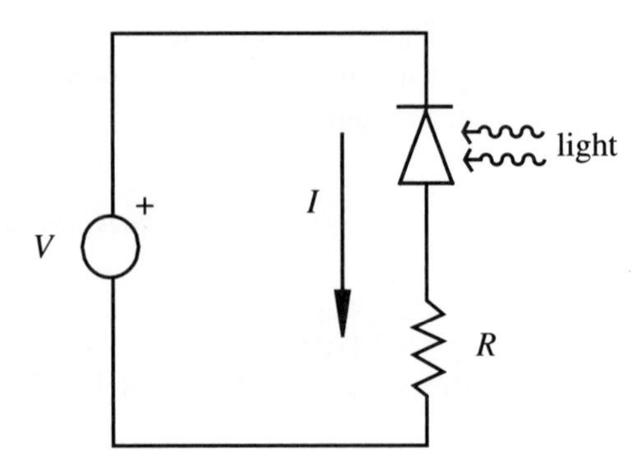


**FIGURE 3.12**
Photodiode light detector circuit.

Earlier we said that a pn junction is sensitive to light. Special diodes, called ***photodiodes,*** are designed to detect photons and can be connected to resistors and used in circuits to sense light as shown in Figure 3.12. Notice that it is the reverse current that flows through the diode when sensing light. It takes a considerable number of photons to provide detectable voltages with these devices, and as we shall see in the transistor section, the photo transistor may be a more sensitive device, although it will be slower to respond. The photodiode is based on quantum effects. If photons excite carriers in a reverse biased pn junction, a very small current proportional to the light intensity will flow. The sensitivity is affected by the wavelength of the light. In Chapter 4, we will show how to use the diode to make a light intensity detector.

### 3.3.4 Analysis of Diode Circuits

Although most of your analyses will include single isolated diodes in circuits, there are situations when you will design a circuit containing multiple diodes. Since the diode is a nonlinear device, one has to be careful not to naively apply the linear circuit analysis methods we have discussed so far.

Dc circuits that contain many diodes may not be easy to analyze by inspection. The following procedure is a straightforward method to determine voltages and currents in these circuits. First, assume current directions for each circuit element. Next, replace each diode with an equivalent open circuit if the assumed current is in a reverse bias direction or a short circuit if it is in the forward bias direction. Then compute the voltage drops and currents in the circuit loops using KVL and KCL. If the sign on a resulting current is opposite to the assumed direction through an element, you have made the wrong assumption and you must change its direction and reanalyze the circuit. Repeat this procedure with different combinations of current directions until there are no inconsistencies between assumed and calculated voltage drops and currents.

▼ ***EXAMPLE 3.3. Analysis of Circuit with More Than One Diode.*** This example illustrates the application of the procedure outlined above to a circuit containing two ideal diodes. In the circuit given below, we want to determine all currents and voltages. We begin by arbitrarily assuming the current directions as shown.

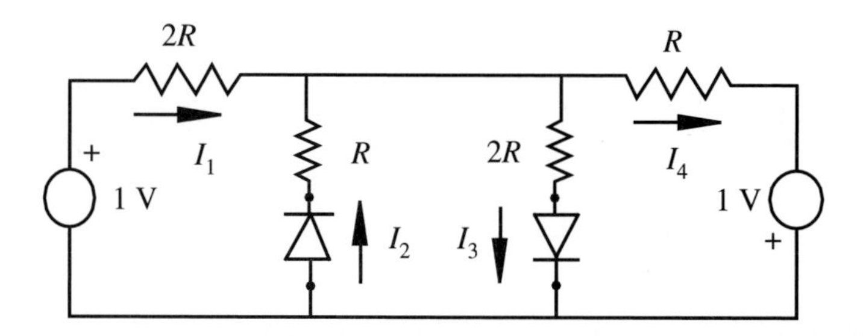


Having assumed the current directions, we replace each diode with a short, since each is assumed to be forward biased. The equivalent circuit is shown below.

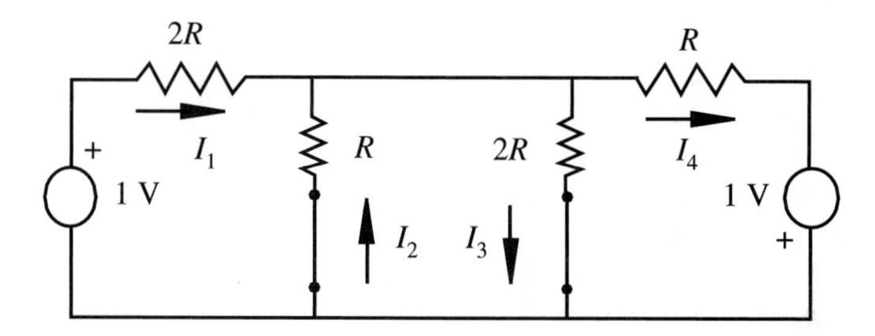


By applying KVL to the loop containing $I_2$ and $I_3$, we find that $I_2 = -2I_3$. We conclude that one of the current directions was incorrectly assumed. Therefore, we need to change one of our initial assumptions. Assume $I_2$ is in the direction opposite to that first chosen. With this assumption, the diode must be replaced with an open circuit since it is reverse biased. Now, the equivalent circuit is as shown next.

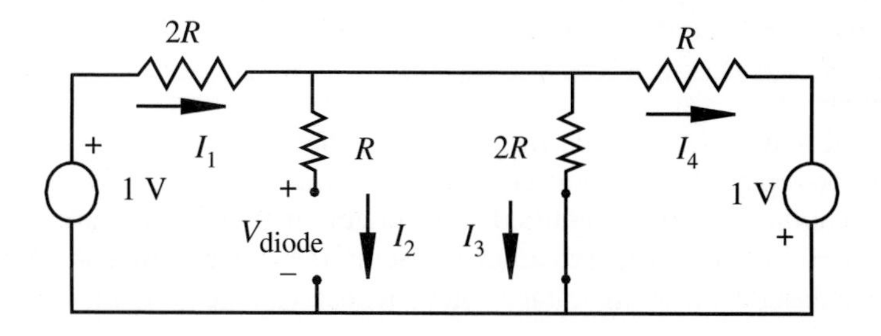

Note that $I_2 = 0$ in this circuit and $V_{diode}$ is the voltage across the diode. By applying KVL to the loop containing $I_3$ and $I_4$, the result is that $I_3 < 0$. Therefore, our assumed direction for $I_3$ is incorrect. We must reverse $I_3$ and replace the diode with an open circuit. The resulting circuit is shown.

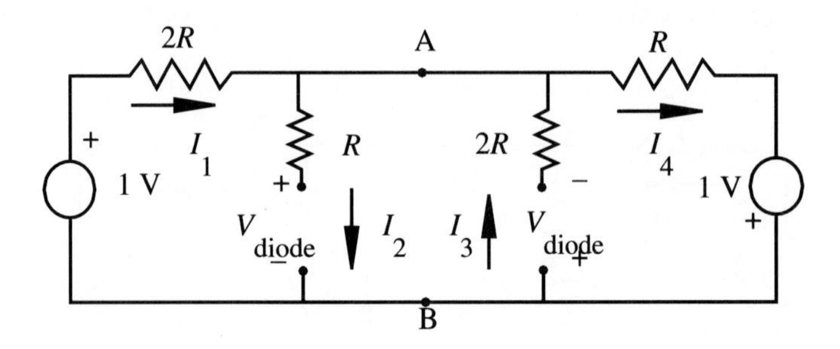


Note that $I_2$ and $I_3$ are both zero in this circuit, and each diode has a nonzero voltage across its terminals. Since the voltage at node A is positive with respect to node B (from KVL around the outside loop), the assumed bias on the right diode is incorrect. Therefore, the next choice for assumed current directions, which is the only combination we have not investigated, is shown below.

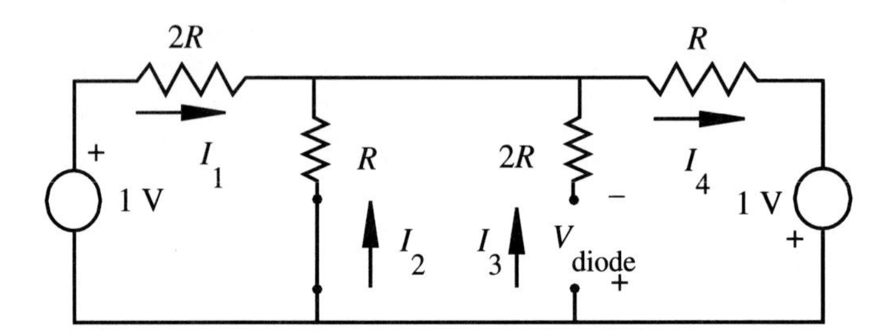


If we analyze this circuit (see Question 3.4), we find that $I_2 > 0$ and $V_{diode} > 0$ as assumed. Therefore, there are no inconsistencies and our results are correct.

In this example, we performed an exhaustive search to check every possible combination of diode biasing. If we had chosen the correct combination earlier in the procedure, by luck or by an educated guess, the exhaustive search would not have been required.

---

This procedure and example have assumed that when a diode is forward biased, it can be replaced by a short circuit, which implies a zero forward bias voltage. The procedure has to be modified to accurately model real diodes. To account for the forward bias voltage, instead of replacing the diode with a short, it would have to be replaced by a small voltage source whose voltage is equal to the forward bias value of the diode.

▼▼▼ ***CLASS DISCUSSION ITEM 3.7. Voltage Limiter.*** The circuit below is called a voltage limiter. Explain why. Sketch some input and output waveforms that illustrate the circuit's behavior.

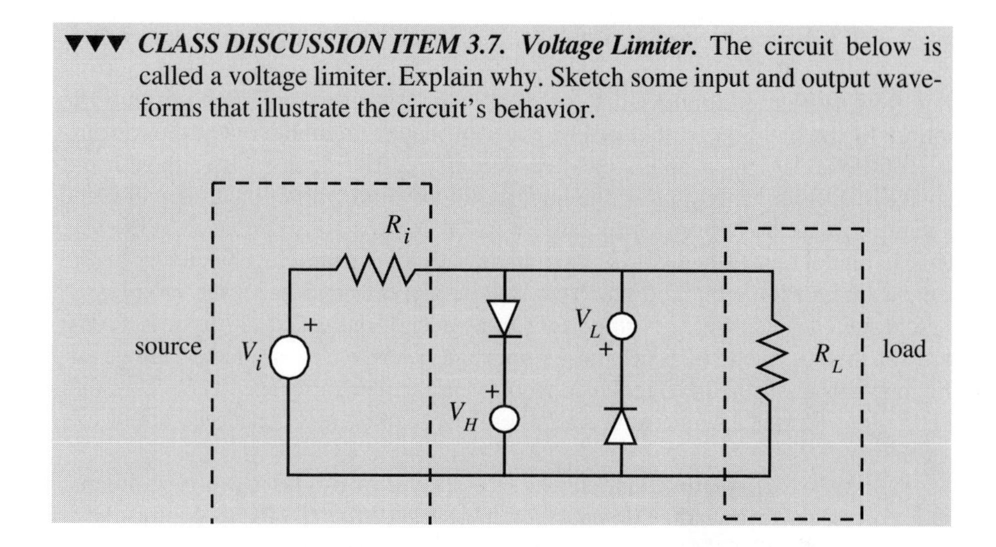


# 3.4 THE BIPOLAR JUNCTION TRANSISTOR

The transistor has truly revolutionized human existence in the ways we go about our everyday lives. We begin by providing the physical foundations necessary to understand the function of transistors.

## 3.4.1 Understanding the Physics of a Bipolar Transistor

In our earlier discussion of semiconductor materials, we took a rather broad approach to the description of the physics of the diode to provide an unintimidating introduction. Now we will move on to other semiconductor devices where more than two layers of doped silicon are used, and whose complexity contributes to their usefulness in electronics and in mechatronic system design. We extend the discussion of the physics with the following subtlety: Semiconductor material regions in a device can be divided into two groups based on how current is conducted through the material. An electronic device in which the current is made up of the charges dominant in the lattice (i.e., electrons in n-type material or holes in p-type material) is called a ***majority carrier device.*** If the current is made up of the charge not dominant in the lattice (i.e., electrons in p-type material or holes in n-type material), the resulting device is called a ***minority carrier device.*** This very important distinction helps explain the operation of a bipolar transistor and also has a large bearing on the device's ability to change state rapidly.

A fish tank can be used to illustrate the concept of charge carriers and their effect on the speed of response of the device. The analogy can be used to illustrate how current is conducted through silicon. If you put an air hose onto the bottom of the fish tank, it takes some time for the air to bubble out at the top surface of the tank. This is analogous to a minority carrier device where the current carrier is different from that of the bulk medium. On the other hand, if you place a water hose at the bottom of the tank, water is immediately displaced at the top. This is analogous to a majority carrier device where the current carrier is the same as that of the bulk medium. This will help you to remember the mobility of the charge carriers, which is important in assessing the speed of the device. We will see that n-type devices are more important because they are faster. However, the mechatronic system designer may not have the switching speed concerns that a computer designer would have and, therefore, can be less concerned about this problem.

We saw earlier that a diode consists of a layer of p-type and n-type silicon, each connected to a lead. A ***bipolar junction transistor (BJT),*** in contrast, is a three-terminal device. In an npn BJT, two layers of n-type silicon sandwich a layer of p-type silicon. In a pnp BJT, two layers of p-type silicon sandwich a layer of n-type silicon. The npn is the more common of the two. Figures 3.13 and 3.14 illustrate the construction of npn and pnp bipolar transistors. Also shown are the circuit symbols and associated voltage and current nomenclature. The middle controlling lead of a BJT is known as the ***base,*** and the other two leads are known as the ***collector*** and ***emitter.*** $V_{CE}$ is the voltage between the collector and emitter, and $V_{BE}$ is the voltage between the base and emitter. The relationships involving the transistor currents and voltages follow:

$$I_E = I_C + I_B \tag{3.9}$$

$$V_{BE} = V_B - V_E \tag{3.10}$$

$$V_{CE} = V_C - V_E \tag{3.11}$$

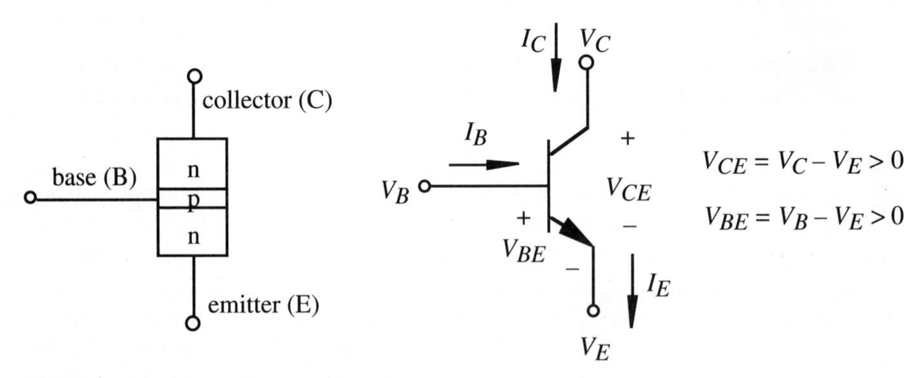


**FIGURE 3.13**
npn Bipolar junction transistor.

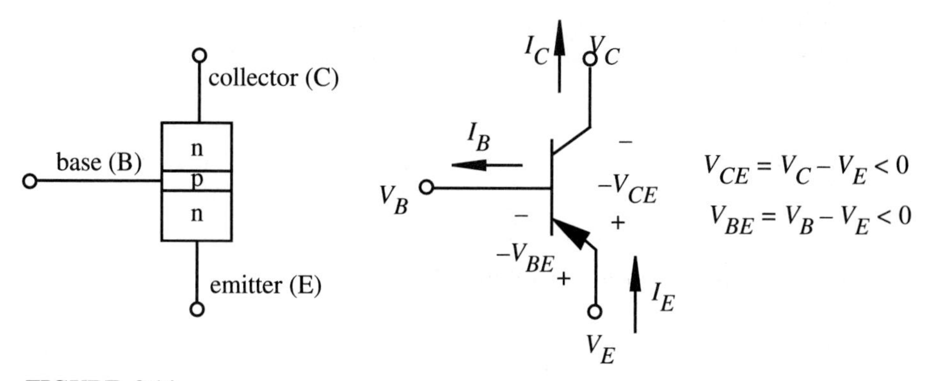


**FIGURE 3.14**
pnp Bipolar junction transistor.

For a basic understanding, a transistor can be thought of as a variable resistor between two leads. A small current injected into a third lead known as the ***base*** can control the magnitude of a much larger current flowing between the other two leads known as the ***collector*** and ***emitter***. The property that a small base current can control a larger collector current is what allows the transistor to act as an amplifier. Also, the base current controls when the collector current switches on or off. Thus a transistor can linearly amplify current or it can simply switch current. This current-switching action is the basis for most digital computers producing a binary number representation. It is important to distinguish the two principal modes of action, switching and amplification, of the transistor. We will focus primarily on switching in mechatronic applications. Amplifiers are the realm of the electrical engineer. The reason an engineer who plans to use a lot of integrated circuits (where you don't have to analyze the function of each of the thousands of transistors) needs to understand transistors is that the transistor will provide a common interface from ICs to the world and will be used as an electronic switch in a number of support components in mechatronic systems.

To understand the physics of the bipolar transistor, consider an npn bipolar transistor. The base and emitter make up a pn junction much like a diode. If the "diode" is forward biased, i.e., the base-to-emitter voltage is 0.7 V, the depletion region shrinks and electrons move from the n region to the p region where the holes exist. The electrons are minority carriers in the p region. The collector is more positive than the base (the base-collector diode is reverse biased), and although some electrons are attracted to the base making up the base current, most of the electrons accelerate toward the higher potential collector. Thus, the smaller base current controls the larger collector current. Therefore, the bipolar transistor is a current amplifier, and this characteristic can be approximated with the following equation:

$$I_C = \beta I_B \tag{3.12}$$

which says that the collector current is equal to the base current times a current amplification factor know as the ***beta*** for the transistor. For typical BJTs, beta is on the order of 100. The problem is that the variation in manufacturing specifications is rather large, and also beta is very temperature dependent. To make linear amplifiers, we have to go to a lot of trouble to obviate these problems. So if you are using discrete transistors as linear amplifiers, you will have to study more electrical engineering. However, as we will see now, there are some important applications that can be easily executed.

### 3.4.2 The Common Emitter Transistor Circuit

If we consider a bipolar transistor with the emitter grounded, an input voltage applied to the base, and the output current measured in the collector, we have a ***common emitter*** circuit as shown in Figure 3.15. The characteristics of the common emitter circuit, and therefore the transistor, are best described by plotting the collector current $I_C$ versus the collector-emitter voltage $V_{CE}$ for different values of base current $I_B$. The resulting family of curves (see Figure 3.16) describe the ***common emitter characteristics*** for the transistor. For a silicon BJT, as the base current is gradually increased, the base-to-emitter diode of the transistor will turn on when $V_{BE}$ is about 0.6 V. At this point $I_C$ will begin to flow and be proportional to $I_B$ ($I_C = \beta I_B$). As $I_B$ is further increased, $V_{BE}$ will slowly increase to 0.7 V but $I_C$ will rise exponentially. As $I_C$ rises, the voltage drop across $R_C$ will increase and $V_{CE}$ will drop toward ground. The collector cannot drop below ground (otherwise, the collector-to-base diode would also be forward biased), so at this point the transistor is said to go into saturation—meaning that the collector current is determined by $R_C$ and the linear relation between $I_C$ and $I_B$ no longer holds.

Hence, the transistor has a ***cutoff region*** (no collector current flows), an ***active region*** (where collector current is proportional to base current), and a ***saturation region*** (where collector current is strictly controlled by the collector circuit, assuming sufficient base current). When designing a transistor switch, we will need to guarantee that the transistor is fully in saturation when it is on to prevent serious heating of the transistor ($I_C V_{CE}$). For BJTs, $V_{CE}$ at saturation is about 0.2 V.

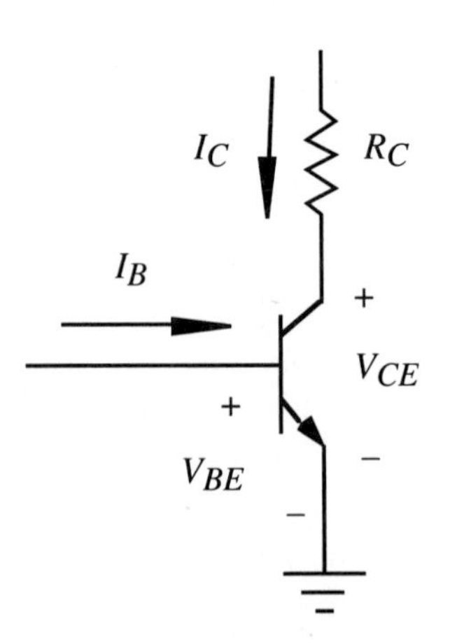


**FIGURE 3.15**
Common emitter circuit.

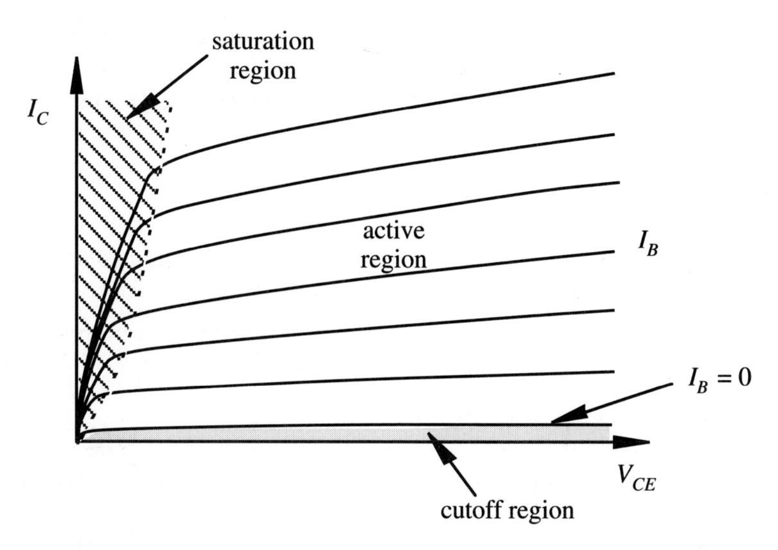


**FIGURE 3.16**
Common emitter characteristics for a transistor.

---

▼ ***EXAMPLE 3.4. Guaranteeing that a Transistor is in Saturation.*** A small signal transistor that is very useful is the 2N3904 manufactured by many companies as a general purpose amplifier and switch. Examine the specifications in a discrete transistor handbook if you have time. You will find a figure displaying the packages and the ratings and electrical characteristics. Here is some of the other information provided:

- Maximum collector current (continuous) = 200 mA
- $V_{CE}$(sat) = 0.2 V
- $h_{FE}$ = beta = 100 (depending on collector current and many other things)

In the circuit below, what minimum input voltage $V_{in}$ is necessary to saturate the transistor?

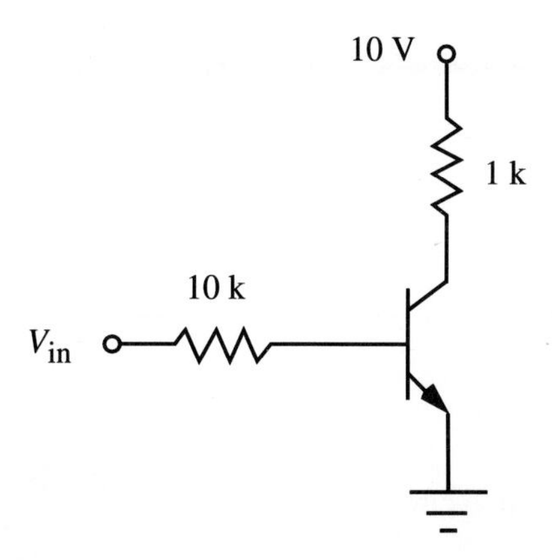

Since $V_{CE}$(sat) for the 2N3904 is 0.2 V, the collector current must be 9.8 mA as the transistor goes into saturation. Since the dc current gain $h_{FE}$ for $I_C = 10$ mA is about 100, $I_B$ must be $I_C/100$ or 0.098 mA. To ensure this base current,

$$I_B = 0.098 \text{ mA} = (V_{in} - 0.7 \text{ V})/10 \text{ k}\Omega$$

This implies $V_{in}$ minimum is at least 1.68 V. Normally you would increase this by a factor to ensure that the transistor is fully saturated for small fluctuations in parameters.

---

### 3.4.3 The Bipolar Transistor Switch

Figure 3.17 illustrates a simple transistor switch circuit. When $V_{in}$ is less than 0.7 V, the *BE* junction of the transistor is not forward biased ($V_{BE} < 0.7$ V) and the transistor does not conduct ($I_C = I_E = 0$). You can therefore assume that the collector-to-emitter circuit can be replaced by a very high impedance or, for all practical purposes, an open circuit. This state is illustrated in Figure 3.18a and is referred to as the cutoff or "OFF" state of the transistor. In cutoff, the output voltage $V_{out}$ will be $V_C$ since there is no current through or voltage drop across $R_C$.

When the *BE* junction is forward biased ($V_{BE} = 0.7$ V), the transistor conducts. Current passes through the *CE* circuit, and $V_{out}$ is close to ground potential (0.2 V for a bipolar transistor). This state is modeled by the forward biased diode illus-

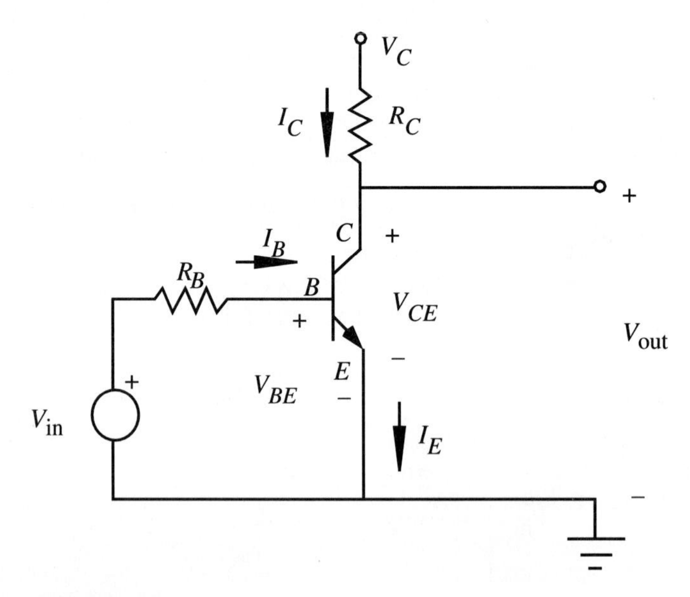


**FIGURE 3.17**
Transistor switch circuit.

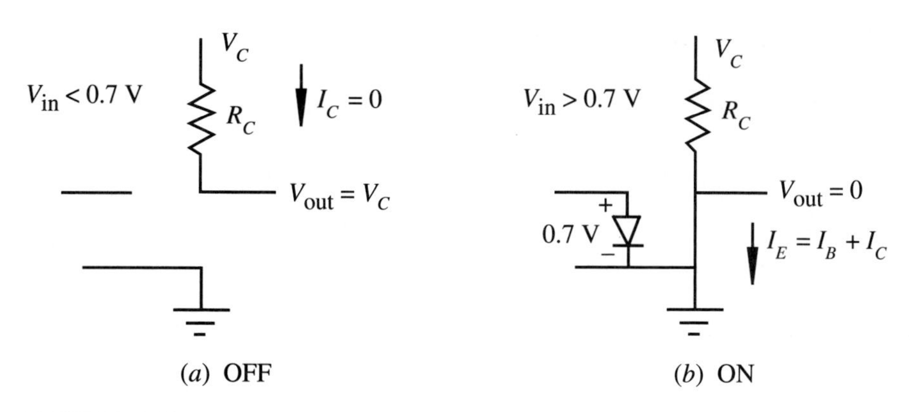


**FIGURE 3.18**
Models for transistor switch states.

trated in Figure 3.18b and is referred to as the saturated or "ON" state of the transistor. We assume that there is enough base current to saturate the transistor. The resistor $R_B$ is required in this circuit to limit the base current since the *BE* junction essentially behaves like a diode. The relationship between the base current and $R_B$ is given by

$$I_B = (V_{in} - V_{BE})/R_B \tag{3.13}$$

When $V_{in} < 0.7$, $I_B = 0$ and $V_{BE} = V_{in}$.

The circuit in Figure 3.17 can serve as a semiconductor switch to turn on or off an LED, electric motor, solenoid, electric light, or some other load (represented by $R_C$ in the figure). These loads may require hundreds of milliamps or many amps to function properly. When the input voltage and current are increased enough to saturate the transistor, large collector currents will flow through the load. The magnitude of the collector current is determined by the load resistance $R_C$ and the collector voltage $V_C$. When the input voltage is below 0.7 V, the transistor will be off and no current will flow through the load. The transistors used in power applications are called ***power transistors*** and are designed to conduct large currents in the collector circuit. Power transistors are the basis for interfacing low output current devices such as ICs and computer ports to other devices requiring large currents.

---

▼▼ ***DESIGN EXAMPLE 3.2. LED Switch.*** Our objective is to turn a dashboard LED on or off with a digital device with an output voltage of either 0 V or 5 V and a maximum output current of 5 mA. The LED requires 40–50 mA to provide a bright display. We will use a transistor switch circuit employing a small signal transistor (e.g., 2N3904 npn) to provide sufficient current to the LED. The required circuit follows.

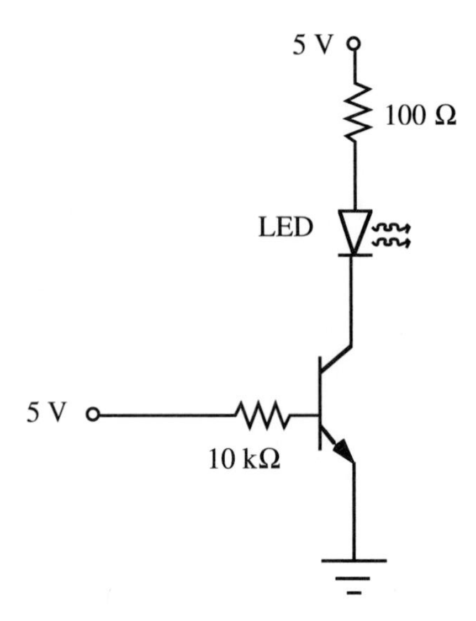


When the digital output is 0 V, the transistor is in cutoff and the LED will be OFF. When the digital output is 5 V (as shown in the figure above), the transistor is in saturation and the base current will be

$$0.43 \text{ mA} = (5 \text{ V} - 0.7 \text{ V})/10 \text{ k}\Omega$$

which is within the specifications. The 100 Ω collector resistance limits the LED current to a value within the desired range for the LED to be bright:

$$48 \text{ mA} = (5 \text{ V} - 0.2 \text{ V})/100 \ \Omega$$

---

Let us summarize the guidelines for making a transistor switch. The collector must be more positive than the base or emitter. To be ON, the base-to-emitter voltage must be 0.7 V. There are maximum values of $I_C$, $I_B$, and $V_{CE}$ that must be calculated and held to specifications. The collector current $I_C$ is independent of base current $I_B$ when the transistor is saturated, but there must be enough base current to ensure saturation. Be careful that the base-to-emitter voltage is not too large, resulting in excess base current; that is why the transistor should not be driven directly from a voltage source.

### 3.4.4 Bipolar Transistor Packages

When you select transistors from a manufacturer for a mechatronic system prototype, you will find that they come in a number of packages as illustrated in Figure 3.19. The small signal transistor packages are normally the TO-92 and TO-226, and the power transistor packages are the TO-220. Surface mount technology is becoming increasingly popular for use on production printed circuit boards, but such devices are less useful for prototyping because of their small size.

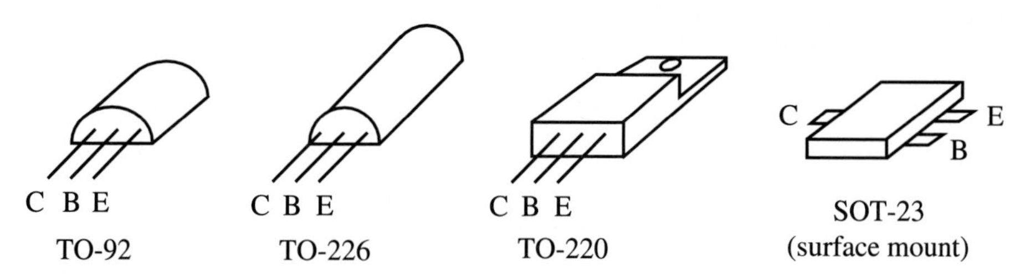


**FIGURE 3.19**
Bipolar transistor packages.

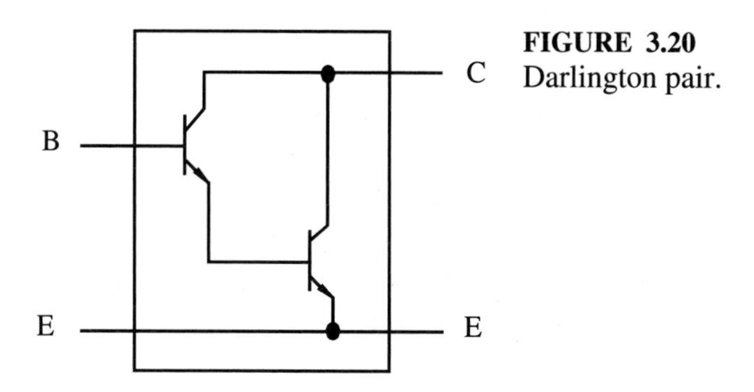


**FIGURE 3.20**
Darlington pair.

### 3.4.5 Darlington Transistor

The schematic in Figure 3.20 represents a transistor known as a ***Darlington pair***, which usually comes in a single package. The advantage of this combination is that the current gain is the product of the two individual transistors and many exceed 10,000. They may often be found in power circuits for mechatronic systems.

### 3.4.6 The Phototransistor and Opto-Isolator

A special class of transistor is the ***phototransistor***, whose junction between the base and emitter acts as a photodiode (see Section 3.3.3). LEDs and phototransistors are often found in pairs where the LED is used to create the light and light in turn biases the phototransistor. The pair can be used to detect the presence of an object that may partially or completely interrupt the light beam between the LED and phototransistor.

An ***opto-isolator*** is composed of an LED and a phototransistor separated by a small gap as illustrated in Figure 3.21. The light emitted by the LED causes current to flow in the phototransistor circuit. This output circuit has a different

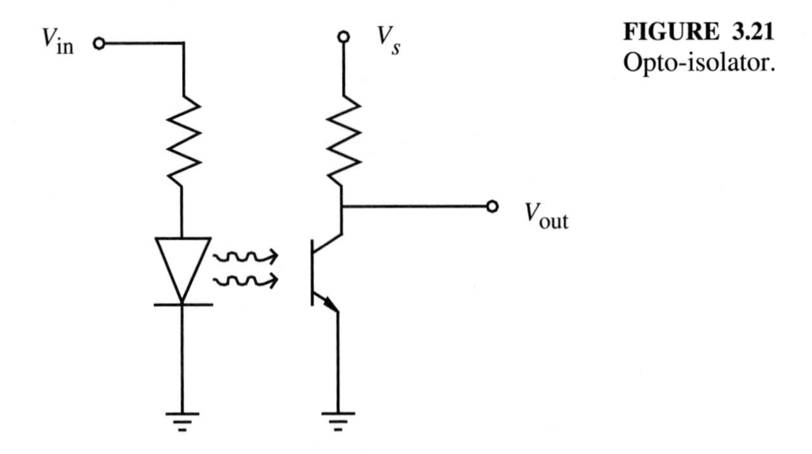


**FIGURE 3.21**
Opto-isolator.

ground reference, and the supply voltage $V_s$ can be chosen to establish a desired output voltage range. The opto-isolator creates a state of electrical isolation between the input and output circuits by transmitting the signal optically rather than through an electrical connection. A benefit of this isolation is that the output is protected from any excessive input voltages that could damage components in the output circuit.

---

▼▼ ***DESIGN EXAMPLE 3.3. Angular Position of a Robotic Scanner.*** This design example will illustrate a use of semiconductor optoelectronic components. Suppose in the design of an autonomous robot, you wish to include a laser scanning device to sweep the environment to detect obstacles. The head of the scanner is rotated through 360 degrees by a dc motor. Your problem here is to track the angular position of the scan head. How could you do this if you want to use the sensed values later in a microprocessor computation?

The solution requires a sensor that provides a digital output, i.e., one that can be handled by a digital computer. We will learn more about digital interfacing in Chapter 6. To keep the solution simple at this point, we choose a device that produces a 5 V or digital output. An LED-phototransistor pair, also known as a ***photo-interrupter***, will be at the heart of the design, which is illustrated in the figure on the next page. The pair, which is readily available in a single package, produces a beam of light that can be broken or interrupted. A slotted disk must be designed to attach to the shaft of the motor driving the scan head and to pass through the gap in the photo-interrupter pair. Each slot in the disk will provide a digital pulse as it interrupts the light beam during rotation.

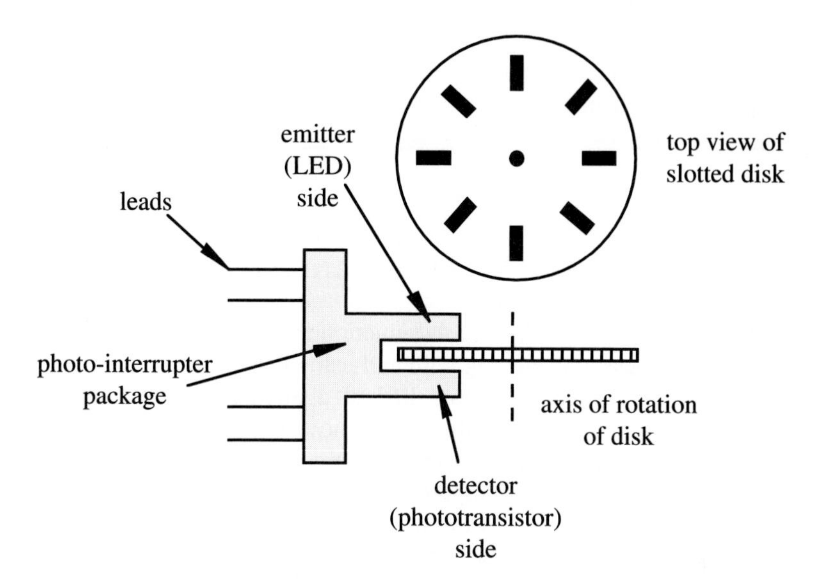


To cause the sensor to function properly, we must add the external components shown in the figure below to provide a digital pulse each time a slot is encountered. The emitter LED is powered with 5 V dc and a current limiting resistor $R_1$. The detector is a phototransistor in the same package to which is added a ***pull-up resistor*** $R_2$ to provide the output.

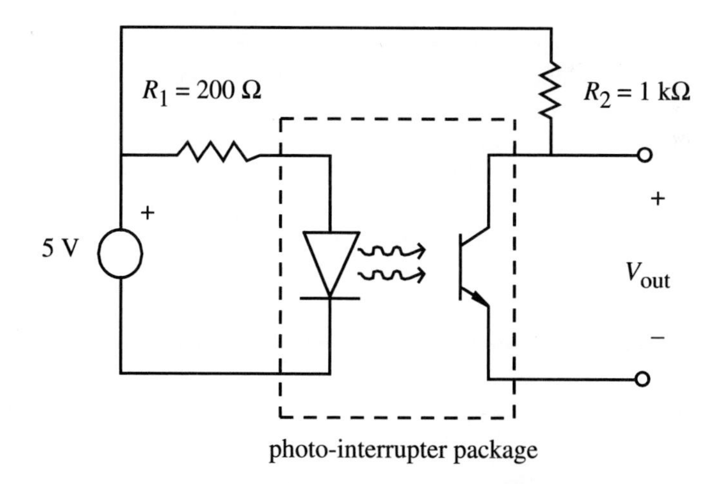


As the slotted disk rotates, light will pass through each slot producing a 0 V output, and then return to 5 V when the segments between the slots interrupt the light. The result is a train of pulses. The number of pulses produced provides the measure of angular rotation as a digital approximation. For example, if the disk had 360 slots, each pulse would correspond to 1 degree of rotation. Providing a zero reference, adding and subtracting pulses, and interfacing to a microprocessor will be left for later when you are prepared to deal with those problems.

## 3.5 FIELD EFFECT TRANSISTORS

Now that we have examined the bipolar junction transistor (BJT) and hopefully enhanced your eagerness in designing with and using discrete semiconductor devices, we move on and make life a little more difficult again. But bear with us, as the additional semiconductor devices we will consider now will further improve your understanding of modern mechatronic systems and tremendously enhance your strength as a designer. We now consider the ***field effect transistor*** or ***FET*** which, unlike the bipolar transistor, controls current through two terminals with an electric field produced by a voltage at a third terminal. The microscopic cross section of a typical FET is shown in Figure 3.22. Current can flow between two terminals of the FET, called the ***drain*** and the ***source,*** but the current is controlled by an electric field produced by the third terminal called the ***gate.*** The silicon substrate (the p-type material in this case) is referred to as the ***base.*** A positive charge on the gate attracts electrons in the substrate, forming a conducting region called a ***channel*** between the source and drain allowing current to flow. The source, gate, and drain of an FET are analogous to a BJT's emitter (source), base (gate), and collector (drain).

There are two kinds of FETs: the ***JFET*** has a pn junction (J) at the gate contact, and the ***MOSFET*** has a metal gate insulated from the semiconductor substrate by a very thin layer of silicon oxide (as illustrated in Figure 3.22). MOS stands for metal oxide semiconductor. Because the silicon oxide layer is an insulator, manufacturers sometimes call MOSFETs *IGFETs,* for *i*nsulated *g*ate *FETs.* Like the BJT, FETs also have two polarities: ***n-channel*** where the conduction is by electrons and ***p-channel*** where the conduction is by holes. The MOSFET illustrated in Figure 3.22 is n-channel. When a positive voltage is applied to the gate, electrons

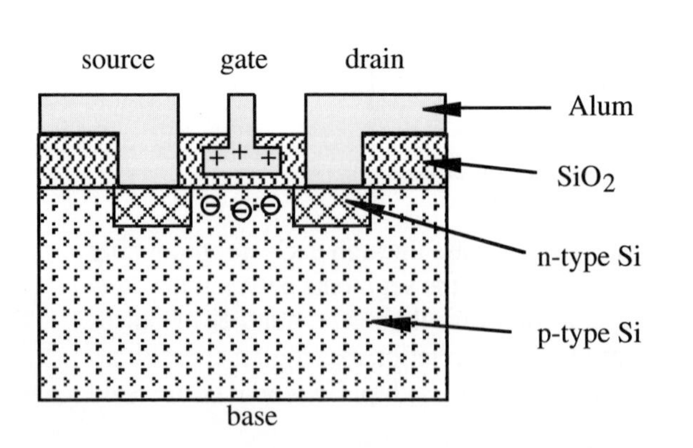


**FIGURE 3.22**
Field effect transistor cross section (n-channel MOSFET).

are attracted to a thin channel adjacent to the gate in the p-type substrate, allowing conduction between the drain and source. This is called an ***enhancement mode*** n-channel MOSFET because the channel is enhanced with electrons by the voltage applied to the gate. Another type of n-channel MOSFET is the ***depletion mode*** variety where the channel is doped (in this case creating a thin n-type channel in the p-type substrate) and a negative voltage is necessary to reduce the conduction between the drain and the source by depleting the electrons in the n-type channel.

The npn BJT is similar to the n-channel MOSFET, and the pnp BJT is similar to the p-channel MOSFET. However, the MOSFETs have an almost nonexistent gate current, which is a major advantage over BJTs. This results in an extremely high input impedance at the gate ($< 10^{14}\ \Omega$) making the FET very versatile and easy to use. It is now the basis for most digital integrated circuits, the smallest of devices, and it is also manufactured in other larger forms for use as fast, high-current switches. Both are very important in mechatronic system design.

Before getting into more detail, let's see how easy it is to use a MOSFET. Figure 3.23 illustrates the analogous use of a BJT and MOSFET to control current in a load. In each case, the switch is used to activate a large current through the load. The MOSFET circuit is simpler than the BJT circuit because with the MOSFET we don't have to worry about biasing the BE junction and providing adequate base current to saturate the transistor. We would have to pick appropriate resistors for the BJT circuit on the left. For the MOSFET circuit on the right, we simply apply the full voltage to the high impedance gate to turn on the current through the load. The drain of the MOSFET drops close to ground when it is on.

Furthermore, continuing to think of the FET as an analog switch, we can build the circuit in Figure 3.24 to pass or block a small signal. When the control voltage $V_{control}$ is large enough to saturate the MOSFET (about +5 V), the input signal is passed ($V_{out} = V_{in}$); but when the control voltage is 0 V, the input signal is blocked ($V_{out} = 0$). This circuit is not possible with a BJT.

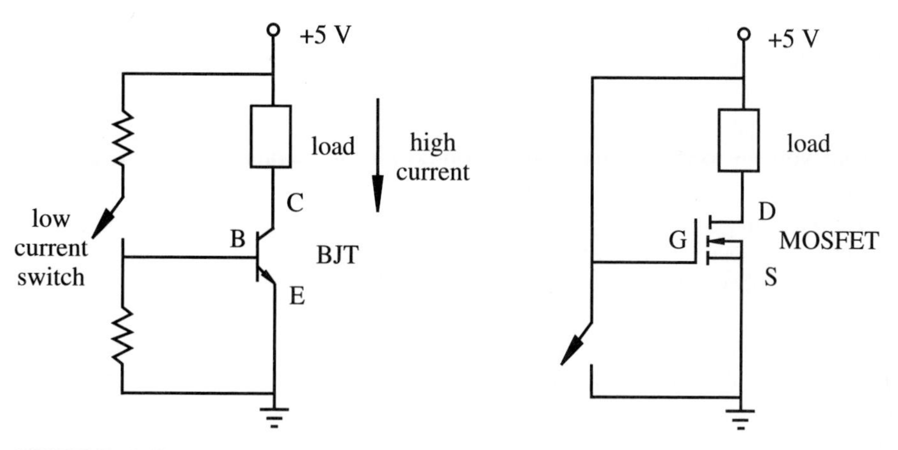


**FIGURE 3.23**
Comparison of BJT and FET switches.

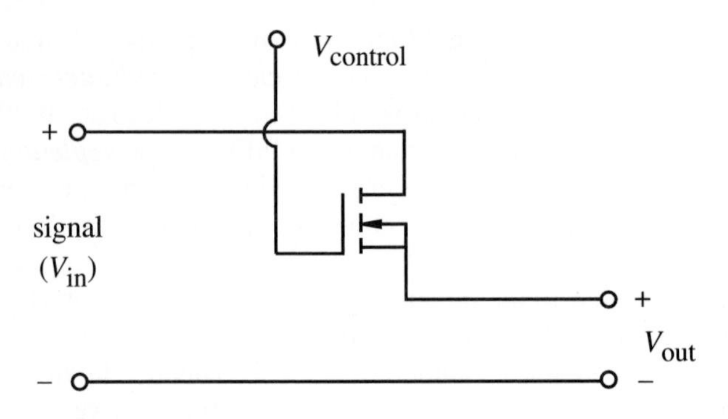


**FIGURE 3.24**
FET analog switch circuit.

### 3.5.1 Symbols Representing Field Effect Transistors

Since you will come across FETs in numerous circuit designs, it is important to recognize the subtleties of the schematic symbols for them. Since FETs (JFETs and MOSFETs) have two different kinds of channel doping and since the base can be p-type or n-type, there are eight potential FET configurations. The symbols for the four most important classes of FETs are shown in Figure 3.25. The terminal designations are G for gate, S for source, D for drain, and B for base or substrate. Some of the principal characteristics of the schematic are:

1. The direction of the gate or substrate arrow distinguishes between p-type (arrow out) and n-type (arrow in).
2. There is a separation shown between the gate and the source in the MOSFET but not in the JFET. The separation represents the insulating layer of the metal oxide in the MOSFET.
3. A broken line between source and drain indicates an enhancement mode device in contrast to a solid line for a depletion mode device. Enhancement mode FETs require a gate voltage to start conduction, and depletion mode FETs require a gate voltage to reduce the conduction. JFETs are available only in depletion mode, but MOSFETs are available in both varieties.
4. The gate line is offset toward the source, so the source side can be easily identified. Many people often show the gate line centered, but in this case there is no way to distinguish the drain from the source unless they are labeled.

The base (substrate) of a MOSFET may be connected to a separate terminal or internally connected to the source (as illustrated in Figures 3.23 and 3.24). If there is a separate substrate lead, it must not be biased more positive than the source or drain for an n-channel device and must not be biased more negative than the source or drain for a p-channel device. It should always be connected to something.

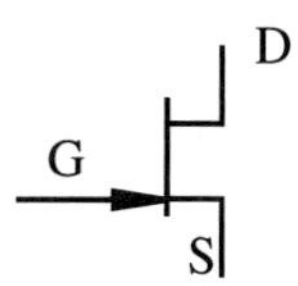


n-channel depletion mode JFET

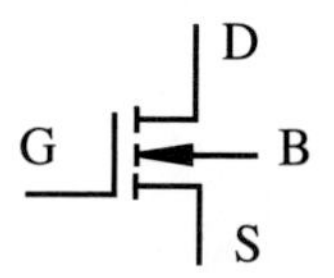


n-channel enhancement mode MOSFET

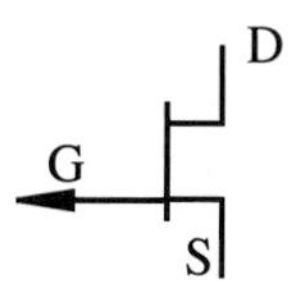


p-channel depletion mode JFET

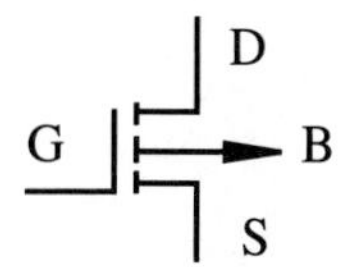


p-channel enhancement mode MOSFET

**FIGURE 3.25**
Field effect transistor schematic symbols.

In summary, the most common FET classes are the two enhancement mode MOSFETs and the two depletion mode JFETs. Since there is a lot of similarity, we have and will focus primarily on the n-channel enhancement mode MOSFET, as this is the version you are most likely to see. Recall that an n-channel enhancement mode MOSFET is analogous to an npn BJT. Conversely, a p-channel enhancement mode MOSFET is analogous to a pnp BJT.

### 3.5.2 Behavior of Field Effect Transistors

We will now consider the operating characteristics of FETs. Using the n-channel enhancement mode MOSFET as our model, we assume that no drain-to-source current ($I_D$) will flow until the gate is positive with respect to the source ($V_{DS} > 0$). Furthermore, due to the insulating nature of the gate, all drain current flows to the source. As the drain-to-source voltage $V_{DS}$ is increased gradually for a fixed gate-to-source voltage $V_{GS}$, the drain current $I_D$ increases and then levels off. Further increases in $V_{DS}$ do not change $I_D$. As shown in Figure 3.26, these characteristics can be summarized as a family of curves. The FET family of curves is similar to that of the BJTs, but the slope in saturation (corresponding to the active region of a BJT) is almost flat. For $V_{DS}$ greater than about a volt, the FET acts like a transconductance amplifier, which means the output current $I_D$ is controlled by input voltage $V_{GS}$ in a roughly square relation. In contrast, the BJT is a current amplifier where $I_C$ is roughly proportional to $I_B$.

Some other facts about FETs also need consideration. Because of the poorer mobility of the hole majority carriers, p-channel MOSFETS are poorer performers than n-channel. MOSFETs are also very susceptible to static discharges, which can destroy their thin gates. This is why you must be properly grounded against static discharges when working with them.

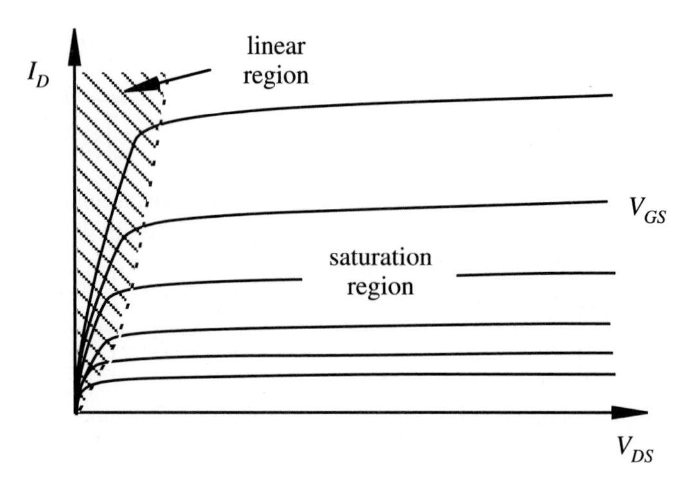


**FIGURE 3.26**
Common source curves for an FET.

FETs are very useful in a variety of mechatronic applications. MOSFETs can be used to make excellent high-current voltage-controlled switches. Special MOSFETs are designed specifically as analog switches, where signals can be gated (blocked or passed) in circuits. They are used in special circuits for driving dc motors and stepper motors. We will find them later in more sophisticated mechatronic design examples and in operational amplifiers and digital ICs. Because of their characteristics they can be designed as current sources using the flat characteristics of the saturation region. MOSFETs are extremely important for achieving high levels of integration such as that found in memory chips and microprocessors. MOSFETs are often fabricated in complementary (n-channel and p-channel) pairs, and the resulting ICs are known as ***complementary metal-oxide-semiconductor* (*CMOS*)** devices. The symmetry of the n-channel and p-channel transistors allows for compact fabrication on a single IC, and it is useful in the internal design of logic ICs.

---

▼▼ ***DESIGN EXAMPLE 3.4. Circuit-to-Switch Power.*** Among the general problems in a mechatronic system design is delivering electrical power to different portions of the system. Electrical power in its many and various forms must be portioned out to various parts either continuously or intermittently. We will return to this problem a few times in this book as we gain experience in the various options to execute this task. Transistors are one of the primary means of controlling electrical current.

An example of a problem you may now handle is the following. Suppose you have a circuit that produces an output that is binary in form, which means its output can be one of two states. For the moment, consider that the output circuit consists of an npn transistor that can be in cutoff or saturation, but with the collector as of yet not connected to anything. As we will see later, this represents the ***open collector output*** of a digital device. All you need to know for now is that the output transistor can be turned on and off. Also, it can sink only a very small current, in the milliamp range. How then can we interface the binary output to control the current to a load that may require a current of many amps? We will consider a solution to this problem employing an n-channel enhancement mode MOSFET power transistor. The solution is shown in the figure below. The equivalent output from the digital device is drawn to the left of the dashed line, and the portion we are designing is to the right.

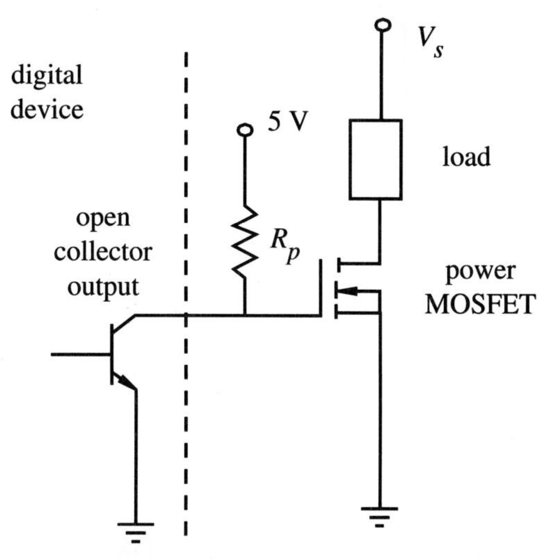


Resistor $R_p$, connected to the collector lead to complete the digital output circuit, is called a ***pull-up resistor*** since it "pulls up" the collector circuit to a dc power source (in this case +5 V). It results in 0 V at the gate when the output transistor is on and 5 V at the gate when it is off. To drive a load with a large current not available from the digital output, we've decided to use a MOSFET power transistor to switch a different power supply $V_s$.

Now if you are given the specific current and voltage requirements for a load to be driven by your digital circuit, you can refer to a discrete semiconductor handbook to select the appropriate MOSFET to do the job. This specific example will arise again and again as we interface from digital or microprocessor systems to sensors and actuators.

▼▼▼ ***CLASS DISCUSSION 3.8. Common Usage of Semiconductor Components.*** Cite specific examples in your experience where and how each of the following electrical components is used:

- Light-emitting diode (LED)
- Diode
- Transistor

## QUESTIONS AND EXERCISES

**3.1.** Sketch the output waveform for $V_o$ in the circuit below on axes as shown below. Assume the diode is ideal.

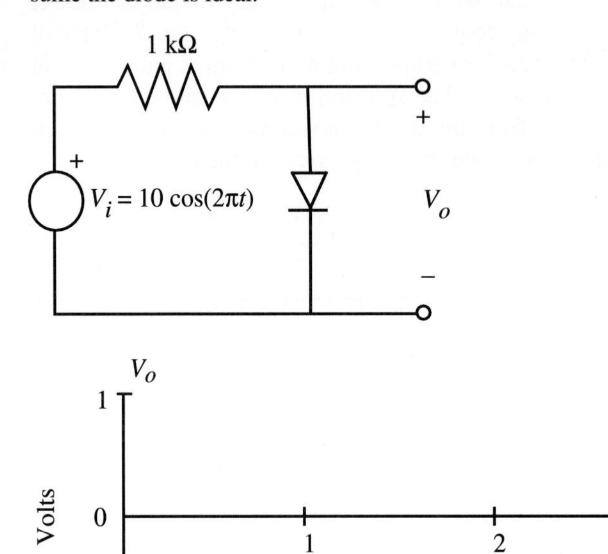


**3.2.** Sketch the output waveform for $V_o$ in the circuit below on axes as shown below. Assume ideal diodes and show your work. Also, explain why this circuit is called a ***full-wave rectifier.***

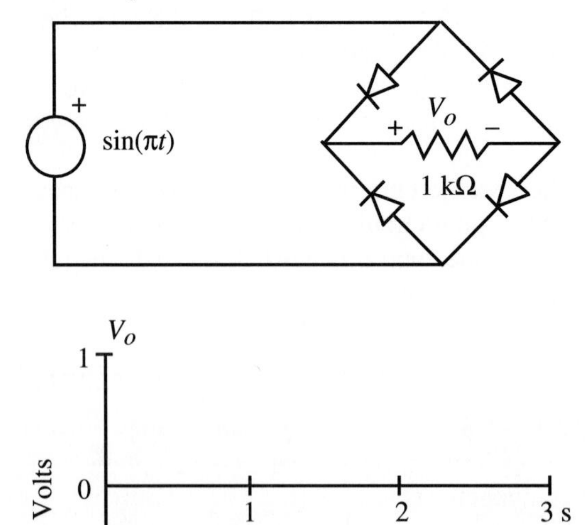

**3.3.** The circuits below are called ***clipping circuits***. Assume ideal diodes and sketch the output voltage $V_o$ for two cycles of a 1 V amplitude sine wave at $V_i$.

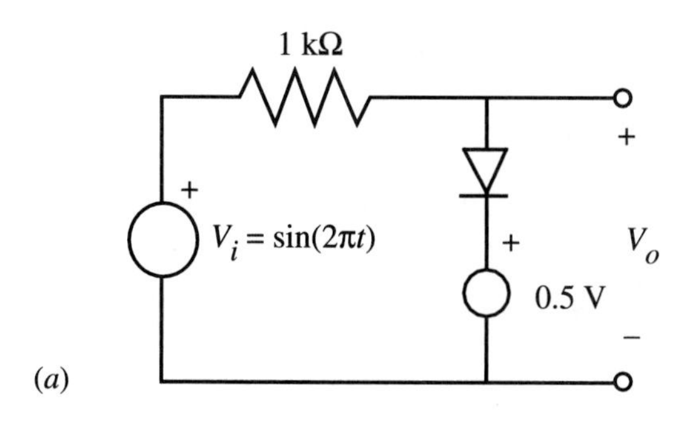


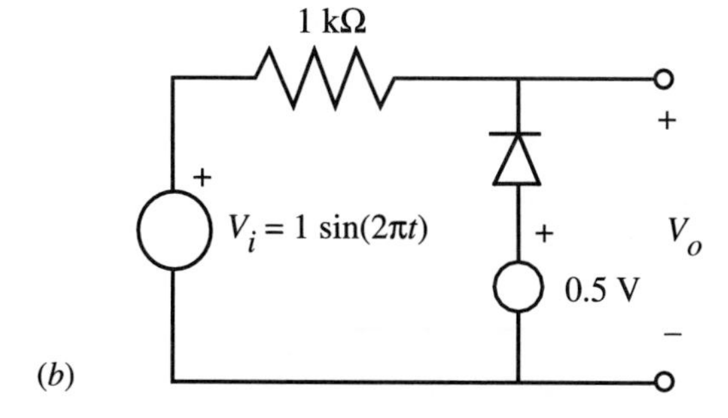


**3.4.** Compute the currents and voltages in the final circuit shown in Example 3.3.

**3.5.** In the circuit below, what minimum steady state voltage $V_{in}$ is required to turn the LED on and keep the transistor fully saturated? Assume that the forward bias voltage for the LED is 2 V and there is a 0.2 V collector-to-emitter voltage drop when the transistor is saturated.

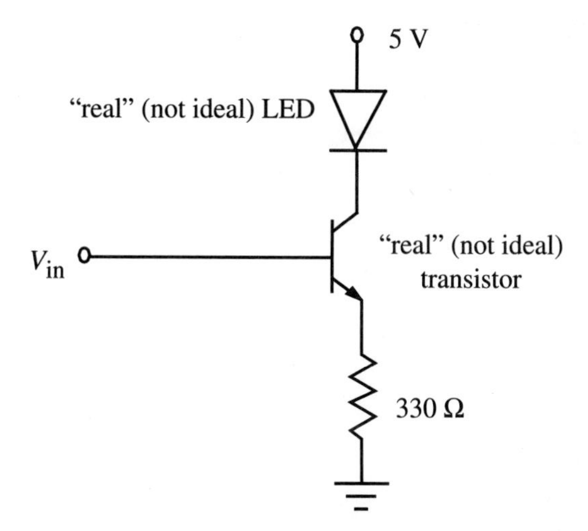

**3.6.** For the circuit below, assuming ideal diodes and given $R = 1\ \text{k}\Omega$ and $V_{in} = 10\sin(\pi t)$ V, plot the output voltage $V_{out}$ on axes with labeled scales.

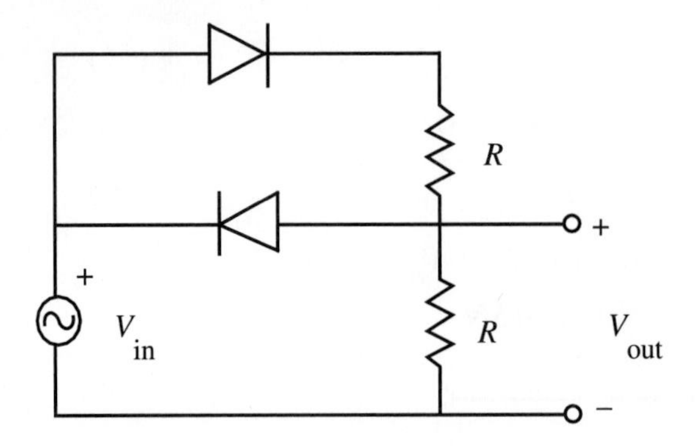


**3.7.** Given $V_{in}$ for the circuit below, sketch the LED on-off curve on a graph similar to that on the next page. Assume that the LED has a forward bias voltage of 1 V.

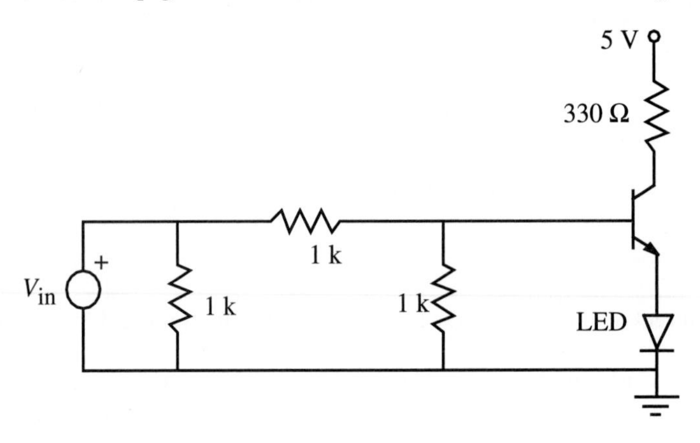


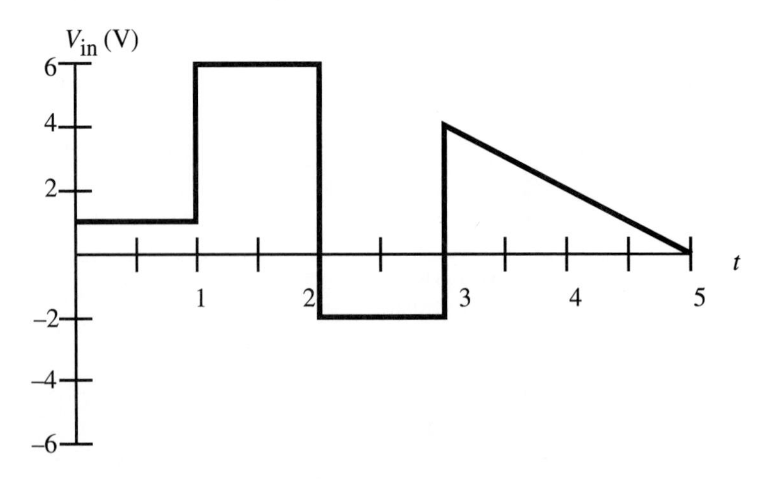

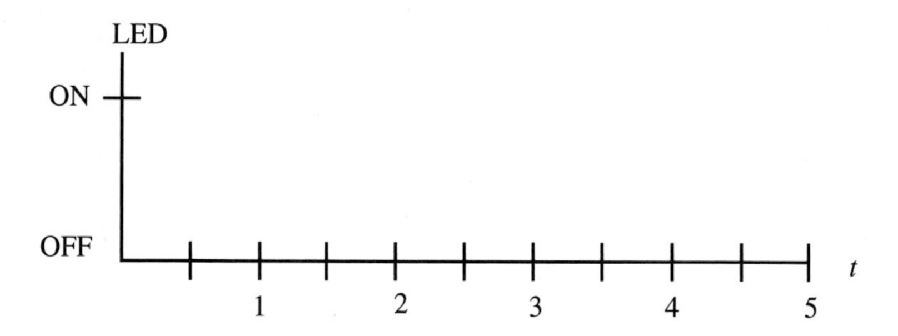


**3.8.** In Design Example 3.4, suppose $V_s = 15$ V. Replace the MOSFET with a BJT power transistor and specify the type (npn or pnp) and any additional components required. Draw the circuit schematic and describe its characteristics.

**3.9.** The output of most CMOS devices looks like:

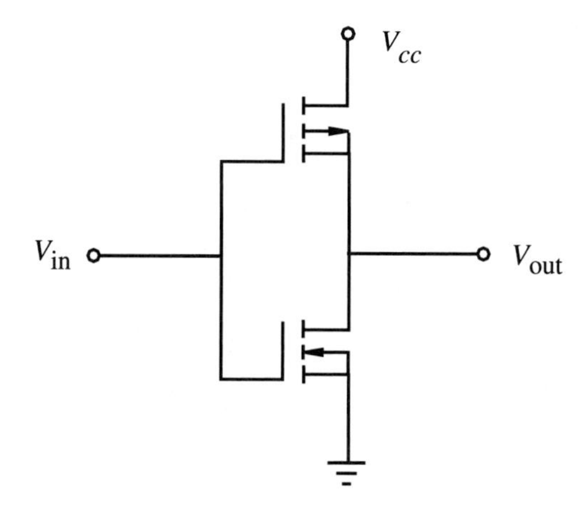


Identify the types of FETs used. What is the value of $V_{\text{out}}$ for $V_{\text{in}} = 5$ V and for $V_{\text{in}} = -5$ V?

## BIBLIOGRAPHY

Gibson, G. and Liu, Y., *Microcomputers for Engineers and Scientists,* Prentice-Hall, Englewood Cliffs, NJ, 1980.

Horowitz, P. and Winfield, H., *The Art of Electronics,* 2nd Edition, Cambridge University Press, New York, 1989.

Johnson, D., Hilburn, J., and Johnson, J., *Basic Electric Circuit Analysis,* 2nd Edition, Prentice-Hall, Englewood Cliffs, NJ, 1984.

Lerner, R. and Trigg, G., *Encyclopedia of Physics,* VCH Publishers, New York, 1991.

McWhorter, G. and Evans, A., *Basic Electronics,* Master Publishing, Inc., Richardson, TX, 1994.

Mims, F., *Engineer's Mini-Notebook: Basic Semiconductor Circuits,* Radio Shack Archer Catalog No. 276-5013, 1986.

Mims, F., *Engineer's Mini-Notebook: Optoelectronics Circuits,* Radio Shack Archer Catalog No. 276-5012A, 1986.

Mims, F., *Getting Started in Electronics,* Radio Shack Archer Catalog No. 276-5003A, 1991.

# 4

# SYSTEM RESPONSE

**OBJECTIVES:** After you read, discuss, study, and apply ideas in this chapter, you will be able to:

- Understand the three characteristics of a good measurement system: amplitude linearity, phase linearity, and adequate bandwidth
- Define the Fourier series representation of a periodic signal and use it to show the components of the spectrum of the signal
- Show the relationship between an instrument's bandwidth and the spectrums of input and output signals
- Understand the dynamic response of zero, first, and second order measurement systems
- Use system response tools such as step and frequency response to analyze and characterize the behavior of measurement and mechatronic systems
- Understand the similarities between mechanical, electrical, and hydraulic systems

## 4.1 MEASUREMENT SYSTEM RESPONSE

The relationship between the desired output of a mechatronic or measurement system and its actual output is the basis of system response analysis. This chapter deals with analysis techniques that characterize and predict how linear systems respond to specific inputs. We will concentrate on measurement systems, which are often integral parts of mechatronic systems.

As we saw in Chapter 1, a measurement system consists of three components: a transducer, a signal processor, and a recorder. A transducer is a device that usually converts a physical quantity into a time-varying voltage, called an ***analog***

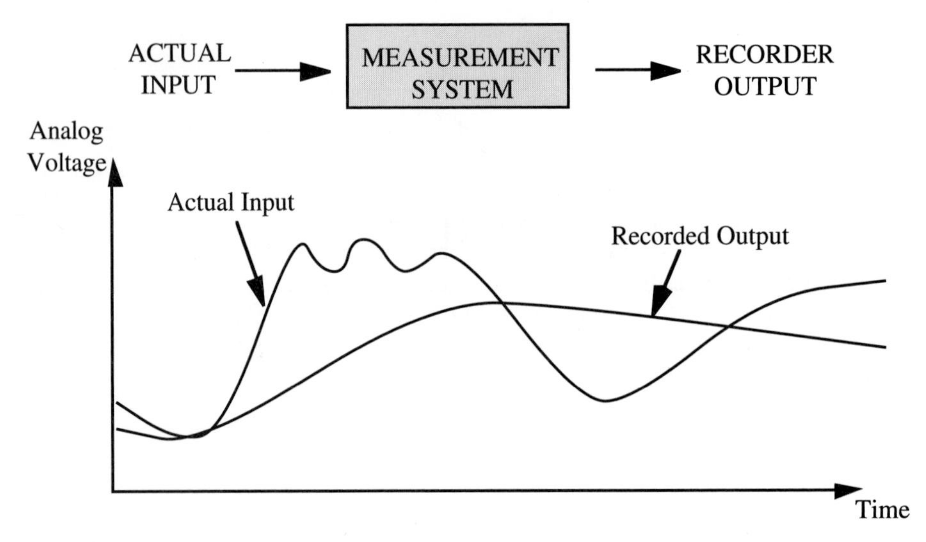


**FIGURE 4.1**
Measurement system input-output.

***signal.*** A signal processor may modify the analog signal, and a recorder provides either a transitory display or storage of the signal. The physical variable we wish to measure is called the input to the measurement system. The transducer transforms the input into a form compatible with the signal processor, which in turn modifies the signal, which then becomes the output of the measurement system. Usually, the recorded output is different from the actual input, as illustrated in Figure 4.1. Generally, we want to have the reproduced output signal match the input as closely as possible unless there is information in the input that we want to eliminate.

Certain conditions must be satisfied to accomplish adequate reproduction of the input. For a measurement system with time-varying inputs, three criteria must be satisfied in order to ensure that we obtain a quality measurement:

1. Amplitude linearity
2. Adequate bandwidth
3. Phase linearity

We will examine each of these criteria in detail in the following sections.

## 4.2 AMPLITUDE LINEARITY

A good measurement system satisfies the criterion of ***amplitude linearity.*** Mathematically, this is expressed as

$$V_{out}(t) - V_{out}(0) = \alpha[V_{in}(t) - V_{in}(0)] \tag{4.1}$$

where $\alpha$ is a constant of proportionality. This means that the output always changes by the same factor times the change in the input. If this does not occur, then the system is not linear with respect to amplitude, and it becomes more difficult to interpret the output. Figure 4.2 displays examples of amplitude linearity and nonlinearity. The first example is linear and $\alpha = 20$. The second two examples are nonlinear since $\alpha$ is not constant. In the third example, the output changes by a factor of 20 on the first pulse and by a factor of 15 on the second pulse.

Normally, a measurement system will satisfy amplitude linearity only over a limited range of input amplitudes. Also, the system usually responds linearly only when the rate of change of the input is within certain limits. This second issue is related to the bandwidth of the system and is addressed in Section 4.4. An ideal measurement system will exhibit amplitude linearity for any amplitude or frequency of the input.

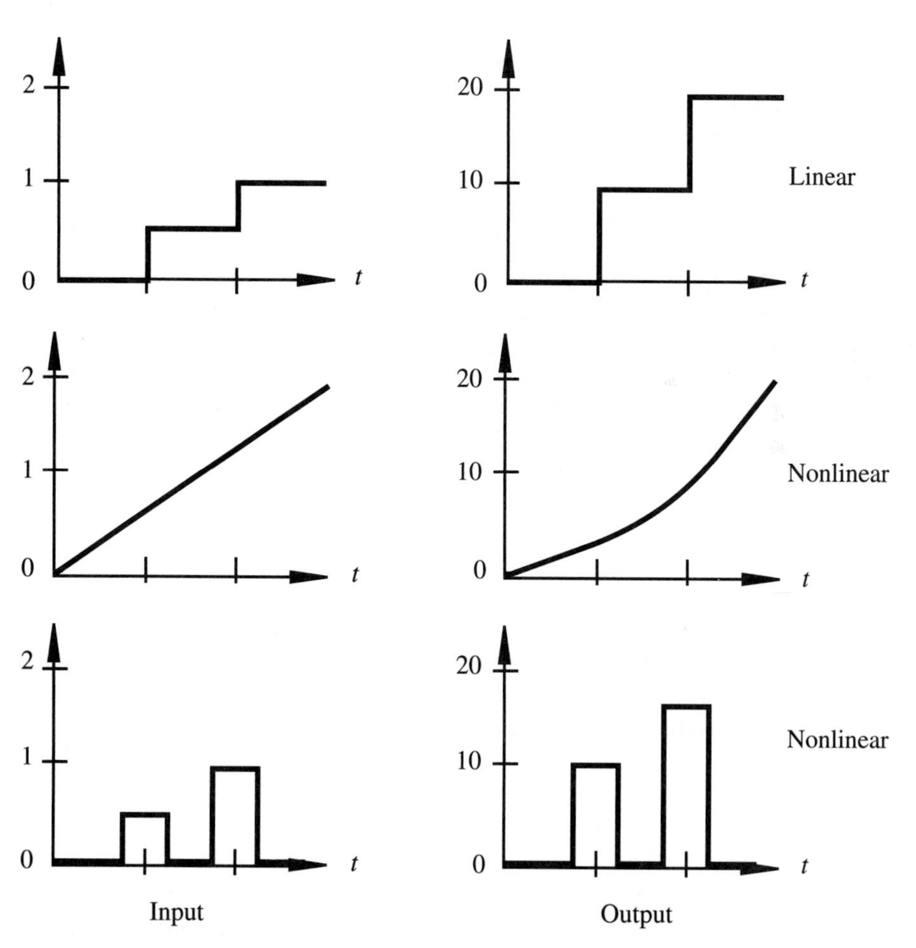


**FIGURE 4.2**
Amplitude linearity and nonlinearity.

## 4.3 FOURIER SERIES REPRESENTATION OF SIGNALS

Before we look at the concepts of bandwidth and phase linearity, which relate to frequency components of the input signal, we first need to review the concept of ***Fourier series*** representation of a signal. The fundamental premise for the Fourier series representation of a signal is that any ***periodic*** waveform can be represented as an infinite series of sine and cosine waveforms of different amplitudes and frequencies. When this infinite series is summed up, it will reproduce the original periodic waveform exactly. What this means is that we can take any complicated but periodic waveform and decompose it into a series of sine and cosine waveforms. In practice, we do not need the entire infinite series since a finite number of the sine and cosine waveforms will adequately represent the original signal.

We define the ***fundamental*** or first ***harmonic*** $\omega_0$ as the lowest frequency component of a periodic waveform. The other sine and cosine waveforms will have frequencies that are an integer number times the fundamental frequency. The second harmonic would be $2\omega_0$, the third harmonic would be $3\omega_0$, and so on. The Fourier series representation of an arbitrary periodic waveform $F(t)$ can be expressed mathematically as

$$F(t) = C_0 + \sum_{n=1}^{\infty} A_n \cos(n\omega_0 t) + \sum_{n=1}^{\infty} B_n \sin(n\omega_0 t) \tag{4.2}$$

where the constant $C_0$ is the dc component of the signal, and the two summations are infinite series of sine and cosine waveforms whose frequencies are integer multiples of the fundamental. The coefficients of the sine and cosine terms are defined by

$$A_n = \frac{2}{T}\int_0^T F(t)\cos(n\omega_0 t)dt \tag{4.3}$$

$$B_n = \frac{2}{T}\int_0^T F(t)\sin(n\omega_0 t)dt \tag{4.4}$$

where $F(t)$ is the waveform being represented. The dc term $C_0$ represents the average value of the waveform over its period; therefore, it can be expressed as

$$C_0 = \frac{A_0}{2} = \frac{1}{T}\int_0^T F(t)dt \tag{4.5}$$

where $A_0$ is given by substituting $n = 0$ into Equation 4.3.

To illustrate the application and meaning of a Fourier series, consider an ideal square wave as an example of a periodic waveform. The square wave illustrated in Figure 4.3 is defined mathematically as

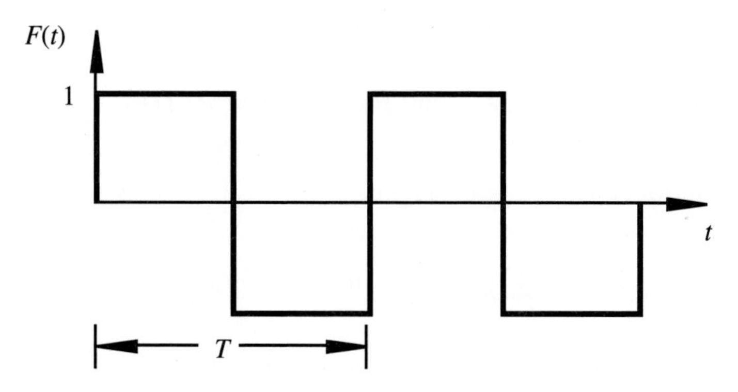


**FIGURE 4.3**
Square wave.

$$F(t) = \begin{cases} 1 & 0 \le t < T/2 \\ -1 & T/2 \le t < T \end{cases} \tag{4.6}$$

where $T$ is the period of the wave. There is a discontinuity at $t = T/2$.

For the square wave defined by Equation 4.6, the coefficients $A_n$, including $A_0$, are zero (see Question 4.1). The coefficients $B_n$ can be found from Equation 4.4:

$$B_n = \frac{2}{T}\left(\int_0^{T/2} \sin(n\omega_0 t)dt - \int_{T/2}^{T} \sin(n\omega_0 t)dt\right) \tag{4.7}$$

Integration results in

$$B_n = \frac{2}{T}\left(-\frac{1}{n\omega_0}[\cos(n\omega_0 t)]_0^{T/2} + \frac{1}{n\omega_0}[\cos(n\omega_0 t)]_{T/2}^{T}\right) \tag{4.8}$$

The fundamental frequency of the square wave $\omega_0$ is related to the period as

$$\omega_0 = \frac{2\pi}{T} \tag{4.9}$$

Using this in evaluating the limits in Equation 4.8 gives

$$B_n = \frac{1}{n\pi}\{-\cos(n\pi) + 1 + 1 - \cos(n\pi)\} \tag{4.10}$$

This can be written as

$$B_n = \frac{2}{n\pi}[1 - \cos(n\pi)] = \begin{cases} \dfrac{4}{n\pi} & n\text{: } odd \\ 0 & n\text{: } even \end{cases} \tag{4.11}$$

Therefore, the Fourier series representation of a square wave of amplitude 1 is

$$F(t) = \frac{4}{\pi}\sin(\omega_0 t) + \frac{4}{3\pi}\sin(3\omega_0 t) + \frac{4}{5\pi}\sin(5\omega_0 t) + \cdots \tag{4.12}$$

or, using an infinite sum of a general term,

$$F(t) = \sum_{n=1}^{\infty} \frac{4}{(2n-1)\pi}\sin[(2n-1)\omega_0 t] \tag{4.13}$$

Figure 4.4 shows the effects of combining the individual harmonics of the signal incrementally. Shown on the left side of the figure are the individual harmonics along with their frequencies and amplitudes. Notice that as the harmonic frequency increases, the amplitudes of the harmonics decrease. Illustrated on the right are the superpositions of the successive harmonics, illustrating how the addition of higher harmonics improves the representation of the square wave. If we take the first, third, and fifth harmonics and add them together, we obtain a waveform that begins to look like a square waveform. As we add additional harmonics to this signal, the quality of the reproduction of the waveform improves and sharp changes are better approximated. If an infinite number of harmonics were used, the result would be a perfect square wave. Waveforms such as a square wave, which have very sharp or rapid changes associated with them (e.g., at discontinuities), require a large number of higher harmonics for good reproduction.

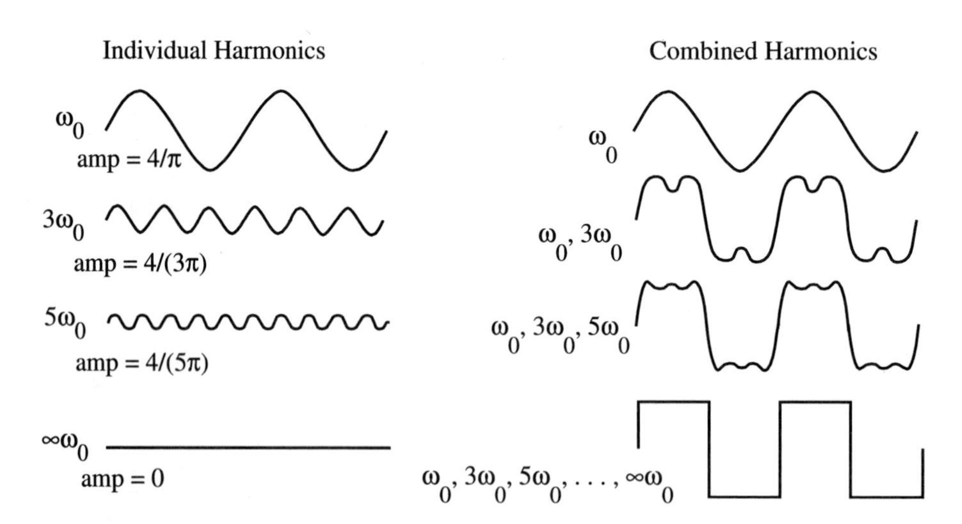


**FIGURE 4.4**
Harmonic decomposition of a square wave.

▼▼▼ ***CLASS DISCUSSION ITEM 4.1. Musical Harmonics.*** Using your knowledge of harmonics, explain each of the following musical phenomena:

- Why does a C on a flute sound different from the same C on a violin?
- Guitar players achieve an effect called "harmonics" by lightly placing a finger on the midpoint of a string before picking. Explain why this changes the quality of the sound produced.
- When striking middle C on a piano while holding the damper pedal down (allowing all strings to ring freely), why do strings other than middle C begin to vibrate? Which strings vibrate with the largest amplitude?

The top half of Figure 4.5 illustrates a plot of the amplitude vs. time of a unity amplitude square wave with a dc offset of 1.5. This plot is known as the ***time-domain*** representation of the signal. The bottom half of the figure is a plot of the Fourier series amplitudes vs. frequency, which is called the signal's ***spectrum.*** Since the harmonics are represented as bars or lines on the graph, it is also called a line spectrum. Note that the dc offset is the zero frequency component. The spectrum is the signal's ***frequency-domain*** representation.

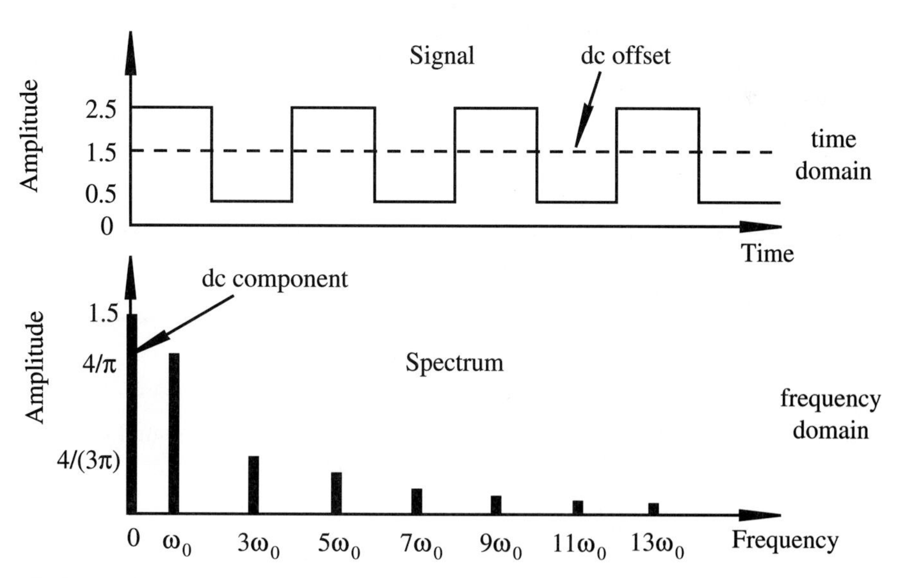


**FIGURE 4.5**
Spectrum of a square wave.

## 4.4 BANDWIDTH AND FREQUENCY RESPONSE

It is important to estimate the spectrum of a signal when choosing a system to measure the signal. Ideally, a measurement system should replicate all frequency components of a signal. Real systems, however, have limitations in their ability to reproduce all frequencies. A scale commonly used to measure the degree of fidelity of a measurement system's reproduction at different frequencies is the ***decibel*** scale. It is a logarithmic scale that allows comparison of the change in amplitude of a component of a signal when it passes through a measurement system. The decibel (dB) is defined as

$$\text{dB} = 20\log_{10}\left(\frac{A_{\text{out}}}{A_{\text{in}}}\right) \tag{4.14}$$

where $A_{\text{in}}$ is the input amplitude and $A_{\text{out}}$ is the output amplitude of a particular harmonic.

The graph shown in Figure 4.6 is an example of a ***frequency response curve*** for a system. This graph is also called a ***Bode Plot.*** It is a plot of the ***amplitude ratio***, $A_{\text{out}}/A_{\text{in}}$, vs. the input frequency. It characterizes how components of an input signal are amplified or attenuated. The term ***bandwidth*** is used to quantify the range of frequencies a system can adequately reproduce. The bandwidth of a system is defined as the range of frequencies where the input of the system is not ***attenuated*** by more than –3 dB. As illustrated in the figure, a system will usually have two frequencies at which the attenuation of the system is –3 dB. They are defined as the low and high ***corner*** or ***cutoff frequencies*** $\omega_L$ and $\omega_H$. These two frequencies define the bandwidth of the system. The –3 dB cutoff value is actually an approximation to the decibel value when the power of the input signal is attenuated to half of its input value:

$$\frac{P_{\text{out}}}{P_{\text{in}}} = \frac{1}{2} \tag{4.15}$$

For this reason, the cutoff frequencies are referred to as the "half-power" points. The power of a sinusoidal signal is proportional to the square of the signal's amplitude, and at the cutoff value,

$$\frac{A_{\text{out}}}{A_{\text{in}}} = \sqrt{\frac{P_{\text{out}}}{P_{\text{in}}}} = \sqrt{\frac{1}{2}} \approx 0.707 \tag{4.16}$$

Thus, at the cutoff frequencies the amplitude of the signal is attenuated by 29.3% (to 70.7% of its original value), which is approximately –3 dB:

$$\text{dB} = 20\log_{10}\sqrt{\frac{1}{2}} \approx -3\text{ dB} \tag{4.17}$$

At first it may seem illogical to define the bandwidth to exclude signal components that exist outside the range of the bandwidth. The half-power points are admittedly somewhat arbitrary, but if applied consistently they allow us to compare a

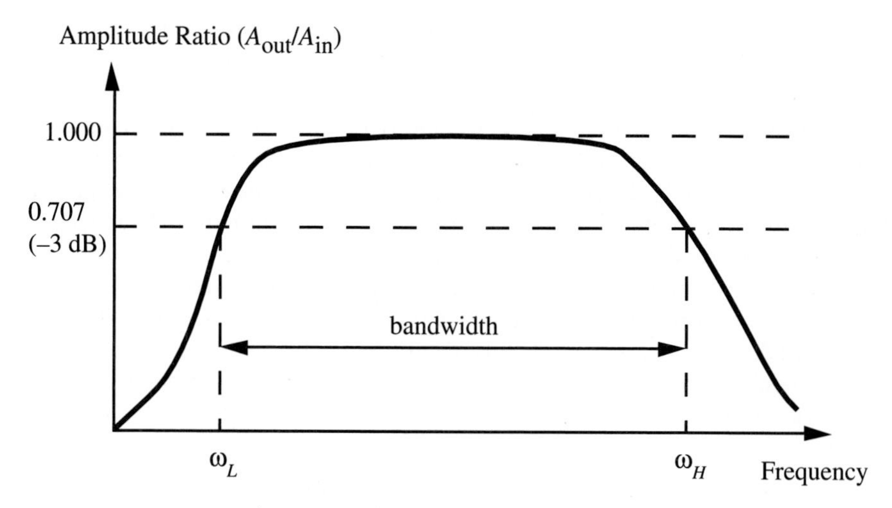


**FIGURE 4.6**
Frequency response and bandwidth.

variety of instruments and system responses. Signal amplitudes of components outside the bandwidth are attenuated by more than 3 dB. Components that lie within the bandwidth, especially those close to the cutoff frequencies, may also be attenuated but by less than 3 dB.

The frequency response of an ideal measurement system has a value of one, extending from zero to infinite frequency. An ideal system reproduces all harmonics in a signal without amplification or attenuation. Real measurement systems, however, will have a limited bandwidth. The bandwidth of a system is influenced by such factors as capacitance, inductance, and resistance in electrical systems, and mass, stiffness, and damping in mechanical systems. Through careful design, these parameters can be selected to result in a desired bandwidth. A properly designed measurement system will reproduce all frequency components in a typical input signal. When it does, the system is said to exhibit high ***fidelity.***

▼▼▼ ***CLASS DISCUSSION ITEM 4.2. Measuring a Square Wave with a Limited Bandwidth System.*** Assume you have a measurement system whose bandwidth is 0 to 5 $\omega_0$. If the input to the system is a square wave with a fundamental frequency of $\omega_0$, describe the difference between the input and the output.

The proper design or selection of a measurement system requires an understanding of measurement system bandwidth and signal spectrum. Figure 4.7 illustrates an input signal spectrum, a measurement system's frequency response, and the resulting spectrum of the output signal, all using the same frequency scale. The measurement system has a limited bandwidth, so not all input signal frequency components are reproduced satisfactorily. Therefore, the output signal will differ from the input signal. When designing or choosing a measurement system for an application, it is important that the bandwidth of the system be large enough to adequately reproduce the important frequency components present in the input signal. A measurement system that does not reproduce high frequencies cannot accurately reproduce signals that have rapid changes associated with them.

To experimentally determine the bandwidth of a measurement system, it is necessary to systematically apply pure sinusoidal signals and to determine the output-to-input amplitude ratio for the desired range of frequencies. The sweep feature on a function generator, which generates a linearly increasing frequency over a selected time, provides a convenient method to perform this for systems with electrical inputs. A method for theoretically determining the frequency response of a system model is presented in Section 4.13, where the method is applied to a second order measurement system.

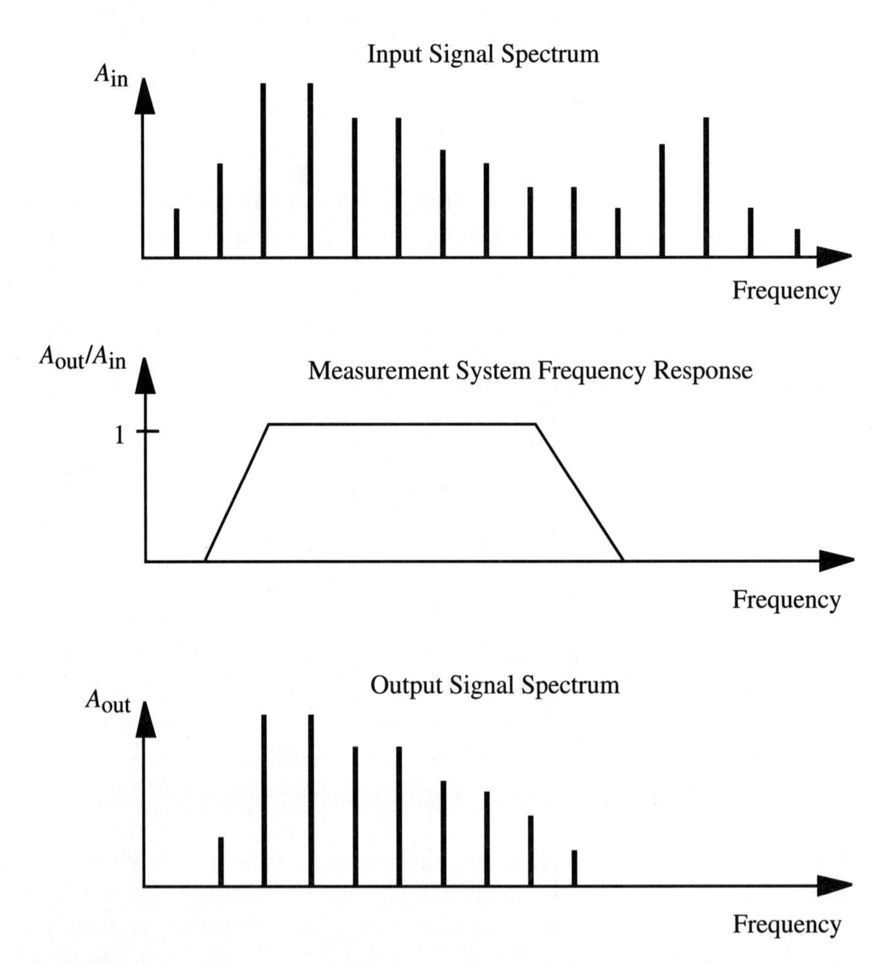


**FIGURE 4.7**
Effect of measurement system bandwidth on signal spectrum.

▼ ***EXAMPLE 4.1. Bandwidth of an Electrical Network.*** The bandwidth of an electrical system can be readily determined analytically with the help of the steady state ac circuit analysis technique presented in Section 2.6. Consider the following circuit as an example.

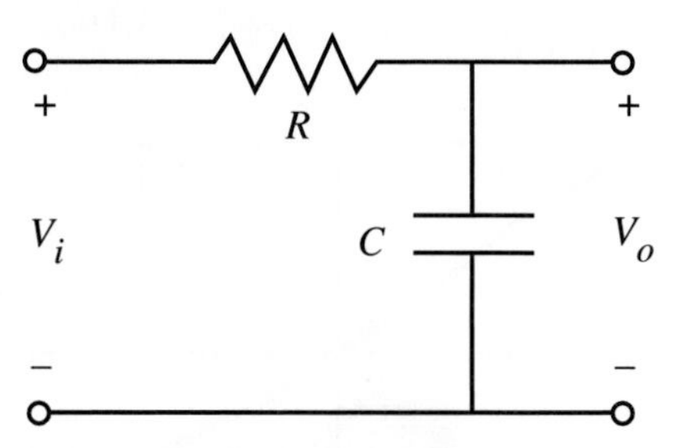


Using the voltage divider rule for the complex impedances of the resistor and capacitor, the output voltage for the network can be written as

$$V_o = V_i \frac{\frac{1}{j\omega C}}{\frac{1}{j\omega C} + R}$$

Thus, the output to input ratio as a function of frequency is

$$\frac{V_o}{V_i} = \frac{1}{j\omega RC + 1}$$

The magnitude of this complex number gives us the amplitude ratio:

$$\left|\frac{V_o}{V_i}\right| = \frac{1}{\sqrt{1 + (\omega RC)^2}}$$

The cutoff frequency $\omega_c$ for the circuit is

$$\omega_c = \frac{1}{RC}$$

because

$$\left|\frac{V_o}{V_i}\right| = \frac{1}{\sqrt{2}} = 0.707$$

when $\omega = \omega_c$. Using $\omega_c$, the amplitude ratio can also be expressed as

$$\left|\frac{V_o}{V_i}\right| = \frac{1}{\sqrt{1 + (\omega/\omega_c)^2}}$$

The frequency response curve representing this relationship is shown on the next page, where $\omega_r = \omega/\omega_c$ and $A_r = |V_o/V_i|$. Note that as $\omega_r$ approaches 0, $A_r$ approaches 1, and as $\omega_r$ approaches $\infty$, $A_r$ approaches 0.

This circuit is called a ***low-pass filter*** since lower frequencies are "passed" to the output with little attenuation, and higher frequencies are significantly attenuated (i.e., not "passed").

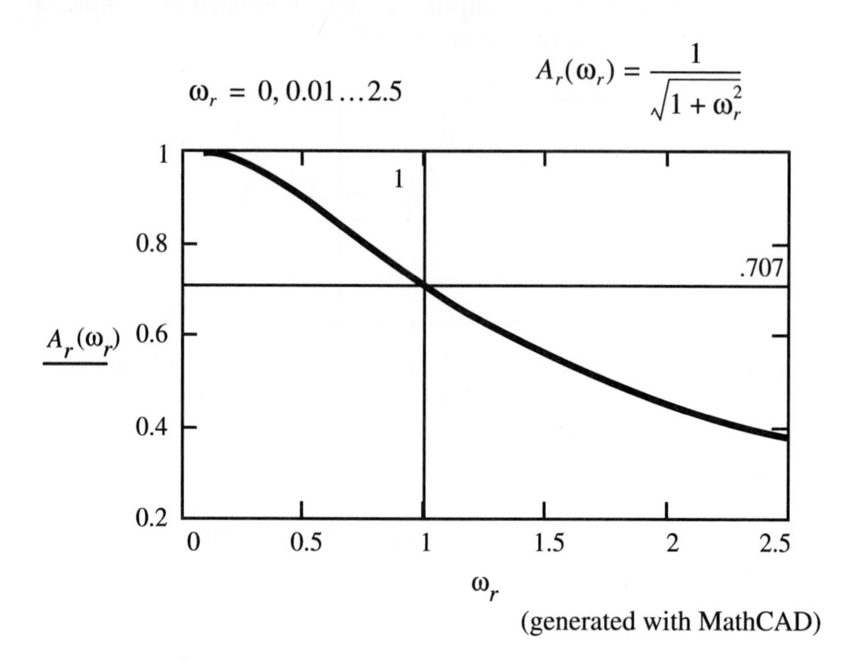


(generated with MathCAD)

If the resistor and capacitor were reversed, the resulting circuit would be called a ***high-pass filter*** since it attenuates low frequencies (see Question 4.8). Two other useful filters are known as the notch filter and band-pass filter. The ***notch filter***, sometimes called the band-reject filter, passes all frequencies except for a narrow band of frequencies that are highly attenuated. A common use for this filter is to eliminate 60 Hz interference found on signal lines. The ***band-pass filter***, on the other hand, passes a narrow band of frequencies and significantly attenuates all others.

---

## 4.5 PHASE LINEARITY

The third important criterion for a good measurement system is ***phase linearity***, which expresses how well a system preserves the phase relationship between frequency components in the input.

Consider the relationship between phase angle and time displacement between two signals as illustrated in Figure 4.8. Signal 2 lags signal 1 since it occurs later on the time axis. The time displacement between the signals $t_d$ is $T/4$, where $T$ is the period of the signals. Since a cycle of a signal corresponds to $2\pi$ radians or 360°, the phase angle between signal 1 and signal 2 is

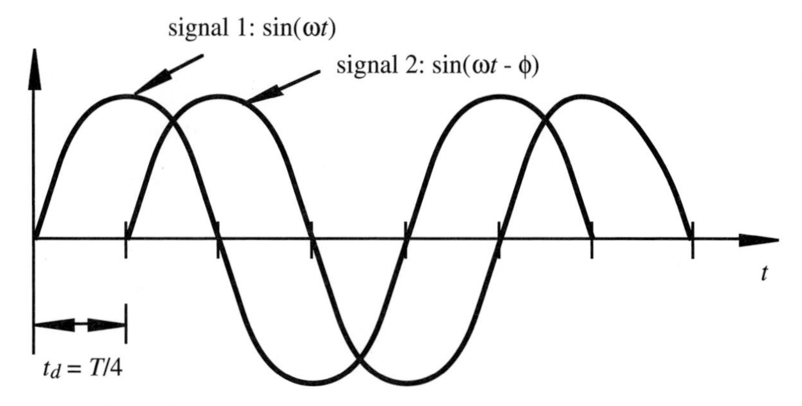


**FIGURE 4.8**
Relationship between phase and time displacement.

$$\phi = 360\ t_d/T \text{ degrees} = 2\pi\ t_d/T \text{ radians} \tag{4.18}$$

So for $t_d = T/4$, the phase angle is 90° or $\pi/2$ radians.

Measurement systems can cause a delay or time displacement between the output and input signals. For a given frequency $f$, where $f = 1/T$, Equation 4.18 can be expressed as

$$\phi = 360f \cdot t_d \text{ (degrees)} = 2\pi f \cdot t_d \text{ (radians)} \tag{4.19}$$

Therefore, for a given time displacement, the phase shift for a signal depends on its frequency. Since an input signal may be composed of many frequency components, it is important that the individual components are all displaced by the same amount of time, otherwise distortion would result with a measurement system. For equal time displacement $t_d$ for all frequency components, the following must hold:

$$\phi = k \cdot f \tag{4.20}$$

where $k$ is a constant equal to $360t_d$ (degrees) or $2\pi t_d$ (radians). Therefore, the phase angle must be linear with frequency for equal time displacement of frequency components. When a system functions in this way, it is said to exhibit phase linearity.

## 4.6
## DISTORTION OF SIGNALS

When a system does not exhibit amplitude linearity, the amplitudes of the output frequency components are attenuated. As a result, the output will suffer from ***amplitude distortion*** as illustrated in Figure 4.9 for a square wave. In the figure, the coefficients of the harmonics have square wave amplitudes that are multiplied by an exponential function whose magnitude decreases (*a*) or increases (*b*) with frequency. Notice the change (distortion) in the square wave.

$$t = 0, 0.01 \,..\, 2 \qquad n = 1 \,..\, 50$$

$$B_n = \frac{4}{\pi \cdot (2 \cdot n - 1)} \exp[-0.1(2n-1)] \qquad C_n = \frac{4}{\pi(2n-1)} \{1 - \exp[-(2n-1)]\}$$

$$F_{\text{high}}(t) = \sum_n B_n \sin[(2n-1)2\pi t] \qquad F_{\text{low}}(t) = \sum_n C_n \sin[(2n-1)2\pi t]$$

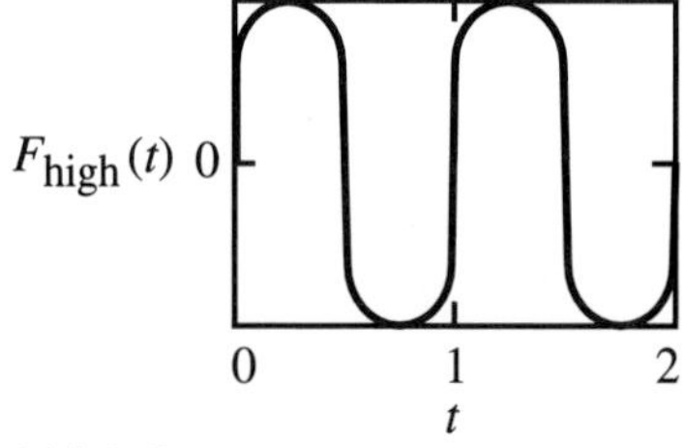


(a) high frequency components attenuated

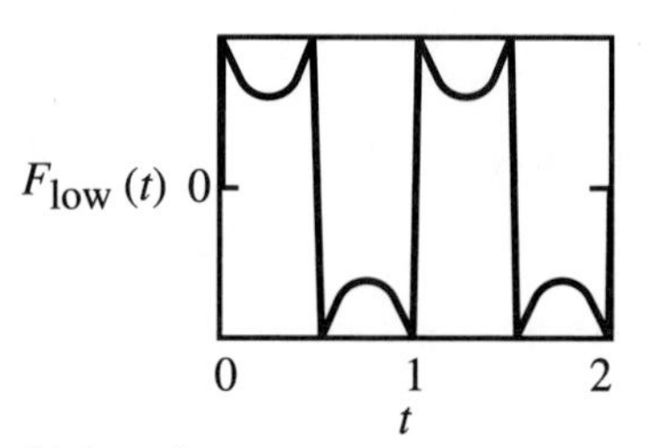


(b) low frequency components attenuated

**FIGURE 4.9**
Amplitude distortion of a square wave (generated with MathCAD).

▼▼▼ ***CLASS DISCUSSION ITEM 4.3. Analytical Attenuation.*** In Figure 4.9, the following exponential terms were used to attenuate amplitude components at either low or high frequencies:

$$e^{-0.1(2n-1)} \quad \text{and} \quad 1 - e^{-(2n-1)}$$

Explain how these exponential terms cause the resulting distortion of the square wave. (Hint: Plot the exponential functions and describe how they change the amplitudes of the components of the square wave spectrum.)

When a system does not exhibit phase linearity, the output frequency components may be of the proper amplitude but are displaced in time with respect to one another. As a result, the output will exhibit ***phase distortion*** as illustrated in Figure 4.10 for a square wave.

A high fidelity measurement system must exhibit amplitude linearity to prevent amplitude distortion, have adequate bandwidth to pass all frequency components contained in an input signal, and exhibit phase linearity to prevent phase distortion.

When designing or analyzing a measurement system, we want to predict the system's performance. Therefore, we need to be able to model the system and express its behavior in mathematical terms. The remainder of the chapter will present systems analysis tools that will let us do this.

$$t = 0, 0.01 \,..\, 2 \qquad n = 1 \,..\, 50$$

$$B_n = \frac{4}{\pi(2n-1)} \qquad dt_n = 0.05\{1 - \exp[-(2n-1)]\}$$

$$F_{\text{lag}}(t) = \sum_n B_n \sin[(2n-1)2\pi(t - dt_n)] \quad F_{\text{lead}}(t) = \sum_n B_n \sin[(2n-1)2\pi(t + dt_n)]$$

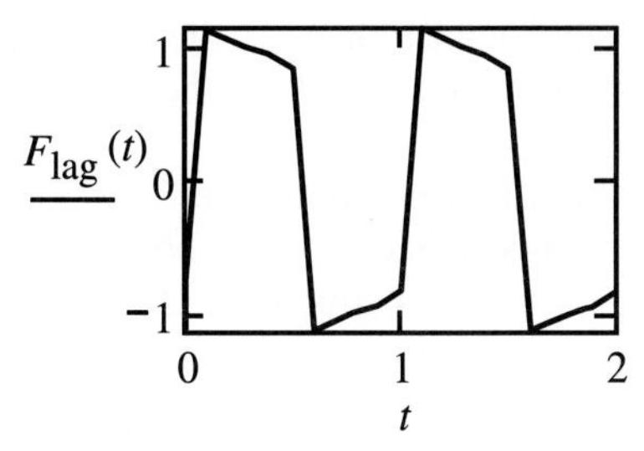


(a) high frequency components lagging

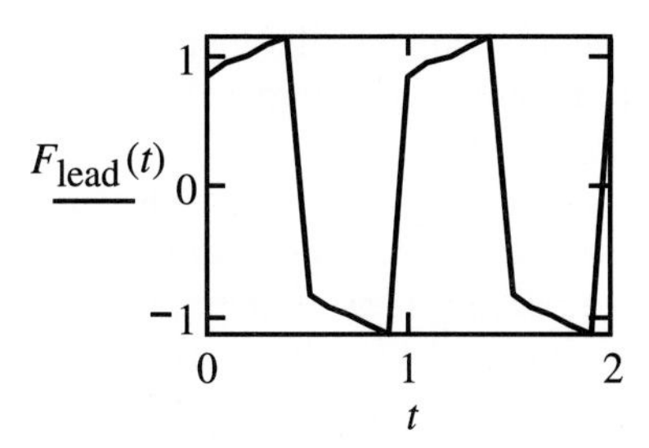


(b) high frequency components leading

**FIGURE 4.10**
Phase distortion of a square wave (generated with MathCAD).

## 4.7 DYNAMIC CHARACTERISTICS OF MEASUREMENT SYSTEMS

As we will see, many measurement systems can be modeled as linear ordinary differential equations with constant coefficients. This is also true of mechatronic systems. These linear models have the general form:

$$\sum_{n=0}^{N} A_n \frac{d^n X_{\text{out}}}{dt^n} = \sum_{m=0}^{M} B_m \frac{d^m X_{\text{in}}}{dt^m} \tag{4.21}$$

where $X_{\text{out}}$ is the output variable, $X_{\text{in}}$ is the input variable, and $A_n$ and $B_n$ are constant coefficients. $N$ defines the ***order*** of the system independent of $M$. Many electromechanical systems exhibit nonlinear behavior and cannot be accurately modeled as linear systems. However, a nonlinear system may often exhibit linear behavior over a specific range of inputs and a linear model can be derived that provides an adequate approximation over this range. This process of modeling a nonlinear system with a linear model is called ***linearization***.

## 4.8 ZERO ORDER MEASUREMENT SYSTEMS

When $N = M = 0$ in Equation 4.21, the model represents a ***zero order*** measurement system whose behavior is described by

$$A_0 X_{\text{out}} = B_0 X_{\text{in}} \tag{4.22}$$

or

$$X_{\text{out}} = \frac{B_0}{A_0} X_{\text{in}} = K X_{\text{in}} \tag{4.23}$$

where $K = B_o / A_o$ is a constant referred to as the ***gain*** or ***sensitivity*** of the system since it represents a scaling between the input and the output. When the sensitivity is high, a small change in the input can result in a significant change in the output. Note that the output of a zero order system follows the input exactly without time delay or distortion.

An example of a zero order measurement system is a potentiometer used to measure displacement. A potentiometer is a variable resistance device whose output resistance changes as an internal wiper moves across a resistive surface. As illustrated in Figure 4.11, it is a variable resistance device which produces an output voltage $V_{\text{out}}$ that is directly proportional to the wiper displacement $X_{\text{in}}$. This is a result of the voltage division rule, which gives

$$V_{\text{out}} = \frac{R_x}{R_p} V_s = \left(\frac{V_s}{L}\right) X_{\text{in}} \tag{4.24}$$

where $R_p$ is the maximum resistance of the potentiometer and $L$ is the maximum amount of wiper travel.

---

▼ ***EXAMPLE 4.2. Assumptions for a Zero Order Potentiometer.*** What approximating assumptions must be made regarding the potentiometer to ensure that it is a zero order measurement system? Hint: Consider inertial and temperature effects.

---

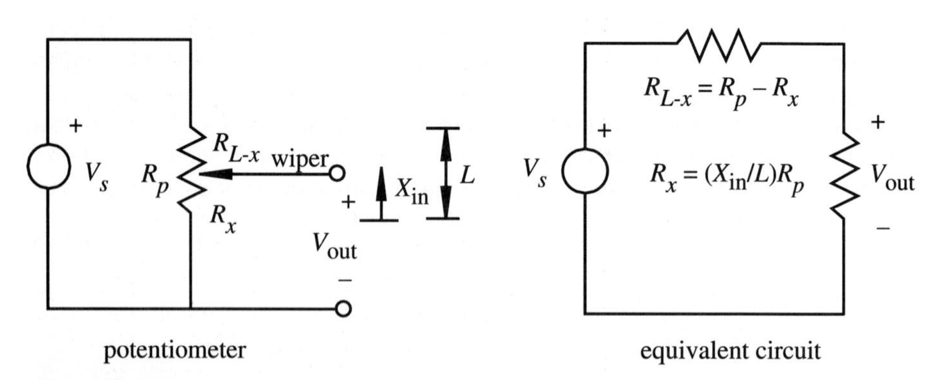


**FIGURE 4.11**
Displacement potentiometer.

## 4.9 FIRST ORDER MEASUREMENT SYSTEMS

When $N = 1$ and $M = 0$ in Equation 4.21, the equation models a ***first order*** measurement system whose behavior is described by

$$A_1 \frac{dX_{out}}{dt} + A_0 X_{out} = B_0 X_{in} \tag{4.25}$$

or

$$\frac{A_1}{A_0}\frac{dX_{out}}{dt} + X_{out} = \frac{B_0}{A_0} X_{in} \tag{4.26}$$

As with the zero order system, the coefficient ratio on the right-hand side is called the sensitivity or static sensitivity defined as

$$K = \frac{B_0}{A_0} \tag{4.27}$$

The coefficient ratio on the left side of Equation 4.26 has a special name and meaning. It is called the ***time constant*** and is defined as

$$\tau = \frac{A_1}{A_0} \tag{4.28}$$

Why we chose this name will become clear in the subsequent analysis. With these definitions, the first order system equation can be written as

$$\tau \frac{dX_{out}}{dt} + X_{out} = K X_{in} \tag{4.29}$$

To characterize how a system responds to various types of inputs, we apply standard inputs to the model, including step, impulse, and sinusoidal functions. Consider the effect of applying a step input to our system. A step input changes instantaneously from zero to a constant value $A_{in}$, and is stated mathematically as

$$X_{in} = \begin{cases} 0 & t < 0 \\ A_{in} & t \geq 0 \end{cases} \tag{4.30}$$

The output of the system in response to this input is called the ***step response*** of the first order system. We can find it by solving Equation 4.29 with the initial condition

$$X_{out}(0) = 0 \tag{4.31}$$

Applying the theory of elementary differential equations where the solution is assumed to be of the form $ce^{\lambda t}$, the ***characteristic equation*** for the homogeneous form of differential equation 4.29 is

$$\tau \lambda + 1 = 0 \tag{4.32}$$

Since the root of this equation is $\lambda = -1/\tau$, the ***homogeneous*** or ***transient solution*** is

$$X_{\text{out}_h} = Ce^{-t/\tau} \tag{4.33}$$

where $C$ is a constant determined later by applying initial conditions. A ***particular*** or ***steady state solution*** is

$$X_{\text{out}_p} = KA_{\text{in}} \tag{4.34}$$

The ***general solution*** is the sum of the homogeneous and particular solutions (Equation 4.33 + Equation 4.34):

$$X_{\text{out}}(t) = X_{\text{out}_h} + X_{\text{out}_p} = Ce^{-t/\tau} + KA_{\text{in}} \tag{4.35}$$

Applying the initial condition (Equation 4.31) to this equation gives

$$0 = C + KA_{\text{in}} \tag{4.36}$$

thus,

$$C = -KA_{\text{in}} \tag{4.37}$$

so the resulting step response is given by

$$X_{\text{out}}(t) = KA_{\text{in}}(1 - e^{-t/\tau}) \tag{4.38}$$

As illustrated in Figure 4.12, this represents an exponential rise in the output toward an asymptotic value of $KA_{\text{in}}$. The rate of the rise depends only on the time constant $\tau$. The smaller the time constant, the faster the response will be.

Note that after one time constant, the output reaches 63.2% of its final value since from Equation 4.38,

$$X_{\text{out}}(\tau) = KA_{\text{in}}(1 - e^{-1}) = 0.632KA_{\text{in}} \tag{4.39}$$

After five time constants, the step response is

$$X_{\text{out}}(5\tau) = KA_{\text{in}}(1 - e^{-5}) = 0.993KA_{\text{in}} \tag{4.40}$$

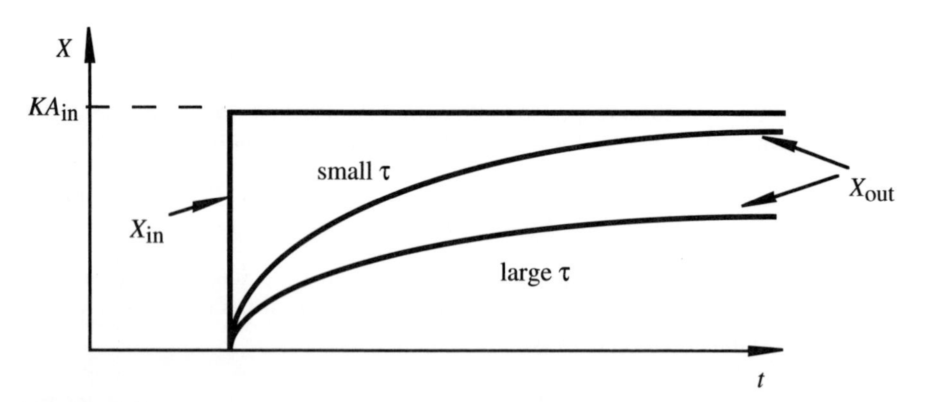


**FIGURE 4.12**
First order response.

Since this value is more than 99% of the steady state value $KA_{in}$, we usually assume that a first order system has reached its steady state value within five time constants.

When designing a first order measurement system, you should look at quantities that affect $\tau$ and reduce them if possible. The larger $\tau$ is, the longer the measurement system takes to respond to an input.

▼▼▼ ***CLASS DISCUSSION ITEM 4.4. Low-Pass Filter Time Constant Relationships.*** Relate the time constant, cutoff frequency, and bandwidth for the low-pass filter circuit shown in Example 4.1 by writing a general first order differential equation for the circuit. What is the dependent variable in the differential equation? Write an expression for the output voltage as a function of time, $V_o(t)$, for a step input voltage of amplitude $A_i$.

## 4.10 EXPERIMENTAL TESTING OF FIRST ORDER MEASUREMENT SYSTEMS

To characterize and evaluate an existing first order measurement system, we need methods to experimentally determine the time constant $\tau$ and the static sensitivity $K$. $K$ may be obtained by static calibration where a known static input is applied and the output is observed. A common method to determine the time constant $\tau$ is to apply a step input to the system and determine the time for the output to reach 63.2% of its final value (see Equation 4.39). An alternative method to determine a value for $\tau$ is presented below.

We can rearrange Equation 4.38, expressing it as

$$\frac{X_{out} - KA_{in}}{KA_{in}} = -e^{-t/\tau} \tag{4.41}$$

Simplifying, we get

$$1 - \frac{X_{out}}{KA_{in}} = e^{-t/\tau} \tag{4.42}$$

If we take the natural logarithm of both sides, we get

$$\ln\left(1 - \frac{X_{out}}{KA_{in}}\right) = -\frac{t}{\tau} \tag{4.43}$$

If we define the left-hand side as $Z$, then

$$Z = -t/\tau \tag{4.44}$$

and a plot of $Z$ vs. $t$ is a straight line with slope

$$\frac{dZ}{dt} = -\frac{1}{\tau} \tag{4.45}$$

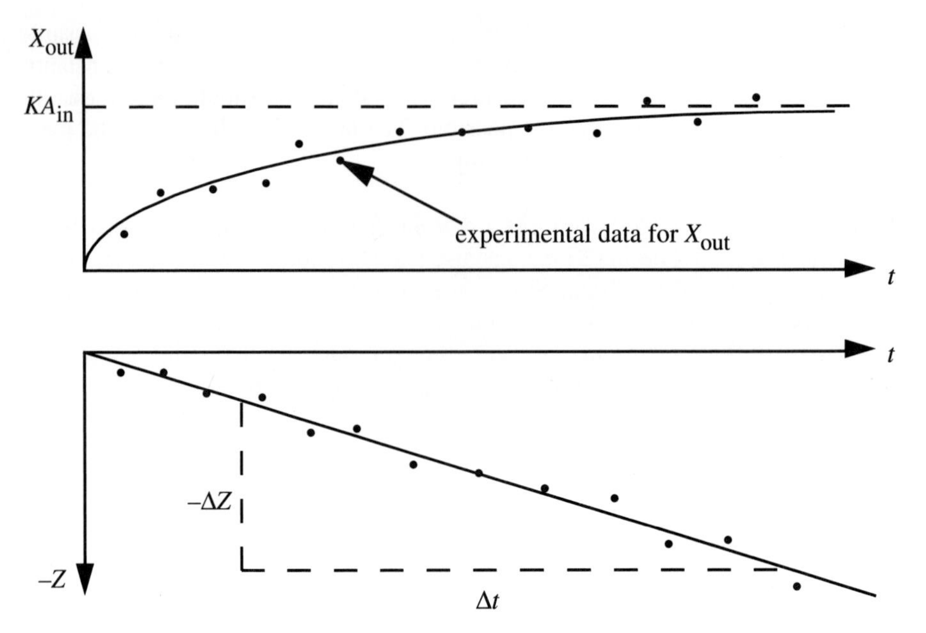


**FIGURE 4.13**
Experimental determination of τ.

Therefore, if we collect experimental data from a step input and plot $Z$ vs. $t$ as illustrated in Figure 4.13, we can determine τ from the slope of the line:

$$\tau = -\frac{\Delta t}{\Delta Z} \tag{4.46}$$

Note that if experimental data for $Z$ vs. $t$ deviates from a straight line, then the system is not first order. In this case, the system is either higher order or nonlinear.

## 4.11 SECOND ORDER MEASUREMENT SYSTEMS

An example of a ***second order*** system, where $N = 2$ in Equation 4.21, is the mechanical spring-mass-dashpot system illustrated in Figure 4.14. Applying Newton's Law of Motion (in this case, $\Sigma F_x = ma_x$) to the free body diagram gives the second order differential equation, which is the mathematical model for the system:

$$m\frac{d^2x}{dt^2} + b\frac{dx}{dt} + kx = F_{ext}(t) \tag{4.47}$$

where $m$ is the mass, $b$ is the damping coefficient, $k$ is the spring constant, and $x$ is the displacement of the mass from the equilibrium (rest) position of the mass measured positively in the downward direction as shown. $F_{ext}(t)$ represents the resul-

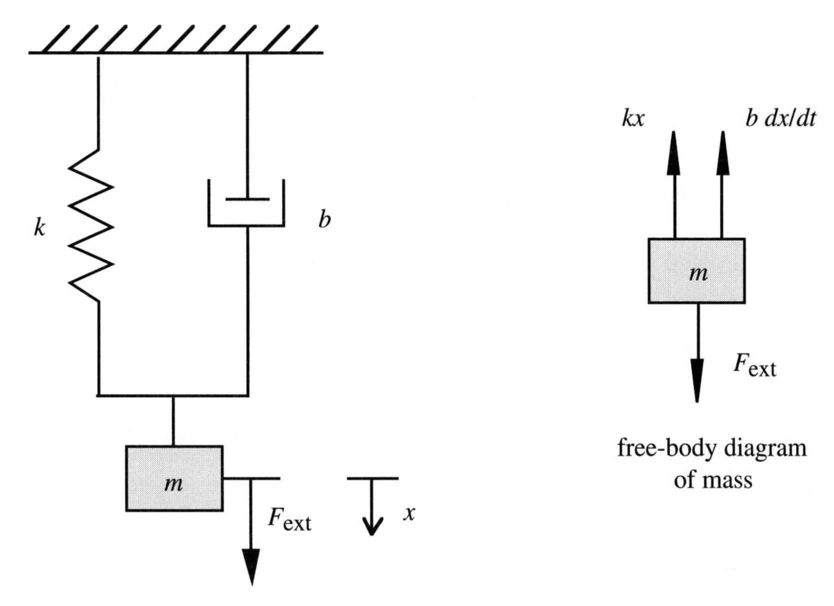


**FIGURE 4.14**
Second order mechanical system and free-body diagram.

tant of all applied external forces (inputs) in the direction of $x$. The weight of the mass is not included in the free body diagram since the displacement $x$ is measured from the equilibrium position. In the equilibrium position, the spring is stretched an amount $\delta$ from its unstretched length such that the effect of gravity is balanced: $k\delta = mg$. Note that the spring force is in the opposite direction from the mass displacement, and the dashpot force is in the opposite direction from the velocity of the mass.

As will be seen in Section 4.14, the governing equations for many second order systems other than mechanical spring-mass-dashpot systems (e.g., mechanical rotational systems and hydraulic systems) have the same form as Equation 4.47.

▼▼▼ ***CLASS DISCUSSION ITEM 4.5. Spring-Mass-Dashpot System in Space.*** Would a spring-mass-dashpot system behave any differently inside a space station orbiting the earth than it does on the surface of the earth? Why or why not? How could you measure the "weight" of an astronaut in the orbiting space station?

A good example of a second order measurement system is the strip chart recorder illustrated in Figure 4.15. The applied force $F_{ext}$ is provided by an electromagnetic coil whose core is attached to the pen carriage. The spring keeps the pen carriage centered in the zero position when the input is zero. The pen carriage bearings and the pen-paper interface result in the damping force. The carriage and the pen constitute the mass to which the forces in the system are applied.

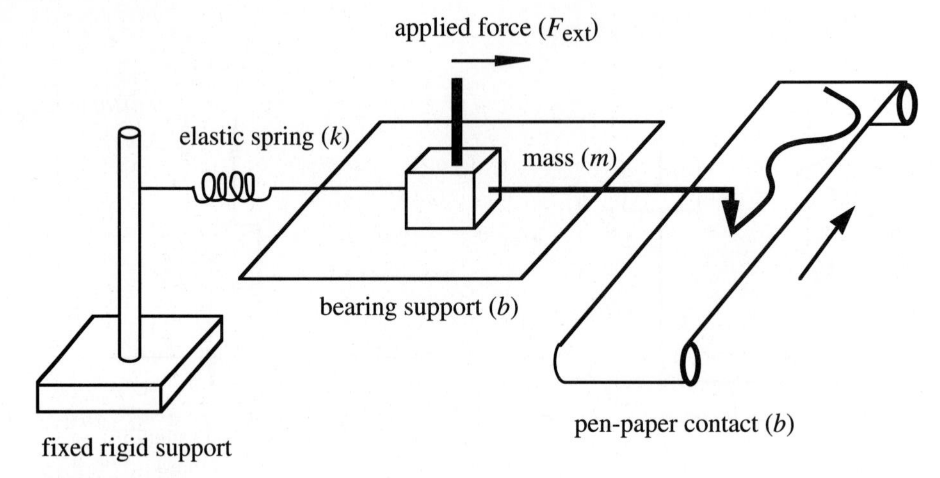


**FIGURE 4.15**
Strip chart recorder as an example of a second order system.

To characterize the unforced response of a second order system, we need to solve the differential equation 4.47 with $F_{ext} = 0$. The characteristic equation found by assuming a solution of the form $x(t) = Ce^{\lambda t}$, where $C$ is a constant, is

$$m\lambda^2 + b\lambda + k = 0 \tag{4.48}$$

This quadratic equation has two roots for $\lambda$:

$$\lambda_1 = -\frac{b}{2m} + \sqrt{\left(\frac{b}{2m}\right)^2 - \frac{k}{m}} \tag{4.49}$$

$$\lambda_2 = -\frac{b}{2m} - \sqrt{\left(\frac{b}{2m}\right)^2 - \frac{k}{m}} \tag{4.50}$$

If there were no damping in the system (i.e., $b = 0$), the roots would be

$$\lambda_1 = j\sqrt{\frac{k}{m}} \tag{4.51}$$

$$\lambda_2 = -j\sqrt{\frac{k}{m}} \tag{4.52}$$

and the corresponding solution would be

$$x_h(t) = A\cos\left(\sqrt{\frac{k}{m}}\,\mathrm{t}\right) + B\sin\left(\sqrt{\frac{k}{m}}\,\mathrm{t}\right) \tag{4.53}$$

where $A$ and $B$ are constants determined from the initial conditions $x(0)$ and $\frac{dx}{dt}(0)$. This motion represents pure ***undamped*** oscillatory motion with radian frequency:

$$\omega_n = \sqrt{\frac{k}{m}} \tag{4.54}$$

This is called the ***natural frequency*** of the system since it is the frequency at which the undamped system would "naturally" oscillate if the spring were stretched and the mass released and allowed to move without any external force ($F_{ext} = 0$).

If there is damping in the system (i.e., $b \neq 0$) and the radicand in Equations 4.49 and 4.50 is zero, the roots are double real roots and the resulting transient homogeneous solution is

$$x_h(t) = (A + Bt)e^{-\omega_n t} \tag{4.55}$$

This represents an exponentially decaying motion. A system with this behavior is said to be ***critically damped*** since it is just on the verge of damped oscillatory motion. The damping constant that results in critical damping is called the ***critical damping constant*** $b_c$. It is the value of $b$ that makes the radicand zero, so

$$b_c = 2\sqrt{km} = 2m\omega_n \tag{4.56}$$

The ***damping ratio*** $\zeta$ for a noncritically damped system is defined as

$$\zeta = \frac{b}{b_c} = \frac{b}{2\sqrt{km}} \tag{4.57}$$

It is a measure of the proximity to critical damping where $\zeta = 1$.

With the definitions of natural frequency and the damping ratio, the roots of the characteristic equations (Equations 4.49 and 4.50) can be written as

$$\lambda_1 = -\zeta\omega_n + \omega_n\sqrt{\zeta^2 - 1} \tag{4.58}$$

$$\lambda_2 = -\zeta\omega_n - \omega_n\sqrt{\zeta^2 - 1} \tag{4.59}$$

If there is damping in the system (i.e., $b \neq 0$) and the radicand in Equations 4.58 and 4.59 is negative (i.e., $\zeta < 1$), the roots are complex conjugates and the resulting transient homogeneous solution is

$$x_h(t) = e^{-\zeta\omega_n t}[A\cos(\omega_n\sqrt{1-\zeta^2}\ \mathrm{t}) + B\sin(\omega_n\sqrt{1-\zeta^2}\ t)] \tag{4.60}$$

This represents an exponentially decaying output. A system with these characteristics is said to be ***underdamped*** since it has less than critical damping ($\zeta < 1$). The frequency of oscillation is

$$\omega_d = \omega_n\sqrt{1-\zeta^2} \tag{4.61}$$

This is called the ***damped natural frequency*** of the system.

If there is damping in the system (i.e., $b \neq 0$) and the radicand in Equations 4.58 and 4.59 is positive (i.e., $\zeta > 1$), the roots are distinct real roots, and the resulting transient homogeneous solution is

$$x_h(t) = Ae^{(-\zeta + \sqrt{\zeta^2 - 1})\omega_n t} + Be^{(-\zeta - \sqrt{\zeta^2 - 1})\omega_n t} \tag{4.62}$$

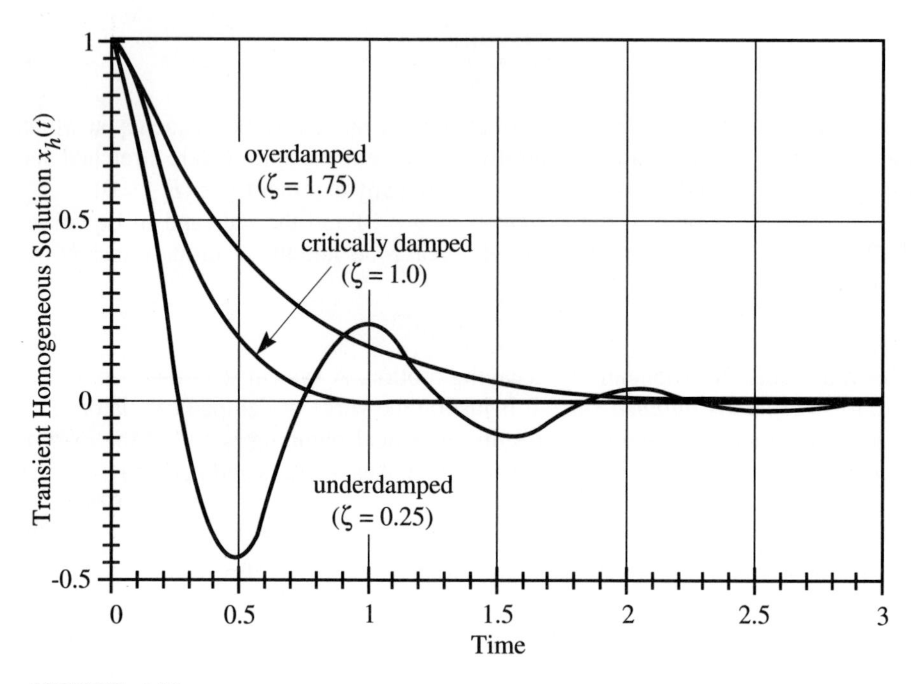


**FIGURE 4.16**
Transient response of second order systems.

This motion represents damped oscillation consisting of sinusoidal motion with exponentially decaying amplitude. A system with these characteristics is said to be ***overdamped*** since its damping exceeds critical damping ($\zeta > 1$).

Examples of transient responses for all three cases of damping (underdamped, critically damped, and overdamped) are illustrated in Figure 4.16. The curves represent unforced motion of a second order system with different amounts of damping, when the system is released from rest ($\dot{x}(0) = 0$) at $x(0) = 1$.

## 4.12 STEP RESPONSE OF SECOND ORDER SYSTEMS

As we found when analyzing a first order system, an important input used to study the dynamic characteristics of a system is a step function. The step response consists of two parts: a transient homogeneous solution $x_h(t)$, which is of the form presented in Section 4.11 for the unforced response, plus a steady state particular solution $x_p(t)$, which is a result of the forcing function. For a step input given by

$$F_{\text{ext}}(t) = \begin{cases} 0 & t < 0 \\ F_o & t \geq 0 \end{cases} \tag{4.63}$$

it is clear from Equation 4.47 that a particular solution is

$$x_p(t) = \frac{F_o}{k} \tag{4.64}$$

The general solution for the step response is then

$$x(t) = x_h(t) + x_p(t) \tag{4.65}$$

where the constants in $x_h(t)$ are determined by applying the initial conditions $x(0)$ and $\frac{dx}{dt}(0)$ to the general solution $x(t)$. As with the unforced case, there are three distinctly different types of response based on the amount of damping in the system, as illustrated in Figure 4.17.

Figure 4.18 illustrates the step response of an underdamped system and several definitions used when describing the step response. The ***steady state value*** is the value the system reaches after all transients dissipate. The ***rise time*** is the time required for the system to first reach 90% of the steady state value. The ***overshoot*** is a measure of the amount the output exceeds the steady state value before settling, usually specified as a percentage of the steady state value. The ***settling time*** is the time required for the system to settle to within an amplitude band whose width is a specified ±percentage of the steady state value. A settling time tolerance band of ±10% is shown in Figure 4.18.

These terms characterize the step response of second order systems with different amounts of damping: Underdamped systems exhibit fast rise times, but they overshoot and may take a while to settle; critically damped systems have average rise times with no overshoot, and they settle the most rapidly; and overdamped systems have no overshoot but have slow rise times.

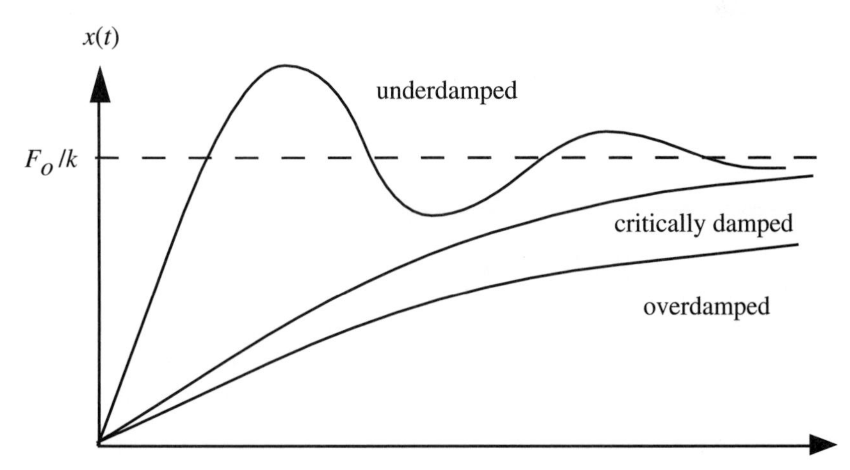


**FIGURE 4.17**
Second order step responses.

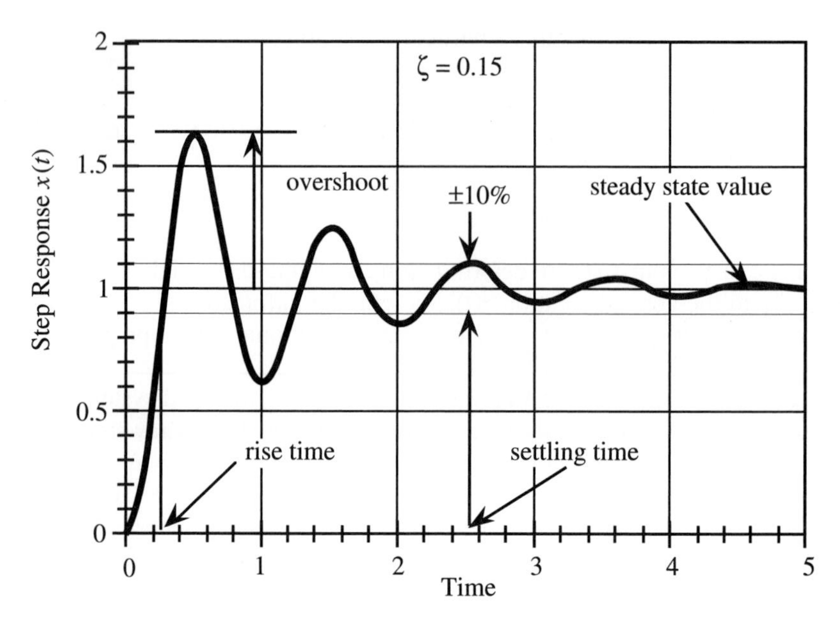


**FIGURE 4.18**
Features of an underdamped step response.

The step response may be useful in identifying the order of a system since the shape of the response curve can help indicate the order. Also, if the system is second order we can learn whether it is underdamped, overdamped, or critically damped.

▼▼▼ ***CLASS DISCUSSION ITEM 4.6. Good Measurement System Response.*** Describe and sketch what you think is a "good" step response of a measurement system. Is overshoot always bad? Does critical damping provide optimal response?

## 4.13 FREQUENCY RESPONSE OF A SYSTEM

In order to determine the frequency response of a linear system, we apply a sinusoidal input and determine the output response for different input frequencies. For the second order system described in Section 4.11, the sinusoidal forcing function input can be represented as

$$F_{\text{ext}}(t) = F_o \sin(\omega t) \tag{4.66}$$

where $F_o$ is the amplitude of the external force and $\omega$ is the input frequency.

When the sinusoidal input is first applied, the system will exhibit a combined transient and steady state response. Once the transients dissipate, the steady state output of the system will be sinusoidal with the same frequency as the input but possibly out of phase with the input. We can represent this steady state output in the following general form:

$$x(t) = X_o \sin(\omega t + \phi) \tag{4.67}$$

where $X_o$ is the output magnitude and $\phi$ is the phase difference between the input and the output.

The procedure to determine analytically the frequency response of a linear system is presented below. The linear system could model a measurement system or a more general mechatronic system of any order. Each step in the procedure is shown in its generic form and then applied to the second order system example.

**Procedure to Determine the Frequency Response of a Linear System Analytically**

1. *Find the Laplace transform of the system differential equation assuming initial conditions are zero: $x(0) = \frac{dx}{dt}(0) = 0$. The Laplace transform converts the differential equation into an algebraic equation that is related to the frequency response of the system.*

   The governing equation for the second order system (Equation 4.47) can be expressed as

$$\frac{1}{\omega_n^2}\ddot{x}(t) + \frac{2\zeta}{\omega_n}\dot{x}(t) + x(t) = \frac{1}{k}F_{\text{ext}}(t) \tag{4.68}$$

   Applying the Laplace transform to both sides of the equation, assuming zero initial conditions, gives

$$\left(\frac{1}{\omega_n^2}s^2 + \frac{2\zeta}{\omega_n}s + 1\right)X(s) = \frac{1}{k}F_{\text{ext}}(s) \tag{4.69}$$

   where $F_{\text{ext}}(s)$ and $X(s)$ are the Laplace transforms of the input forcing function and the output displacement, respectively.

2. *Find the **transfer function** of the system, which is the ratio of the output and input Laplace transforms.*

   For this system, the transfer function is

$$G(s) = \frac{X(s)}{F_{\text{ext}}(s)} = \frac{1/k}{\left(\frac{1}{\omega_n^2}s^2 + \frac{2\zeta}{\omega_n}s + 1\right)} \tag{4.70}$$

3. *To simulate a harmonic input, replace s with jω in the transfer function. This yields the frequency response behavior of the system.*

   For this system,

$$G(j\omega) = \frac{1/k}{\left[1-\left(\frac{\omega}{\omega_n}\right)^2\right] + j\left(2\zeta\frac{\omega}{\omega_n}\right)} \tag{4.71}$$

4. *Find the desired amplitude ratio between the output and input by determining the magnitude of the complex transfer function:*

$$\text{mag}(G(j\omega)) = \|G(j\omega)\| \tag{4.72}$$

   For this system, taking the magnitude of the denominator, which is the square root of the sum of the squares of the real and imaginary components, and moving the $k$ term to the left side of the equation, results in the following dimensionless ratio:

$$\frac{X_o}{F_o/k} = \frac{1}{\left\{\left[1-\left(\frac{\omega}{\omega_n}\right)^2\right]^2 + 4\zeta^2\left(\frac{\omega}{\omega_n}\right)^2\right\}^{1/2}} \tag{4.73}$$

5. *Find the phase angle φ between the output and the input by determining the argument or complex phasor angle of the complex transfer function:*

$$\phi = \arg(G(j\omega)) = \angle G(j\omega) \tag{4.74}$$

   From Equation 4.71, recalling that the argument of a complex number is the arctangent of the ratio of its imaginary and real parts, and that the argument of a ratio is the difference of the numerator and denominator arguments, the phase angle for this system is found as

$$\phi = 0 - \tan^{-1}\left\{\frac{2\zeta\frac{\omega}{\omega_n}}{\left[1-\left(\frac{\omega}{\omega_n}\right)^2\right]}\right\} = -\tan^{-1}\left(\frac{2\zeta}{\frac{\omega_n}{\omega} - \frac{\omega}{\omega_n}}\right) \tag{4.75}$$

A plot of the frequency response of the second order system as a function of the damping ratio $\zeta$ is shown in Figure 4.19. When the input frequency $\omega$ equals the natural frequency $\omega_n$ (i.e., $\omega/\omega_n = 1$), ***resonance*** occurs. $\omega_n$ is often called the resonant frequency. For small damping ratios the output amplitude becomes quite large at the resonant frequency. For input frequencies less than the natural frequency (i.e., $\omega/\omega_n < 1$), the amplitude ratio is close to one. The damping ratio $\zeta$ affects the range of frequencies for which this is true. Recall that for a good measurement system, we require the amplitude ratio to be constant (within 3 dB) over the range of frequencies we wish to measure. Be sure to familiarize yourself with Figure 4.19 and its implications.

A plot of the phase angle as a function of frequency ratio and damping ratio is shown in Figure 4.20. The phase angle is negative, implying that the output lags the

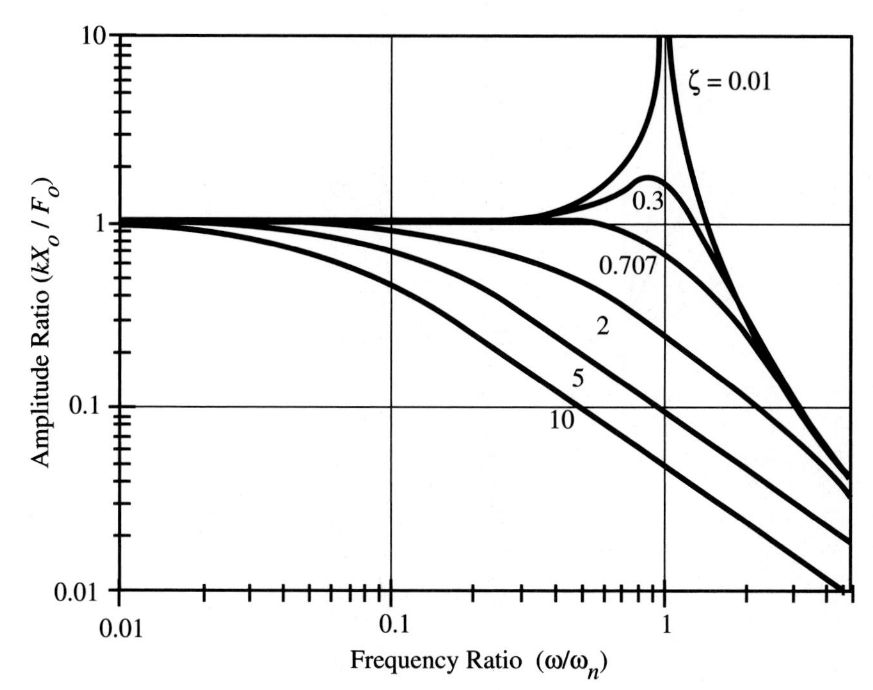


**FIGURE 4.19**
Second order system amplitude response.

input, and the magnitude ranges from 0°, which represents no phase difference, to 180°, which represents a half-cycle phase difference. As $\omega$ approaches the natural frequency $\omega_n$, the output lags the input by 90° (a quarter of a cycle). When the damping in a system is very small (e.g., $\zeta = 0.01$), the amplitude effects at resonance are very pronounced, and as the excitation frequency passes through the natural frequency ($\omega_n$), the phase shift changes abruptly by 180°. A damping ratio of 0.707 provides the best approximation of phase linearity for a second order system.

A very large number of mechanical systems can be approximated by second order linear ordinary differential equations. Also, complex systems can often be reduced to second order systems for a first approximation analysis. Therefore, all engineers should feel comfortable with their analysis.

The output of second order measurement systems always differs from the input in both amplitude and phase. In designing measurement systems, the goal is to minimize these effects when possible. As a general rule of thumb, the choice of 0.707 for the damping ratio is optimal since it results in the best combination of amplitude linearity and phase linearity over the widest range of frequencies.

In general, whenever a periodic input is applied to a second order measurement system, resonance, attenuation, and phase shift may occur, distorting the output. In order to predict whether or not a system will respond adequately, it is important to be able to determine the system's frequency response characteristics.

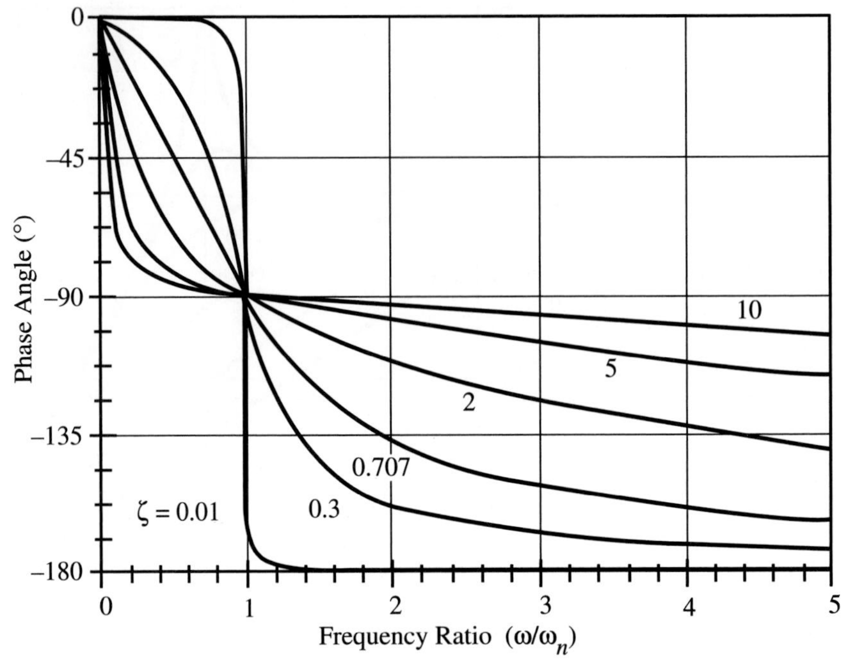


**FIGURE 4.20**
Second order phase response.

▼▼▼ ***CLASS DISCUSSION ITEM 4.7. Slinky Frequency Response.*** Using a Slinky with a mass attached to one end, demonstrate the frequency response characteristics illustrated in Figures 4.19 and 4.20.

---

▼▼ ***DESIGN EXAMPLE 4.1. Automobile Suspension Selection.*** A design team has been asked to select a spring and shock for the front suspension for a new automobile. Their task is to choose a suspension that will provide the best response to varying road conditions. Since many manufacturers specify their products in English units, the design team has agreed to perform all analyses using the English system.

After preliminary analyses and after meetings with other design teams, the suspension team has narrowed their choices down to three alternative suspension designs. The three alternatives are characterized by the following spring constants $k$ and damping constants $b$:

| Alternative | $k$ (lbf/in.) | $b$ (lbf sec/in.) |
|---|---|---|
| 1 | 500 | 10 |
| 2 | 200 | 20 |
| 3 | 120 | 10 |

To evaluate the alternatives under different road conditions, the design team decides to simulate bumps or pot holes as a ground force step input of magnitude $F_0$, and regular bumps or waviness in the road as a sinusoidal ground force input of magnitude $B_0$ and frequency $\omega$. To subject each alternative to similar input displacement levels, the following fixed ratios will be used in the comparison:

$$F_0/k = 6 \text{ in.} \quad B_0/k = 3 \text{ in.}$$

Each alternative is subjected to the following three frequencies to simulate different speeds or bump spacing:

$$\omega_1 = 10 \text{ rad/sec} \quad \omega_2 = 20 \text{ rad/sec} \quad \omega_3 = 30 \text{ rad/sec}$$

The team decides on a single degree of freedom system that models only a single wheel and neglects the response of the other wheels. The resulting model is a second order system described by Equation 4.47, where $m$ is the mass supported ("sprung") by the suspension and $F_{ext}$ is the ground force. The weight of the entire vehicle is 2000 lbf; therefore, the sprung mass supported by a single wheel is assumed to be 1/4 of the total mass:

$$m = 500 \text{ lbm}$$

Note that the wheel mass and tire dynamics are not included in this simplified model (see Discussion Item 4.8).

Using Equations 4.54 and 4.57, the natural frequency $\omega_n$ and the damping constant $\zeta$ are calculated for each alternative:

| Alternative | $\omega_n$ | $\zeta$ |
|---|---|---|
| 1 | 19.6 | 0.20 |
| 2 | 12.4 | 0.62 |
| 3 | 9.6 | 0.40 |

Since $\zeta < 1.0$ for each alternative, the suspensions will exhibit an underdamped step response described by Equation 4.60.

The step response and sinusoidal response analyses were performed in MathCAD. The input variables (for Alternative 1) are defined as

$$k = 500\,\frac{\text{lbf}}{\text{in}} \qquad b = 10\,\text{lbf}\,\frac{\text{sec}}{\text{in}} \qquad m = 500\text{lb}$$

$$F_0 = (6\text{in.})k \qquad B_0 = (3\text{in})k$$

$$\omega_1 = 10\,\frac{\text{rad}}{\text{sec}} \qquad \omega_2 = 20\,\frac{\text{rad}}{\text{sec}} \qquad \omega_3 = 30\,\frac{\text{rad}}{\text{sec}}$$

$$t = 0\text{sec}, 0.01\text{sec} \ldots 2\text{sec}$$

The step response is defined by (see Question 4.15)

$$\omega_n = \sqrt{\frac{k}{m}} \qquad \zeta = \frac{b}{2\sqrt{km}} \qquad \omega_d = \omega_n\sqrt{1-\zeta^2}$$

$$x(t) = \frac{F_0}{k}[1 - e^{-\zeta\omega_n t}\cos(\omega_d t)]$$

and the sinusoidal response is defined by

$$X_0(\omega) = \frac{\frac{B_0}{k}}{\sqrt{\left[1-\left(\frac{\omega}{\omega_n}\right)^2\right]^2 + 4\zeta^2\left(\frac{\omega}{\omega_n}\right)^2}} \qquad \phi(\omega) = -\text{angle}\left[\left(\frac{\omega_n}{\omega} - \frac{\omega}{\omega_n}\right), 2\zeta\right]$$

$$x(t, \omega) = X_0[(\omega)\sin(\omega t + \phi(\omega))]$$

The results for the three alternatives follow.

**ALTERNATIVE 1:**

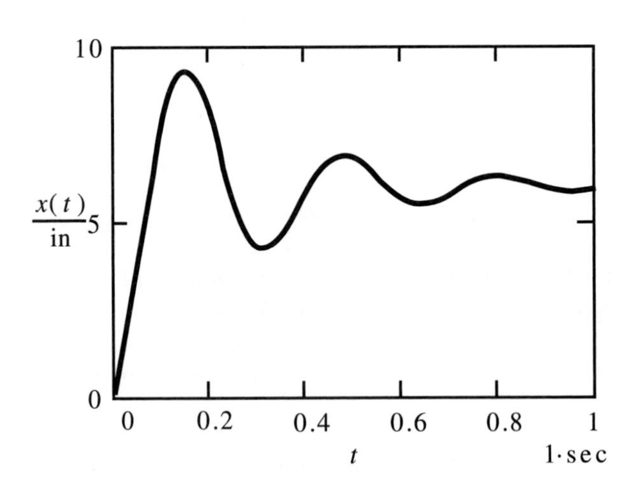

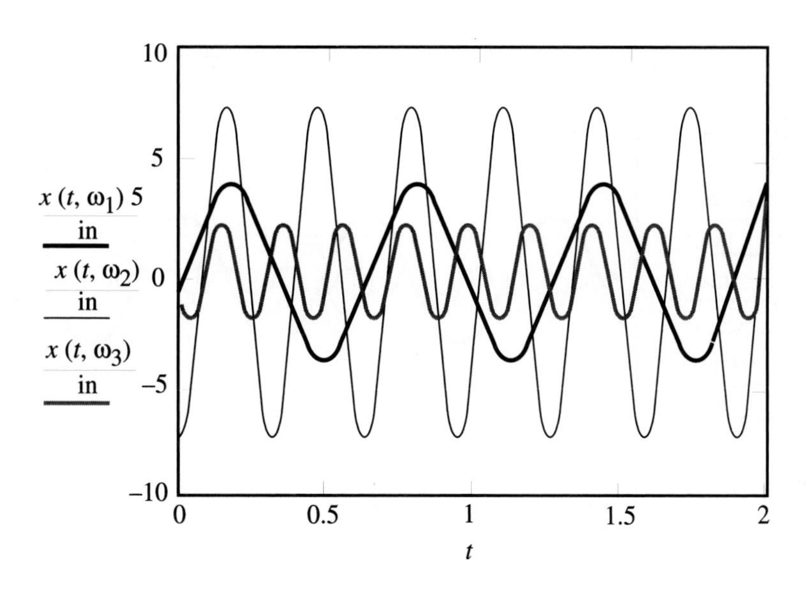


**ALTERNATIVE 2:**

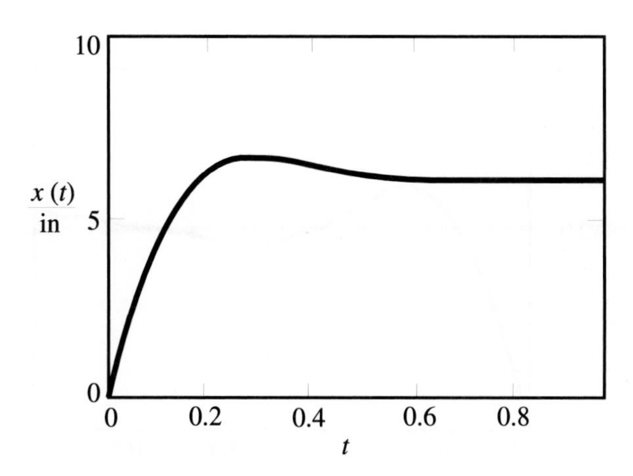

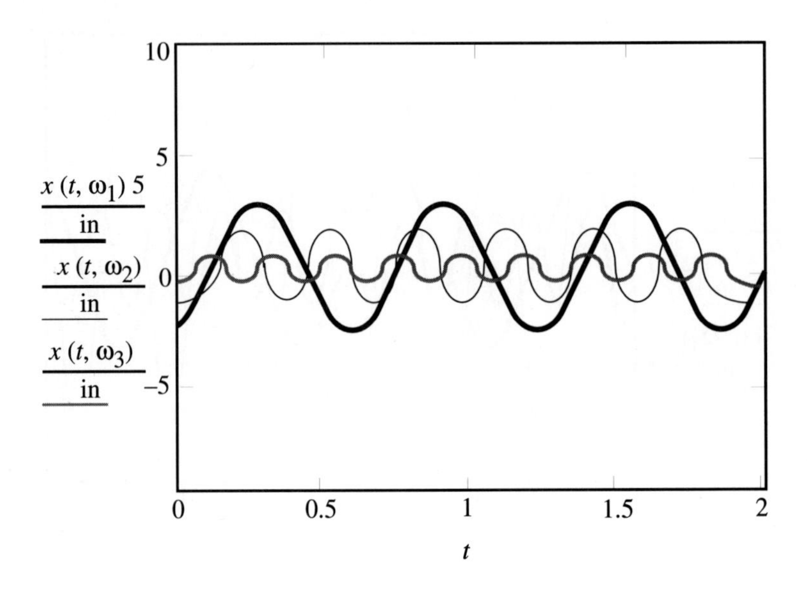


**ALTERNATIVE 3:**

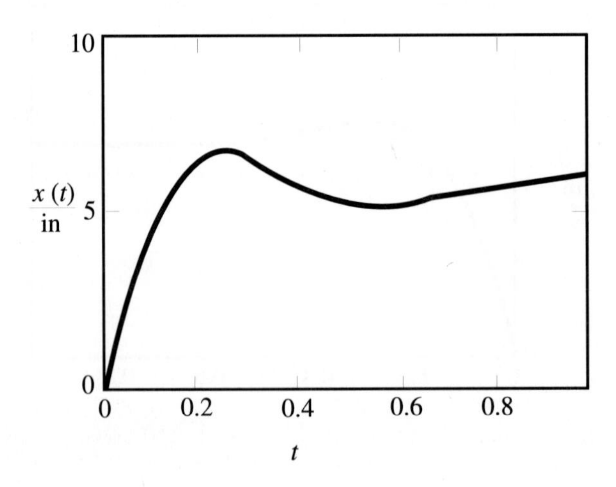

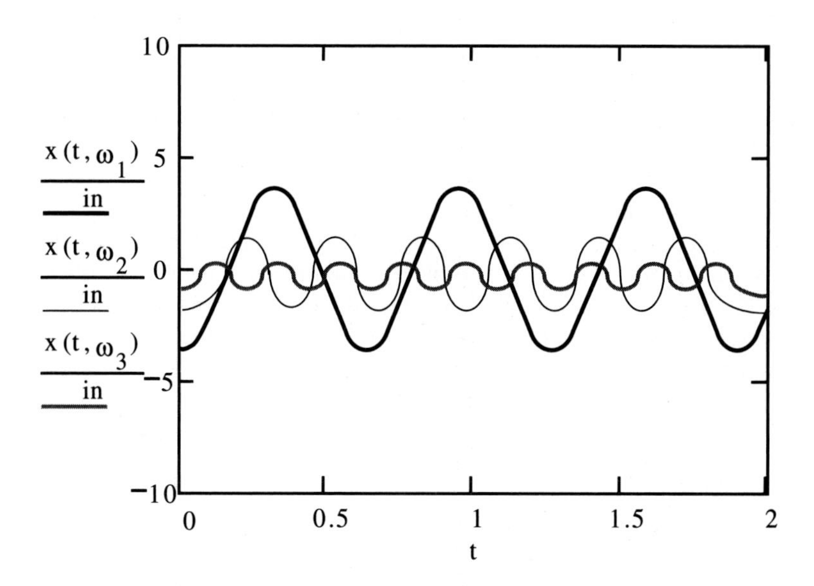


A good suspension is one that provides fast response (short settling time) for good handling, without excessive bounce or large displacement (oscillation) for rider comfort. From the results above, it appears that Alternative 2 is the best of the three designs. The step response settles fast with little bounce and the sinusoidal responses have low amplitude.

▼▼▼ ***CLASS DISCUSSION ITEM 4.8. Suspension Design Results.*** Were the step response and sinusoidal response graphs required to draw conclusions about the three alternatives? Also, why was the amplitude of the sinusoidal response excessive for Alternative 1 and least for Alternative 2? What would it take to include the wheel mass and tire dynamics in the system model?

## 4.14 SYSTEM MODELING AND ANALOGIES

Any linear system—whether it be mechanical, electrical, hydraulic, thermal, or some combination—can be modeled by an ordinary linear differential equation that relates the output response of the system to the input. These differential equations are all very similar mathematically and only differ in the constants that appear in front of the derivative terms. These constants represent the physical parameters of the system. For all second order linear systems, there are analogies between the different systems. For instance, a resistor in an electrical system is analogous to a dashpot or damper in a mechanical system or to a valve or flow restriction in a hydraulic system. Similarly, a mass or an inertia in a mechanical system is analogous to inductance in an electrical system and fluid inertance in a hydraulic system. The generic terms used to describe the analogous system parameters and variables are ***effort***, ***flow***, ***displacement***, ***momentum***, ***resistance***, ***capacitance***, and ***inertia***. Table 4.1 summarizes the generic quantities along with specific analogies for mechanical, electrical, and hydraulic systems. Also included are equations relating variables for energy storage and energy dissipation elements.

The only problematic analogy in the table is hydraulic resistance, since this quantity is usually not constant. It is typically a nonlinear function of flow rate and geometry. Analogies can also be extended to other phenomena such as heat transfer, where temperature difference is the "effort," heat flow is the "flow," the amount of heat transferred is the "displacement," heat capacity due to mass and specific heat is "capacitance," and thermal resistance due to convection and conduction is "resistance." However, heat transfer and many other phenomena cannot be ade-

**TABLE 4.1**
**Second order system modeling analogies**

| Generic quantity | Mechanical translation | Mechanical rotation | Electrical | Hydraulic |
|---|---|---|---|---|
| Effort ($E$) | Force ($F$) | Torque ($T$) | Voltage ($V$) | Pressure ($P$) |
| Flow ($F$) | Speed ($v$) | Angular speed ($\omega$) | Current ($i$) | Volumetric flow rate ($Q$) |
| Displacement ($q$) | Displacement ($x$) | Angular displacement ($\theta$) | Charge ($q$) | Volume ($\forall$) |
| Momentum ($p$) | Linear momentum ($p = mv$) | Angular momentum ($h = J\omega$) | Flux linkage ($I = N\Phi = Li$) | Momentum/area ($\Gamma = IQ$) |
| Resistor ($R$) | Damper ($b$) | Rotary damper ($B$) | Resistor ($R$) | Resistor ($R$) |
| Capacitor ($C$) | Spring ($1/k$) | Torsion spring ($1/k$) | Capacitor ($C$) | Tank ($C$) |
| Inertia ($I$) | Mass ($m$) | Moment of inertia ($J$) | Inductor ($L$) | Inertance ($I$) |
| Inertia energy storage (special case) | $F = \dot{p}$ ($F = ma$) | $T = \dot{h}$ ($T = J\alpha$) | $V = \dot{\lambda}$ ($V = L\, di/dt$) | $P = \dot{\Gamma}$ ($P = I\, dQ/dt$) |
| Capacitor energy storage | $F = kx$ | $T = k\theta$ | $V = (1/C)q$ | $P = (1/C)\forall$ |
| Dissipative | $F = bv$ | $T = B\omega$ | $V = Ri$ | $P = RQ$ |

quately modeled by second order linear differential equations, so there are no direct analogies for the relations given in the table.

Figure 4.21 illustrates an example of electrical, mechanical, and hydraulic system analogies along with their governing equations. Each of these systems is described by differential equations of exactly the same form. The only differences are the constant parameters that describe the analogous elements. Any quantity in one system is directly analogous to a quantity in each of the other two systems. For example, the capacitor in the electrical system is analogous to the spring in the mechanical system and to the tank in the hydraulic system, and the current through the capacitor ($I_2 = I_1 - I_3$) is analogous to the rate of compression of the spring ($v_2 = v_1 - v_3$) in the mechanical system and to the flow rate into the tank ($Q_2 = Q_1 - Q_3$) in the hydraulic system.

Knowledge of these analogies helps one gain a deeper understanding and intuitive feel for how different systems respond. It also provides a framework to model and analyze a great variety of systems in a generic form. An understanding of the characteristics and analysis of one type of system can directly apply to any other system.

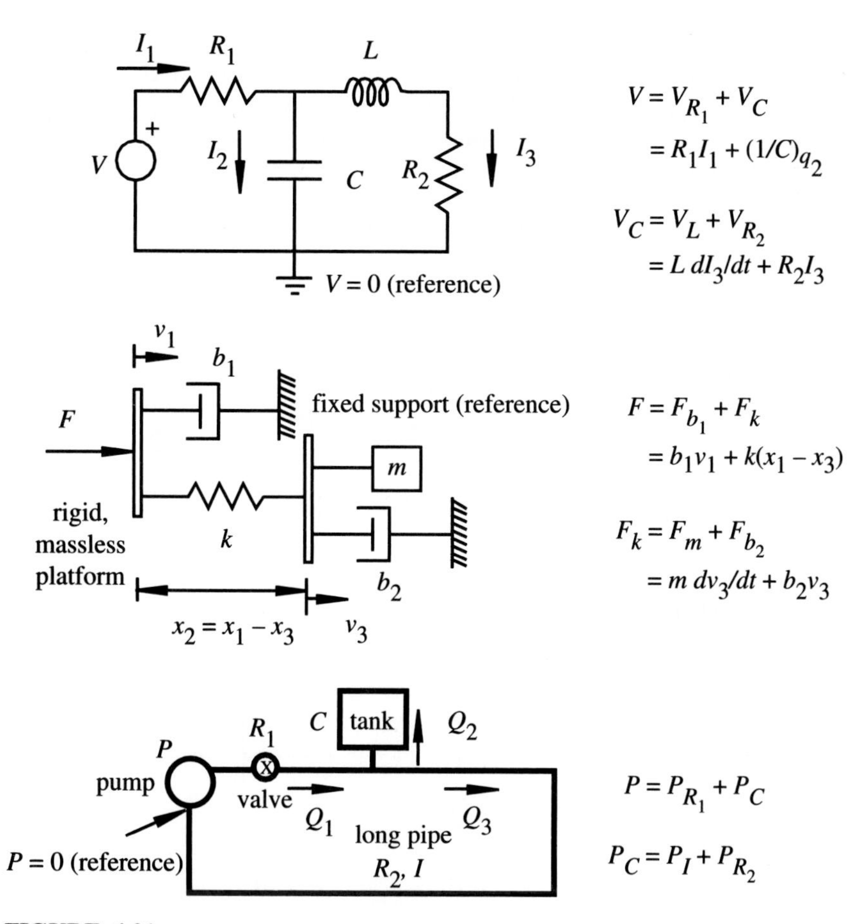


**FIGURE 4.21**
Example of system analogies.

In the past, before the advent of digital computers, modeling analogies were essential in the application of analog computers to simulating and analyzing the behavior of various types of systems. The procedure was to develop an electrical system analogous to the physical system of interest; build the analogous electrical system with the aid of an analog computer, which provided scaling, integration, differentiation, mixing, and other functions; and perform simulations by providing inputs in the form of voltage signals resulting in voltage or current outputs analogous to quantities of interest in the physical system being simulated. As an example, the behavior of the mechanical system shown in Figure 4.21 could be studied by building the electrical circuit in the figure with $C = s/k$, $R_1 = s \cdot b_1$, $R_2 = s \cdot b_2$, and $L = s \cdot m$, where $s$ is some constant scaling factor. Predicting the force on and speed of the mass $m$ given some applied force $F(t)$ would be a simple matter of applying $V(t) = s \cdot F(t)$ volts to the circuit and monitoring the voltage across and the current through the inductor. The force and speed that would be experienced by the mass in the actual mechanical system would be $F_m = (V_L/s)$ and $v_m = (I_L/s)$. Before the availability of numerical integration simulation software on digital computers, this was the only method for simulating the behavior of physical systems.

▼▼▼ ***CLASS DISCUSSION ITEM 4.9. Initial Condition Analogy.*** What is the electrical analogy for an initial displacement of a spring in a mechanical system? In other words, what is the corresponding initial condition in the electrical system?

Presented below is a procedure that facilitates converting from a system in one form to an analogous system in some other form. As an illustrative example, the mechanical system shown in Figure 4.22 will be converted to its analogous electrical system.

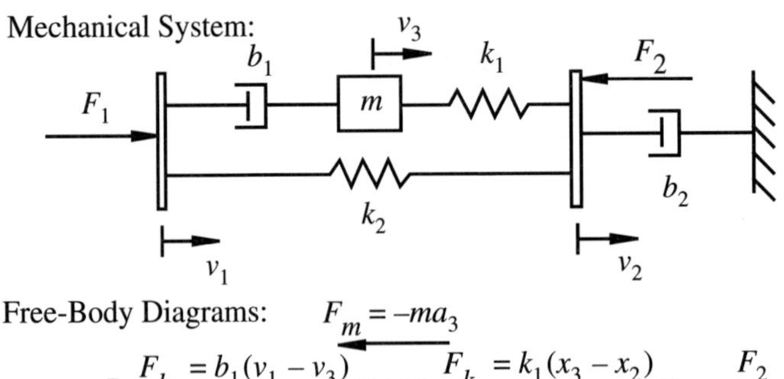


Equations of Motion:

$$F_1 - F_{b_1} - F_{k_2} = 0 \qquad F_{b_1} - F_{k_1} = F_m \qquad F_{k_1} + F_{k_2} - F_2 - F_{b_2} = 0$$

**FIGURE 4.22**
Mechanical system analogy example.

**Steps in Converting from One System to an Analogous System**

1. *Label the flows for each element in the system and its analogy.*

   The algebraic sign given to a flow depends on the direction of the effort on the element. For the example in Figure 4.22, the applied load $F_1$ is in the direction of speed $v_1$, so its flow is $v_1$, but applied load $F_2$ is in the opposite direction to $v_2$, thus its flow is $-v_2$. For the spring and damper flows, since the forces are drawn assuming compression, the flow for $b_1$ is $(v_1 - v_3)$ and the flow for $k_1$ is $(v_3 - v_2)$. The quantities for the analogous system are obtained by direct substitutions with quantities based on Table 4.1. The flows and analogies for the entire example are shown in the table below.

| Mechanical element | Mechanical flow | Electrical flow | Electrical element |
|---|---|---|---|
| $F_1$ | $v_1$ | $I_1$ | $V_1$ |
| $b_1$ | $v_1 - v_3$ | $I_1 - I_3$ | $R_1$ |
| $m$ | $v_3$ | $I_3$ | $L$ |
| $k_1$ | $v_3 - v_2$ | $I_3 - I_2$ | $C_1$ |
| $k_2$ | $v_1 - v_2$ | $I_1 - I_2$ | $C_2$ |
| $b_2$ | $v_2$ | $I_2$ | $R_2$ |
| $F_2$ | $-v_2$ | $-I_2$ | $V_2$ |

2. *Formulate the system equations at each node using the equations of motion ($\Sigma F = ma$) for mechanical systems or KVL loop equations for electrical systems.*

   For the example, see the free-body diagrams and system equations in Figure 4.22. Note that for the massless, rigid platforms the equation is of the form $\Sigma F = 0$ since they are assumed to have negligible mass, and for the mass $m$ the form is $\Sigma F = ma$.

3. *Pick an element to start with and begin constructing the schematic for the analogy using the flows for guidance. It is best to start with an element that is embedded in the system (its flow affects many other elements), although any element would suffice.*

   For the example, start with inductor $L$ (mass $m$) and begin recursively branching to other elements that involve its flow ($I_3$). Note that the FBD or KVL equations and the element flow relationships need to be enforced as the schematic is constructed. The beginning of the analog electrical circuit construction would be as shown in Figure 4.23.

4. *Verify the graph with the analogous system equations.*

   For the example, the KVL loop equations indeed are the same form as the free-body diagram equations of motion (see Figures 4.22 and 4.24).

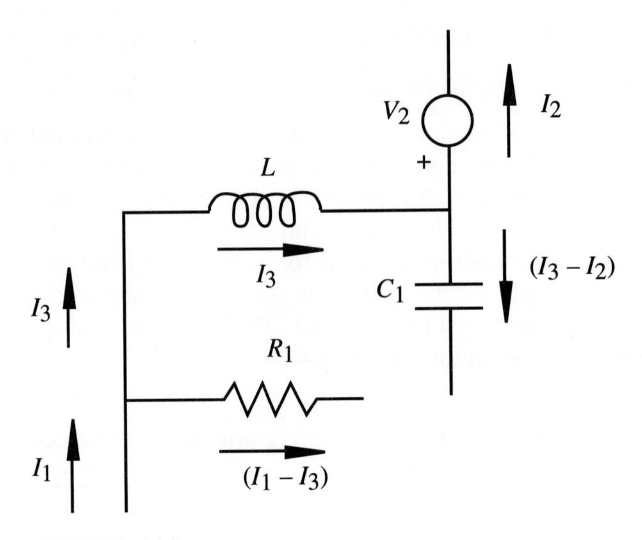


**FIGURE 4.23**
Beginning the analog schematic.

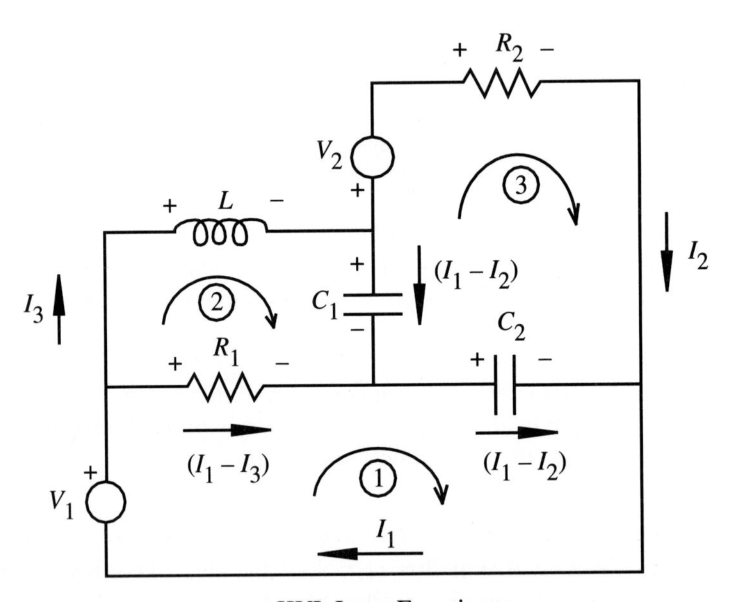


KVL Loop Equations:

① $V_1 = V_{R_1} + V_{C_2}$ ② $V_L = V_{R_1} - V_{C_1}$ ③ $V_2 = V_{C_1} + V_{C_2} - V_{R_2}$

**FIGURE 4.24**
Electrical system analogy example.

▼▼▼ ***CLASS DISCUSSION ITEM 4.10. Measurement System Physical Characteristics.*** Inertia, capacitance, and damping are almost always present in a measurement system. Sometimes these characteristics are desirable; other times they are undesirable. Think of examples of measurement systems containing mechanical, electrical, or hydraulic components and discuss the advantages and disadvantages of changing the physical characteristics of the system. Also describe how these changes might be

## QUESTIONS AND EXERCISES

**4.1.** Show that the Fourier series coefficients $A_n$ for the square wave $F(t)$ defined by Equation 4.6 are zero.

**4.2.** The discrete Fourier series representation of a half-sine-wave pulse train waveform ($V(t)$) is mathematically represented as

$$V(t) = \frac{1}{\pi} + \frac{\sin(2\pi t)}{2} - \frac{2}{\pi}\left[\frac{\cos(4\pi t)}{1 \cdot 3} + \frac{\cos(8\pi t)}{3 \cdot 5} + \frac{\cos(12\pi t)}{5 \cdot 7} + \cdots\right]$$

Using a computer plotting application, plot three cycles of $V(t)$ displaying:

(*a*) Dc component + fundamental (1st harmonic).

(*b*) Dc component + 1st 10 harmonics (hint: some harmonics have zero amplitude).

(*c*) Dc component + as many harmonics as you think are necessary to provide a good reproduction of the waveform.

**4.3.** Given the measurement system frequency response below,

(*a*) What is the bandwidth of the measurement system?

(*b*) If the input signal ($V_{in}$) is a 2 V, peak-to-peak, square wave of period 1 s, what will the steady state output signal ($V_{out}(t)$) from the measurement system be? Use the square wave Fourier series representation given in Equation 4.13 for $V_{in}(t)$.

(*c*) Plot the resulting output on a computer for $t = 0$ to 2 *s*.

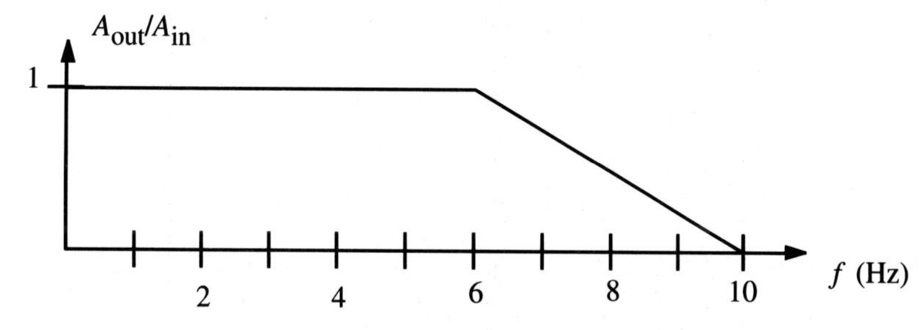

**4.4.** For an $RC$ low-pass filter with $R = 1\ \text{k}\Omega$ and $C = 0.01\ \mu\text{F}$,

(*a*) What is the bandwidth of the system?

(*b*) If the input is a 2 V, peak-to-peak, square wave of period 1 s, what will the filter output be? Use the square wave Fourier series representation given in Equation 4.13 for $V_{in}(t)$.

(*c*) Plot the resulting output on a computer for $t = 0$ to 2 s.

**4.5.** What is the bandwidth (in Hz) of a system with the frequency response below?

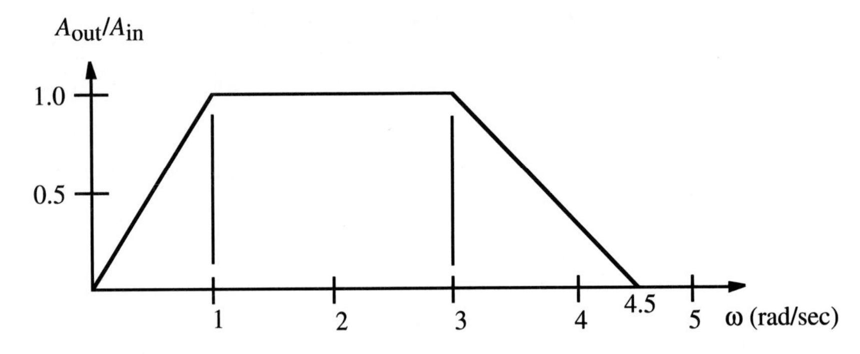


**4.6.** If a signal is given by

$$F(t) = \sum_{n=1}^{5} n \sin(n\omega t)$$

where $\omega = 1$ rad/sec,

(*a*) What is the range of frequencies of the signal?

(*b*) Plot the spectrum of $F(t)$.

(*c*) Assuming the signal is being measured by a measurement system with the frequency response curve shown in Question 4.5, plot the spectrum of the resulting measured output of the system.

**4.7.** Using the Fourier series representation of a square wave in Equation 4.13, plot a square wave using 20 harmonics. Then plot it with the first three harmonics attenuated by 1/4. Finally, plot it with the first three harmonics unattenuated and the next 17 attenuated by 1/4. What do you conclude about the influence of the amplitudes of low and high harmonics?

**4.8.** Determine and plot the frequency response curve for a high-pass filter (see Example 4.1).

**4.9.** For a spring-mass-damper system with $F_{ext}(t) = 20 \sin(0.75t)$ N, $m = 10$ kg, $k = 12$ N/m, and $b = 10$ Ns/m, what is the equation for the steady state sinusoidal response of $x(t)$?

**4.10.** Convert the translational mechanical system on the next page to an analogous electrical system.

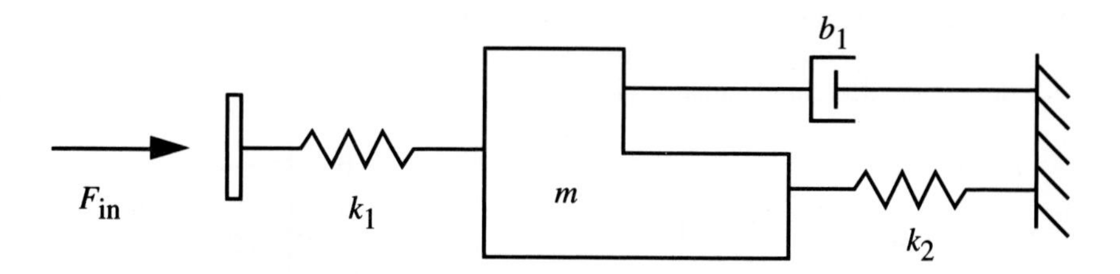


**4.11.** Convert the following electrical system to an analogous hydraulic system.

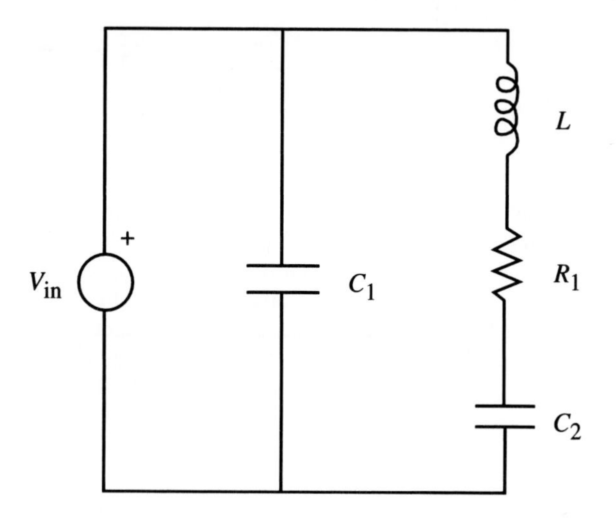


**4.12.** For the translational mechanical system below, draw free-body diagrams for each node, list the flow for each element, and construct (draw) an analogous electrical system with all elements and flows labeled.

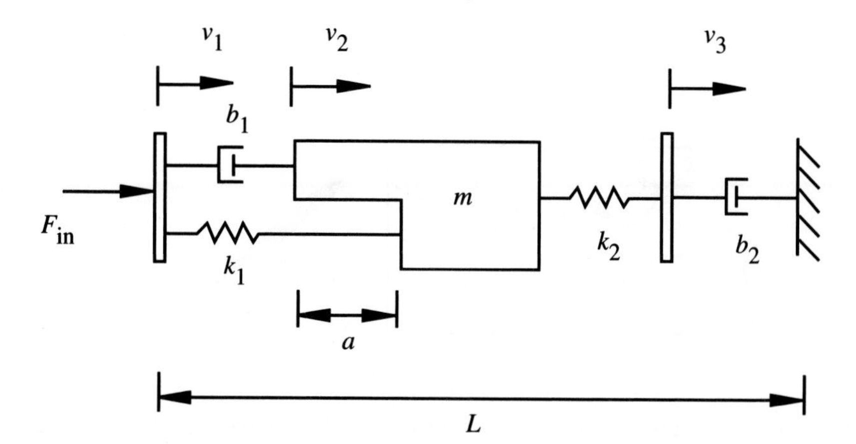


**4.13.** For the three-degrees-of-freedom mechanical system on the next page, draw free-body diagrams for each of the components, carefully showing all external forces and moments. Be sure to show the reference positions for the position variables on the free-body diagrams. Directly from the component free-body diagrams, write down the differential equations of motion.

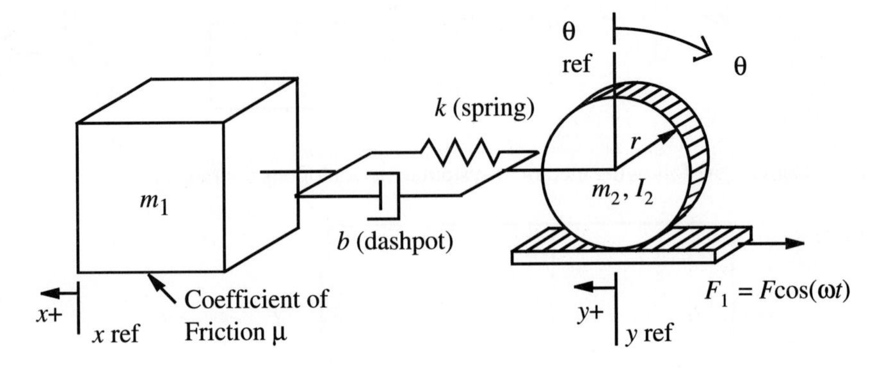


**4.14.** A cheap, mechanically actuated, strip chart recording system consisting of a plotter head of mass $m$ driven by an elastic belt of stiffness $k$ is illustrated below. The input displacement $x_{in}$ is coupled mechanically to the pen head. The spring constant $k$ is 32.4 N/m and the mass of the pen head is 0.10 kg. Derive the differential equation of motion given the displacement input. Assume that the damping constant between the pen and paper is 10 Ns/m. Using the procedure in Section 4.13, determine the steady state output displacement magnitude and phase angle for each steady state input below. Explain the differences between the output and input displacements and phase angles.

(a) $x_{in} = 0.05\sin(10t)$

(b) $x_{in} = 0.05\sin(1000t)$

(c) $x_{in} = 0.05\sin(10{,}000\,t)$

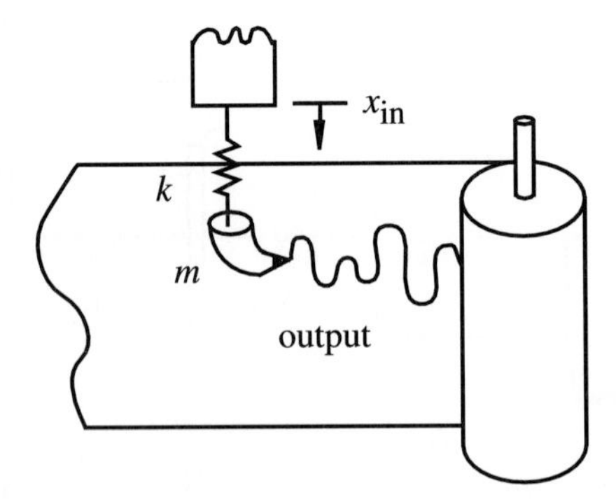


**4.15.** Derive the expression for the step response $x(t)$ in Design Example 4.1 from Equation 4.60.

## BIBLIOGRAPHY

Beckwith, T., Marangoni, R., and Lienhard, J., *Mechanical Measurement,* 5th Edition, Addison-Wesley, Reading, MA, 1993.

Doebelin, E., *Measurement Systems Application and Design,* McGraw-Hill, New York, 1990.

Figliola, R. and Beasley, D., *Theory and Design for Mechanical Measurements,* John Wiley, New York, 1995.

Holman, J. P., *Experimental Methods for Engineers,* McGraw-Hill, New York, 1994.

Ross, S., *Introduction to Ordinary Differential Equations,* 3rd Edition, John Wiley, New York, 1980.

Shearer, J., Murphy, A., and Richardson, H., *Introduction to System Dynamics,* Addison-Wesley, Reading, MA, 1971.

Thomson, W., *Theory of Vibration with Applications,* 2nd Edition, Prentice-Hall, Englewood Cliffs, NJ, 1981.

# 5

# ANALOG SIGNAL PROCESSING USING OPERATIONAL AMPLIFIERS

OBJECTIVES: After you read, discuss, study, and apply ideas in this chapter, you will be able to:

- Understand the input-output characteristics of a linear amplifier
- Understand how to use the model of an ideal operational amplifier
- Design an inverting amplifier, noninverting amplifier, summer, difference amplifier, instrumentation amplifier, integrator, differentiator, and sample and hold amplifier
- Understand the characteristics and limitations of a "real" operational amplifier

## 5.1 INTRODUCTION

Since electrical circuits occur in virtually all mechatronic and measurement systems, it is essential that engineers develop a basic understanding of the acquisition and processing of electrical signals. Usually these signals come from transducers, which convert physical quantities (e.g., temperature, strain, displacement, flow rate) into currents or voltages, usually the latter. The transducer output is usually described as an ***analog signal***, which is continuous and time varying.

Often the signals from transducers are not in the form we would like them to be. They may:

- Be too small, usually in the millivolt range
- Be too "noisy," usually due to electromagnetic interference

- Contain the wrong information, usually due to poor transducer design or installation
- Have a dc offset, usually due to the transducer and instrumentation design

Many of the problems listed above can be remedied, and the desired signal information can be extracted through appropriate analog signal processing. The simplest and most common form of signal processing is amplification where the magnitude of the voltage signal is increased. Other forms include signal inversion, differentiation, integration, addition, subtraction, and comparison.

Analog signals are very different from digital signals, which are discrete, taking only a finite number of states or values. Since computers and microprocessors require digital signals, any application involving computer measurement or control requires analog to digital (A/D) conversion. This chapter covers the basic elements of analog signal processing including the design and analysis of signal processing circuits. The operational amplifier is an integrated circuit that is used as a building block in many of these circuits. Chapter 6 focuses on digital circuits, and Chapter 7 deals with converting analog signals into a format that can be processed by digital devices such as computers.

## 5.2 AMPLIFIERS

People have spent their lives studying and writing about amplifiers, so we can't expect to do justice to the subject in a few pages. However, we will look at the salient features of amplifiers and determine how we may design one using integrated circuits.

Ideally, an ***amplifier*** increases the amplitude of a signal without affecting the phase relationship of different components of the signal. When choosing or designing an amplifier, we must consider size, cost, power consumption, input impedance, output impedance, gain, and bandwidth. Physical size depends on the components used to construct the amplifier. Prior to the 1960s vacuum tube amplifiers were common, but they were heavy power consumers with significant heat dissipation. Portable units were large and heavy and required frequent battery replacement. The advent of ***solid state*** technology, where charge carriers move through a solid semiconductor material, has replaced the vacuum tube technology, where bulky tubes enclosed a gas at low pressure through which electrons flowed. Today, solid state transistors and integrated circuits have dramatically changed amplifier design, resulting in small, cool-running amplifiers. They are relatively light power consumers easily made portable with rechargeable batteries.

Generally, we model the amplifier as a two port device, with an input and output voltage referenced to ground, as illustrated in Figure 5.1. The voltage ***gain*** of an amplifier is defined as the ratio of the output and input voltage amplitudes:

$$A_v = \frac{V_{\text{out}}}{V_{\text{in}}} \tag{5.1}$$

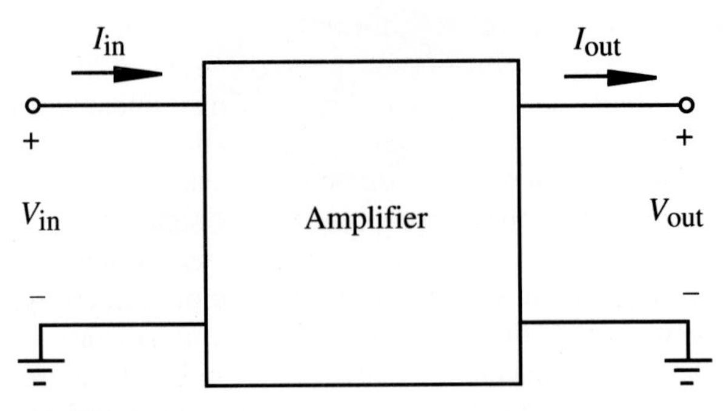


**FIGURE 5.1**
Amplifier model.

Normally we want an amplifier to exhibit amplitude linearity where the gain is constant for all frequencies. However, amplifiers may be designed to intentionally amplify only certain frequencies, resulting in a filtering effect. In such cases, the output characteristics are governed by the amplifier's bandwidth and associated cutoff frequencies.

The input and output impedances of an amplifier, $Z_{in}$ and $Z_{out}$, are found by measuring the ratio of the respective voltage and current:

$$Z_{in} = V_{in}/I_{in} \tag{5.2}$$

$$Z_{out} = V_{out}/I_{out} \tag{5.3}$$

For the operational amplifiers described in the next section, $Z_{in}$ is larger than 100 KΩ and $Z_{out}$ is a few ohms or less.

## 5.3 OPERATIONAL AMPLIFIERS

The ***operational amplifier***, or ***op amp***, is a low-cost and versatile integrated circuit consisting of many internal transistors, resistors, and capacitors. It can be combined with external discrete components to create a wide variety of signal processing circuits. The op amp is the basic building block for:

- Amplifiers
- Integrators
- Summers
- Differentiators
- Comparators
- A/D and D/A converters
- Active filters
- Sample and hold amplifiers

We will consider most of these applications in subsequent sections. The op amp derives its name from the fact that it can perform so many different operations.

## 5.4
## IDEAL MODEL FOR THE OPERATIONAL AMPLIFIER

Figure 5.2 shows the schematic symbol and terminal nomenclature for an ideal op amp. It is a differential input, single output amplifier that is assumed to have infinite gain. The ∞ symbol is sometimes used in the schematic to denote the infinite gain and the assumption that it is an ideal op amp. The voltages are all referenced to a common ground. The op amp is an ***active device*** requiring connection to an external power supply, usually plus and minus 15 volts. The external supply is not normally shown on circuit schematics. Since the op amp is an active device, output voltages and currents can be larger than the values applied to the inverting and noninverting terminals.

As illustrated in Figure 5.3, an op amp circuit usually includes ***feedback*** from the output to the negative (inverting) input. This so-called ***closed loop*** configuration results in the stabilization of the amplifier and control of the gain. When feedback is absent in an op amp circuit, the op amp is said to be in an ***open loop*** configuration. This configuration results in considerable instability due to the infinite gain, and it is seldom used. The utility of feedback will become evident in the examples presented in the following sections.

Figure 5.4 illustrates an ideal model that can aid in analyzing circuits containing op amps. This model is based on the following assumptions that describe an ideal op amp:

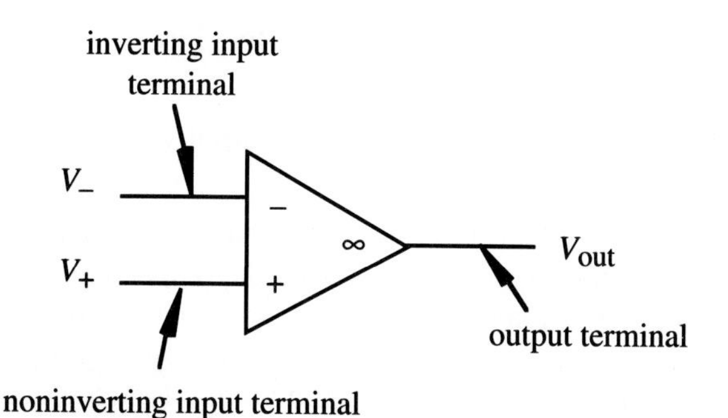


**FIGURE 5.2**
Op amp terminology and schematic representation.

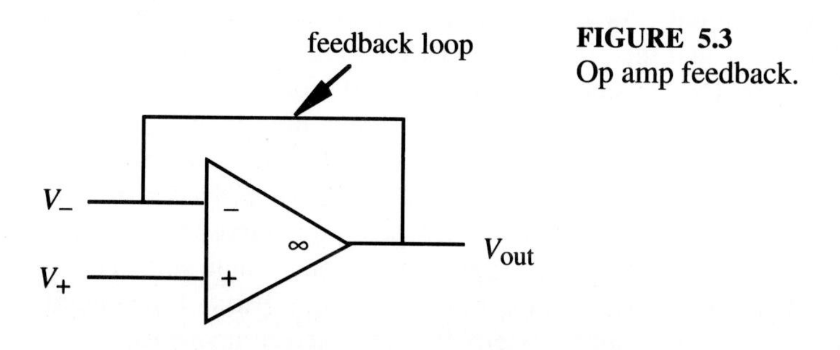


**FIGURE 5.3**
Op amp feedback.

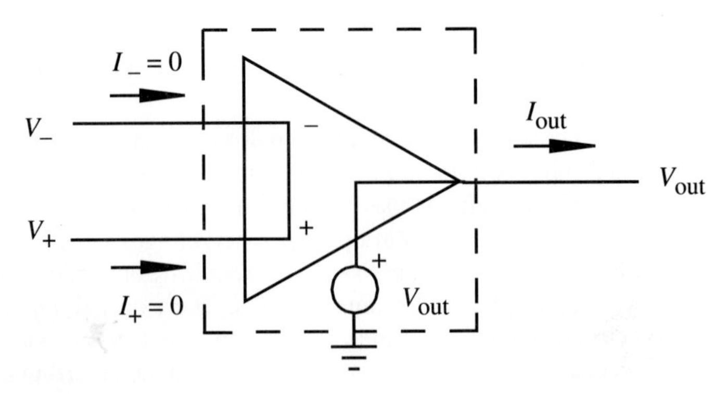


**FIGURE 5.4**
Op amp equivalent circuit.

1. It has infinite impedance at both inputs, hence no current is drawn from the input circuits. Therefore,

$$I_+ = I_- = 0 \tag{5.4}$$

2. It has infinite gain. As a consequence, the difference between the input voltages must be zero; otherwise, the output would be infinite. This is denoted in Figure 5.4 by the shorting of the two inputs. Therefore,

$$V_+ = V_- \tag{5.5}$$

   Even though we indicate a short between the two inputs, we assume no current may flow through this short.

3. It has zero output impedance. Therefore, the output voltage does not depend on the output current.

Note that $V_{out}$, $V_+$, and $V_-$ are all referenced to a common ground. Also, for stable linear behavior, there must be feedback between the output and the inverting input.

The assumptions above may at first seem to stretch common sense, but they do provide a close approximation to the behavior of a real op amp. With the aid of this ideal model, we only need Kirchoff's laws and Ohm's Law to completely analyze op amp circuits.

Actual op amps are usually packaged in 8-pin dual in-line package (DIP) integrated circuit (IC) chips. 741 is the designation for a general purpose op amp produced by many IC manufacturers. It is illustrated in Figure 5.5 with its pin configuration (pin-out). As with all ICs, one end of the chip is marked with an indentation or spot, and the pins are numbered counterclockwise and consecutively starting with 1 at the left side of the marked end. For a 741 series op amp, pin 2 is the inverting input, pin 3 is the noninverting input, pins 4 and 7 are for the external power supply, and pin 6 is the op amp output. Pins 1, 5, and 8 are not normally connected. Figure 5.6 illustrates the internal design of a 741 IC available from National Semiconductor. Notice that the circuits are composed of transistors, resistors, and capacitors that are easily manufactured on a single silicon chip. The most valuable details for the user are the input and output parts of the circuit that have characteristics that might affect externally connected components.

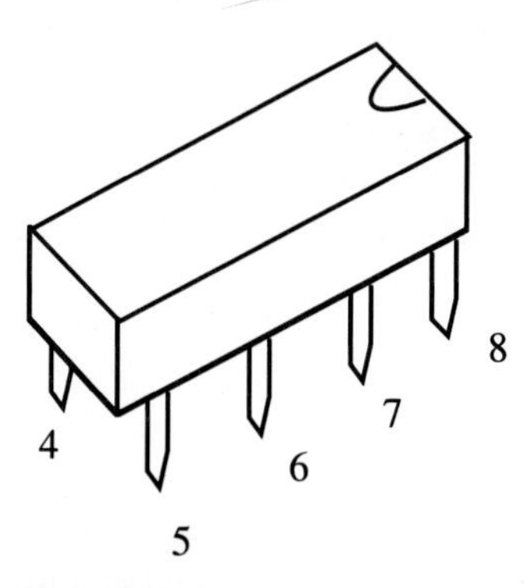


| | | | |
|---|---|---|---|
| 1 | x | x | 8 |
| 2 | – | +15 V | 7 |
| 3 | + | $V_{out}$ | 6 |
| 4 | –15 V | x | 5 |

**FIGURE 5.5**
741 op amp pin-out.

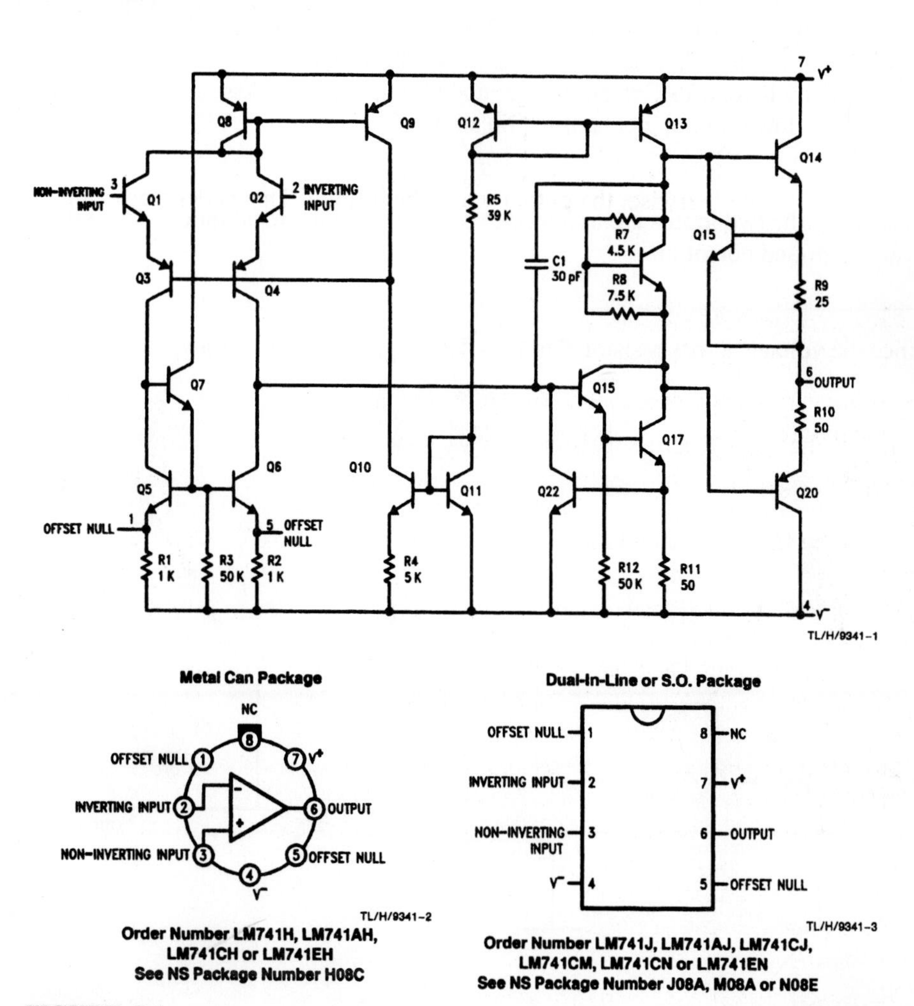


**FIGURE 5.6**
741 internal design. *(Courtesy of National Semiconductor, Santa Clara, CA)*

There are many different op amp designs available from IC manufacturers. The input impedances, bandwidth, and power ratings can vary significantly. Also, some require only a single output power supply. Although the 741 is widely used, another common op amp is the TL071 manufactured by Texas Instruments. Its pin configuration is identical to the 741, but because it has FET inputs, it has a larger input impedance (10 MΩ) and a wider bandwidth.

## 5.5 INVERTING AMPLIFIER

An ***inverting amplifier*** is constructed by connecting two external resistors to an op amp as shown in Figure 5.7. As the name implies, this circuit inverts and amplifies the input voltage. Notice that the resistor $R_F$ forms the feedback loop. This feedback loop always goes to the "–" or inverting input of the op amp, implying negative feedback. We now use Kirchoff's laws and Ohm's Law to analyze this circuit. First, we replace the op amp with its ideal model shown within the dashed box in Figure 5.8.

Applying Kirchoff's Current Law at node C and utilizing Assumption 1 that no current can flow into the inputs of the op amp,

$$i_{in} = -i_{out} \tag{5.6}$$

Also, since the two inputs are assumed to be shorted in the ideal model, C is effectively at ground potential:

$$V_C = 0 \tag{5.7}$$

Since the voltage across resistor $R$ is $V_{in} - V_C = V_{in}$, from Ohm's Law,

$$V_{in} = i_{in}R \tag{5.8}$$

and since the voltage across resistor $R_F$ is $V_{out} - V_C = V_{out}$,

$$V_{out} = i_{out}R_F \tag{5.9}$$

Substituting Equation 5.6 into Equation 5.9 gives

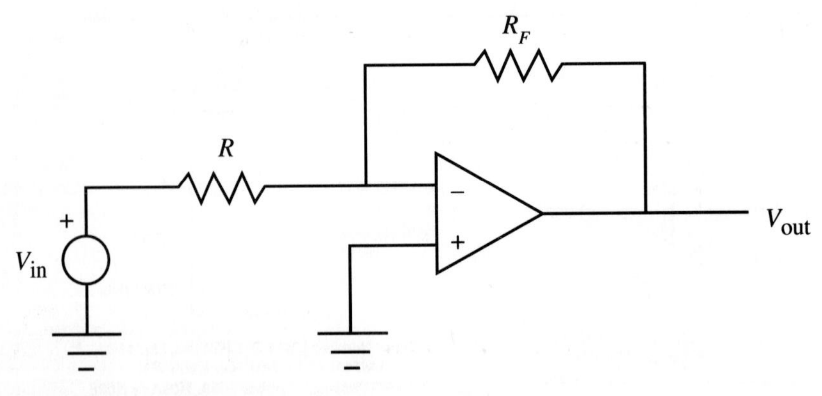


**FIGURE 5.7**
Inverting amplifier.

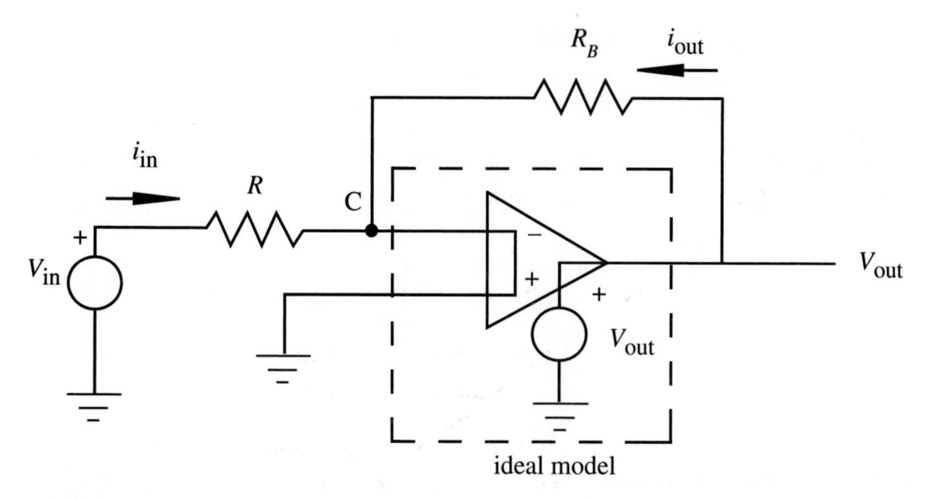


**FIGURE 5.8**
Equivalent circuit for inverting amplifier.

$$V_{out} = -\, i_{in} R_F \tag{5.10}$$

Dividing Equation 5.10 by Equation 5.8 yields the input-output relationship:

$$\frac{V_{out}}{V_{in}} = -\frac{R_F}{R} \tag{5.11}$$

Therefore, the voltage gain of the amplifier is determined simply by the external resistors $R_F$ and $R$, and it is always negative. The reason this circuit is called an inverting amplifier is that it reverses the polarity of the input signal. This results in a phase shift of 180° for periodic signals. For example, if the square wave $V_{in}$ shown in Figure 5.9 is connected to an inverting amplifier with a gain of –2, the output $V_{out}$ will be inverted and amplified, resulting in a larger amplitude signal 180° out of phase with the input.

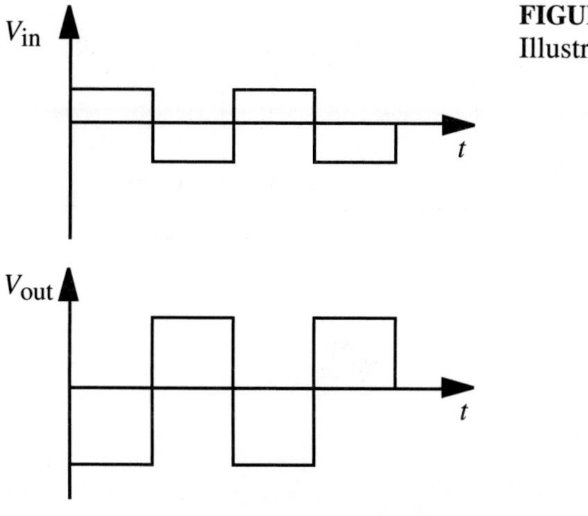


**FIGURE 5.9**
Illustration of inversion.

▼▼▼ ***CLASS DISCUSSION ITEM 5.1. Kitchen Sink in OP Amp Circuits.***
Consider the op amp circuit shown below:

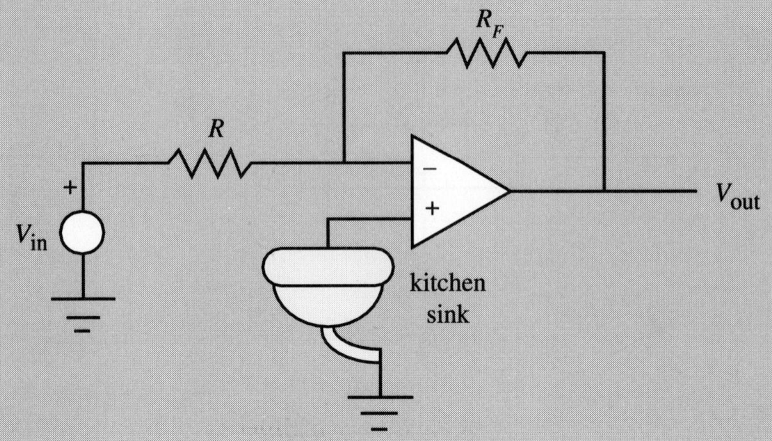


What is the effect of the kitchen sink at the noninverting input of the op amp?

## 5.6 NONINVERTING AMPLIFIER

The schematic of a ***noninverting amplifier*** is shown in Figure 5.10. As the name implies, this circuit amplifies the input voltage without inverting the signal. Again we may apply Kirchoff's laws and Ohm's Law to determine the voltage gain of this amplifier. As before, we replace the op amp with the ideal model shown in the dashed box in Figure 5.11.

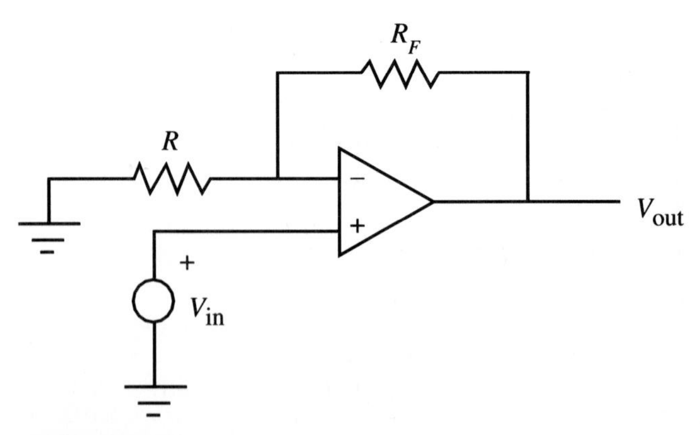


**FIGURE 5.10**
Noninverting amplifier.

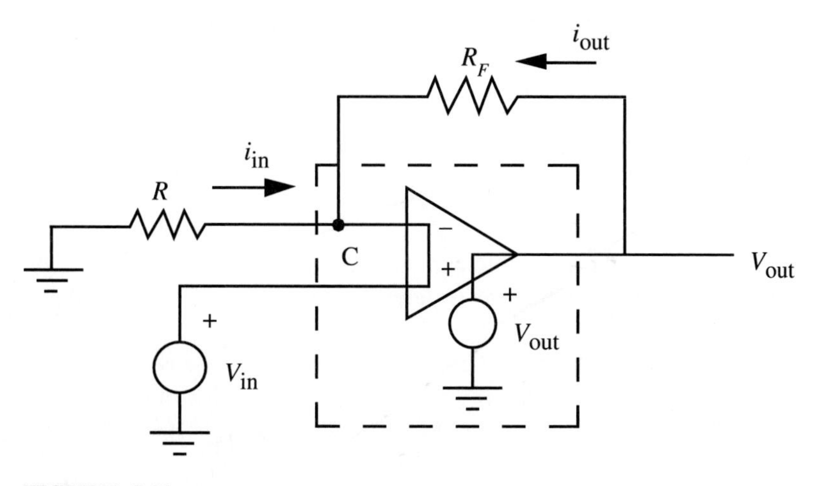


**FIGURE 5.11**
Equivalent circuit for noninverting amplifier.

The voltage at node C is $V_{in}$ since the inverting and noninverting inputs are at the same voltage. Therefore, applying Ohm's Law to resistor $R$,

$$i_{in} = \frac{-V_{in}}{R} \tag{5.12}$$

and applying it to resistor $R_F$,

$$i_{out} = \frac{V_{out} - V_{in}}{R_F} \tag{5.13}$$

Solving Equation 5.13 for $V_{out}$ gives

$$V_{out} = i_{out}R_F + V_{in} \tag{5.14}$$

Applying KCL at node C gives

$$i_{in} = -i_{out} \tag{5.15}$$

so Equation 5.12 can be written as

$$V_{in} = i_{out}R \tag{5.16}$$

Using Equations 5.14 and 5.16, the voltage gain can be written as

$$\frac{V_{out}}{V_{in}} = \frac{i_{out}R_F + V_{in}}{V_{in}} = \frac{i_{out}R_F + i_{out}R}{i_{out}R} = 1 + \frac{R_F}{R} \tag{5.17}$$

Therefore, the noninverting amplifier has a positive gain greater than or equal to one. This is useful in isolating one portion of a circuit from another by transmitting a scaled voltage without drawing appreciable current. If we let $R_F = 0$ and $R = \infty$, the resulting circuit can be represented as shown in Figure 5.12. This circuit is known as a ***buffer*** or ***follower*** since $V_{out} = V_{in}$. It has a high input impedance and low output impedance. This circuit is useful in applications where you need to

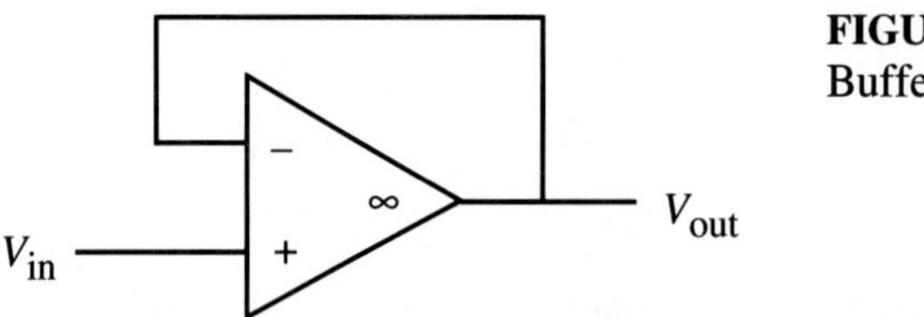


**FIGURE 5.12**
Buffer or follower.

couple to a voltage signal without loading the source of the voltage. The high input impedance of the op amp effectively ***isolates*** the source from the rest of the circuit.

▼▼▼ *CLASS DISCUSSION ITEM 5.2.* ***Positive Feedback.*** Ideally, what would happen to the output of the buffer amplifier shown in Figure 5.12 if feedback was from the output to the noninverting input instead?

▼▼▼ *CLASS DISCUSSION ITEM 5.3.* ***Example of Positive Feedback.*** A good example of positive feedback is the effect Jimi Hendrix used to achieve when he would move his guitar close to the front of his amplifier speaker. Describe the effect of this technique and describe what is going on physically.

## 5.7 SUMMER

The ***summer*** op amp circuit shown in Figure 5.13 is used to add analog signals. By analyzing the circuit with

$$R_1 = R_2 = R_F \tag{5.18}$$

we can show (see Question 5.1) that

$$V_{out} = -(V_1 + V_2) \tag{5.19}$$

Thus the circuit output is the negative sum of the inputs.

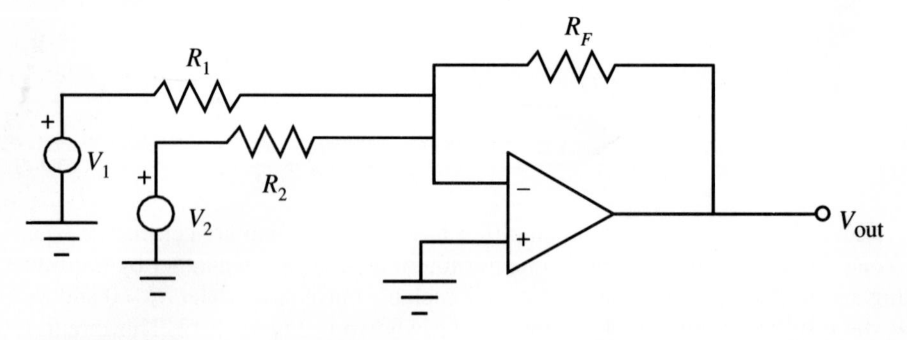


**FIGURE 5.13**
Summer circuit.

## 5.8
## DIFFERENCE AMPLIFIER

The ***difference amplifier*** circuit shown in Figure 5.14 is used to subtract analog signals. In analyzing this circuit we will use the principle of ***superposition,*** which states that whenever multiple inputs are applied to a linear system (e.g., an op amp circuit) we can analyze the circuit and determine the response for each of the individual inputs independently. The sum of the individual responses is equivalent to the overall response to the multiple inputs. Specifically, when the inputs are ideal voltage sources, to analyze the response due to one source, the other sources are shorted. If some inputs are current sources, they are successively replaced by open circuits.

The first step in analyzing the circuit in Figure 5.14 is to replace $V_2$ with a short circuit effectively grounding $R_2$. As shown in Figure 5.15, the result is an inverting amplifier (see Figure 5.7 and Class Discussion Item 5.1). Therefore, from Equation 5.11 the output due to input $V_1$ is:

$$V_{\text{out}_1} = -\frac{R_F}{R_1}V_1 \tag{5.20}$$

The second step in analyzing the circuit in Figure 5.14 is to replace $V_1$ with a short circuit effectively grounding $R_{1,}$ as shown in Figure 5.16a. This circuit is equivalent to the circuit shown in Figure 5.16b where the input voltage is

$$V_3 = \frac{R_F}{R_2 + R_F}V_2 \tag{5.21}$$

since $V_2$ is divided between resistors $R_2$ and $R_F$.

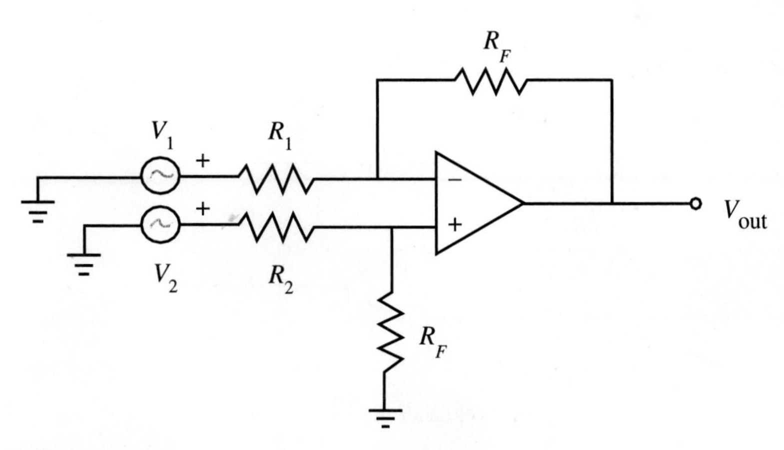


**FIGURE 5.14**
Difference amplifier circuit.

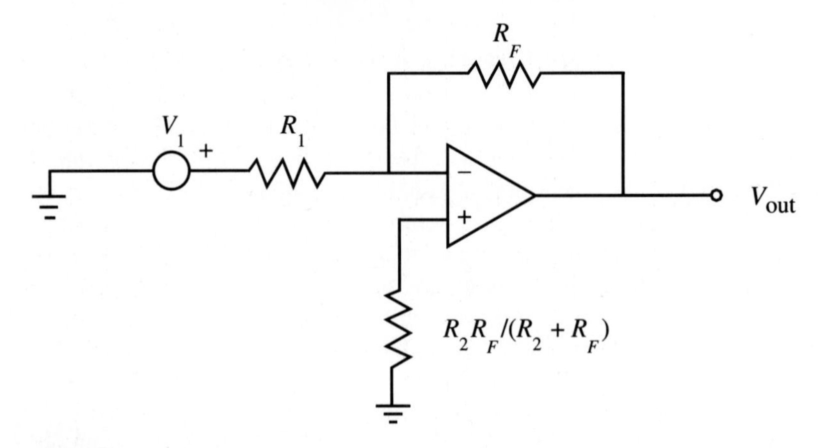


**FIGURE 5.15**
Difference amplifier with $V_2$ shorted.

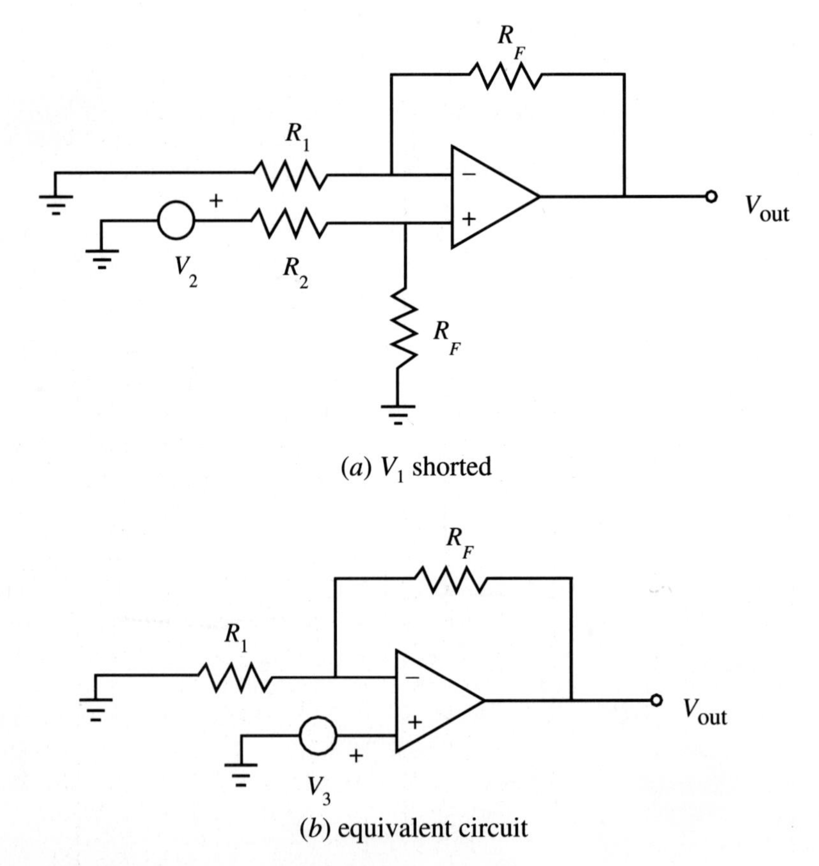


**FIGURE 5.16**
Difference amplifier with $V_1$ shorted.

The circuit in Figure 5.16b is a noninverting amplifier (see Figure 5.10). Therefore, the output due to input $V_2$ is given by Equation 5.17:

$$V_{\text{out}_2} = \left(1 + \frac{R_F}{R_1}\right)V_3 \tag{5.22}$$

By substituting Equation 5.21, this equation can be written as

$$V_{\text{out}_2} = \left(1 + \frac{R_F}{R_1}\right)\left(\frac{R_F}{R_2 + R_F}\right)V_2 \tag{5.23}$$

The principle of superposition implies that the total output $V_{\text{out}}$ is the sum of the outputs due to the individual inputs:

$$V_{\text{out}} = V_{\text{out}_1} + V_{\text{out}_2} = -\left(\frac{R_F}{R_1}\right)V_1 + \left(1 + \frac{R_F}{R_1}\right)\left(\frac{R_F}{R_2 + R_F}\right)V_2 \tag{5.24}$$

When $R_1 = R_2 = R$, the output voltage is an amplified difference of the input voltages:

$$V_{\text{out}} = \frac{R_F}{R}(V_2 - V_1) \tag{5.25}$$

This result can also be obtained using the op amp rules, KCL, and Ohm's Law (see Question 5.3).

## 5.9
## INSTRUMENTATION AMPLIFIER

The difference amplifier presented in the last section may be satisfactory for low impedance sources, but its input impedance is too low for high output impedance sources. Furthermore, if the input signals are very low level and include noise, the difference amplifier is unable to extract a satisfactory difference signal. The solution to this problem is the ***instrumentation amplifier.*** It has the following characteristics:

- Very high input impedance.
- Large ***common mode rejection ratio (CMRR)***. The CMRR is the ratio of the difference mode gain to the common mode gain. The ***difference mode gain*** is the amplification factor for the difference between the input signals, and the ***common mode gain*** is the amplification factor for the average of the input signals. For an ideal difference amplifier, the common mode gain is zero, implying an infinite CMRR. When the common mode gain is nonzero, the output will be nonzero when the inputs are equal and nonzero. It is desirable to minimize the common mode gain to suppress signals such as noise that are common to both inputs.
- Capability to amplify low-level signals in a noisy environment, often a requirement in differential output sensor signal conditioning applications.
- Consistent bandwidth over a large range of gains.

Instrumentation amplifiers are commercially available as monolithic ICs (e.g., Analog Devices 524 and 624 and National Semiconductor LM 623). A single external resistor is used to set the gain. This gain can be higher and is more stable than gains achievable with a simple difference amplifier.

An instrumentation amplifier can also be constructed with inexpensive discrete op amps and precision resistors as illustrated in Figure 5.17. We will analyze this circuit in two parts. The two op amps on the left provide a high impedance amplifier stage where each input is amplified separately. This stage involves a moderate CMRR. The outputs $V_3$ and $V_4$ are supplied to the op amp circuit on the right, which is a difference amplifier with a potentiometer $R_5$ used to maximize the overall CMRR.

We first apply KCL and Ohm's Law to the left portion of the circuit to express $V_3$ and $V_4$ in terms of $V_1$ and $V_2$. Using the assumptions and rules for an ideal op amp, it's clear that the current $I_1$ passes through both feedback resistors $R_2$ and $R_1$. Applying Ohm's Law to the feedback resistors gives

$$V_3 - V_1 = I_1 R_2 \tag{5.26}$$

$$V_2 - V_4 = I_1 R_2 \tag{5.27}$$

Applying Ohm's Law to $R_1$ gives

$$V_1 - V_2 = I_1 R_1 \tag{5.28}$$

To express $V_3$ and $V_4$ in terms of $V_1$ and $V_2$, we eliminate $I_1$ by solving Equation 5.28 for $I_1$ and substituting it into Equations 5.26 and 5.27. The results are

$$V_3 = \left(\frac{R_2}{R_1} + 1\right)V_1 - \frac{R_2}{R_1}V_2 \tag{5.29}$$

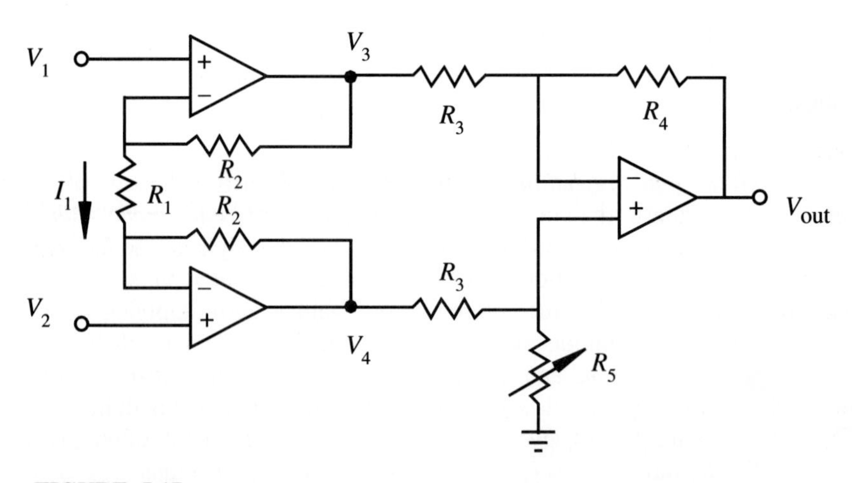


**FIGURE 5.17**
Instrumentation amplifier.

$$V_4 = -\frac{R_2}{R_1}V_1 + \left(\frac{R_2}{R_1} + 1\right)V_2 \tag{5.30}$$

By analyzing the right portion of the circuit, it can be shown that (see Question 5.4 )

$$V_{out} = \frac{R_5(R_3 + R_4)}{R_3(R_3 + R_5)}V_4 - \frac{R_4}{R_3}V_3 \tag{5.31}$$

We can substitute the expressions for $V_3$ and $V_4$ from Equations 5.29 and 5.30 into Equation 5.31 to express the output voltage $V_{out}$ in terms of the input voltages $V_1$ and $V_2$. Assuming $R_5 = R_4$, the result is

$$V_{out} = \left[\frac{R_4}{R_3}\left(1 + 2\frac{R_2}{R_1}\right)\right](V_2 - V_1) \tag{5.32}$$

A design objective for the instrumentation amplifier is to maximize the CMRR by minimizing the common mode gain. For a common mode input, $V_1 = V_2$, Equation 5.32 yields an output voltage $V_{out} = 0$. Hence, the common mode gain is zero and the CMRR is infinite if $R_5 = R_4$. In practice, the resistances will never match exactly. Also, if temperature varies within the discrete circuit, resistance mismatches will be further exaggerated. By using a potentiometer for $R_5$, the designer can minimize the mismatch between $R_5$ and $R_4$, resulting in a maximum CMRR.

The problems of resistor matching with discrete components are avoided by using a monolithic instrumentation amplifier constructed with laser-trimmed resistors. These amplifiers have a very high CMRR not usually obtainable with discrete components. In addition, the gain is programmable by selecting an appropriate external resistor $R_1$.

## 5.10 INTEGRATOR

If the feedback resistor of the inverting op amp is replaced by a capacitor, the result is an ***integrator*** circuit. It is shown in Figure 5.18. Referring to the analysis for the inverting amplifier, Equation 5.9 is replaced by the relationship between voltage and current for a capacitor:

$$\frac{dV_{out}}{dt} = \frac{i_{out}}{C} \tag{5.33}$$

Integrating gives

$$V_{out}(t) = \frac{1}{C}\int_0^t i_{out}(\tau)d\tau \tag{5.34}$$

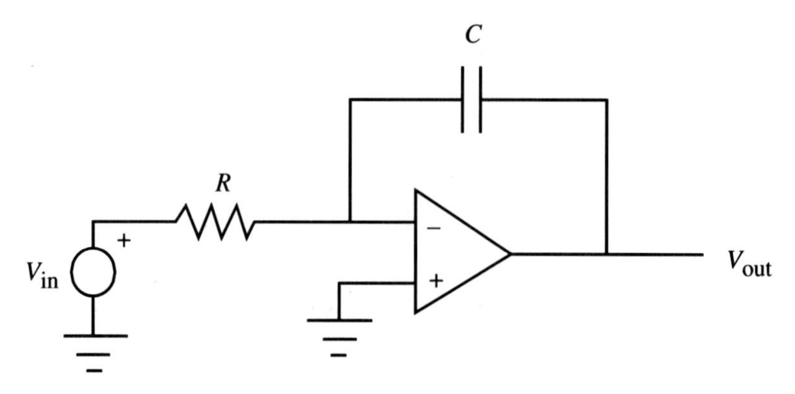


**FIGURE 5.18**
Ideal integrator.

where $\tau$ is a dummy variable of integration. Since $i_{out} = -i_{in}$ and $i_{in} = V_{in} / R$,

$$V_{out}(t) = -\frac{1}{RC}\int_0^t V_{in}(\tau)d\tau \tag{5.35}$$

Therefore, the output signal is a scaled integral of the input signal.

▼▼▼ ***CLASS DISCUSSION ITEM 5.4. Integrator Inputs.*** If a dc voltage is applied as an input to an ideal integrator, how does the output change over time? What is the output given a sinusoidal input? If the integrator is made with a real op amp, what happens to the output in contrast to the ideal integrator?

A more practical integrator circuit is shown in Figure 5.19. The resistor $R_s$ placed across the feedback capacitor is called a ***shunt resistor***. Its purpose is to limit the low-frequency gain of the circuit. This is necessary due to the fact that even a small dc offset at the input would be integrated over time, eventually saturating the op amp. Note that the integrator is only useful when the scaled integral always remains below the maximum output voltage for the op amp (see Section 5.14). As a good rule of thumb, $R_s$ should be greater than $10R_1$.

Because of the impedance and frequency response of the feedback circuit $R_s$ and $C$, the circuit in Figure 5.19 acts as an integrator only for higher frequencies. At low frequencies, the circuit behaves as an inverting amplifier because the impedance of the feedback loop is effectively $R_s$ since the impedance of $C$ is large at low frequencies.

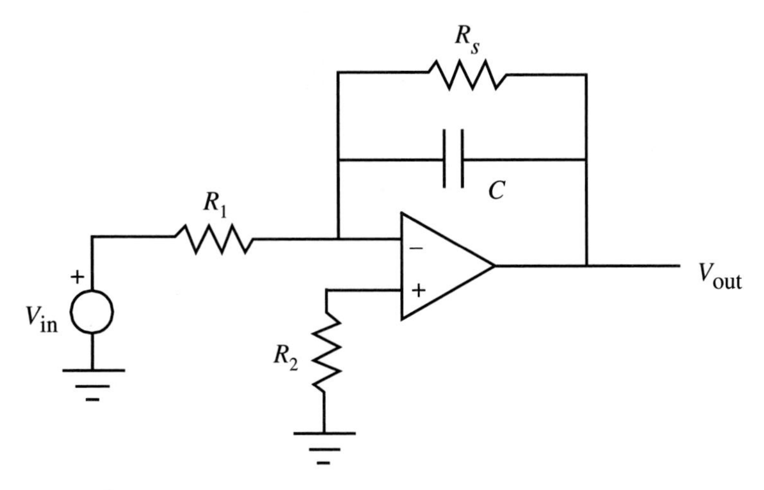


**FIGURE 5.19**
Improved integrator.

Any dc offset due to the input bias currents (see paragraph B in Section 5.14.1) is minimized by $R_2$, which should be chosen to approximate the parallel combination of the input and shunt resistors:

$$R_2 = \frac{R_1 R_s}{R_1 + R_s} \tag{5.36}$$

The reason for this is that the input bias current flowing into the inverting terminal is a result of the currents through $R_1$ and $R_s$, and the input bias current flowing into the noninverting terminal flows through $R_2$. If the voltages generated by the bias currents are the same, they will have no net effect on the output.

## 5.11 DIFFERENTIATOR

If the input resistor of the inverting op amp is replaced by a capacitor, the result is a ***differentiator*** circuit. It is shown in Figure 5.20. Referring to the analysis for the inverting amplifier, Equation 5.8 is replaced by the relationship between voltage and current for a capacitor:

$$\frac{dV_{in}}{dt} = \frac{i_{in}}{C} \tag{5.37}$$

Since $i_{in} = -i_{out}$ and $i_{out} = V_{out} / R$,

$$V_{out} = -RC\frac{dV_{in}}{dt} \tag{5.38}$$

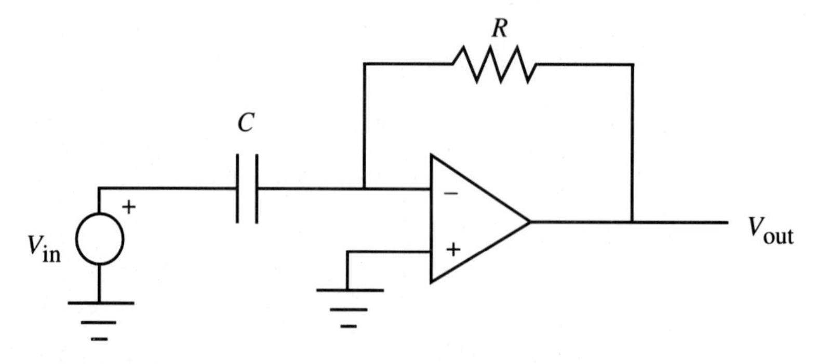


**FIGURE 5.20**
Differentiator.

Therefore, the output signal is a scaled derivative of the input signal.

Differentiation is a signal processing method that tends to accentuate the effects of noise whereas integration smooths signals over time.

▼▼▼ *CLASS DISCUSSION ITEM 5.5. **Differentiator Improvements.*** Recommend possible improvements to the differentiator circuit in Figure 5.20. Consider the effects of high-frequency noise in the input signal.

▼▼▼ *CLASS DISCUSSION ITEM 5.6. **Integrator and Differentiator Applications.*** Think of various applications for the integrator and differentiator circuits. Consider how a differential equation could be solved with an analog computer. Also consider how to convert between sawtooth and square wave function generator outputs and how to process position and speed sensor signals.

## 5.12 SAMPLE AND HOLD CIRCUIT

The ***sample and hold*** circuit is used extensively in analog to digital conversion, where a signal value must be stabilized while it is converted to a digital representation. The circuit that performs this is illustrated in Figure 5.21. It consists of a voltage-holding capacitor and a voltage follower. With switch S closed,

$$V_{out}(t) = V_{in}(t) \tag{5.39}$$

When the switch is opened, the capacitor $C$ holds the input voltage corresponding to the last sampled value, since negligible current is drawn by the follower. Therefore,

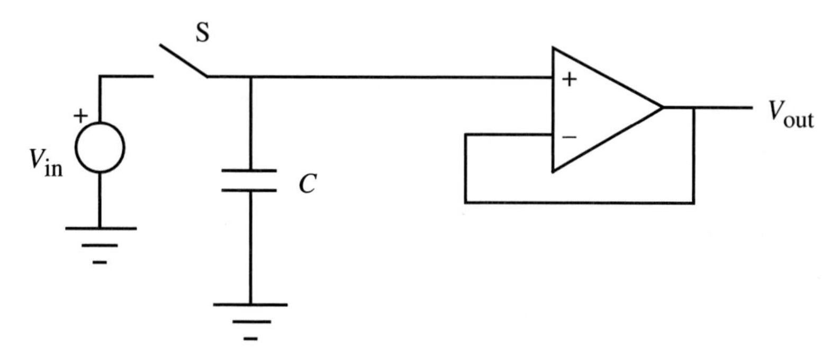


**FIGURE 5.21**
Sample and hold circuit.

$$V_{out}\,(t - t_{sampled}) = V_{in}\,(t_{sampled}) \tag{5.40}$$

where $t_{sampled}$ is the time when the switch was last opened. Often, an op amp buffer is also used on the $V_{in}$ side of the switch to minimize current drain from the input voltage source $V_{in}$.

The type of capacitor used for this application is important. A low-leakage capacitor such as a polystryene or polypropylene would be a good choice. An electrolytic capacitor would be a poor choice because of its high leakage. This leakage would cause the output voltage value to drop during a "hold."

## 5.13 COMPARATOR

The ***comparator*** circuit illustrated in Figure 5.22 is used to determine whether one signal is greater than another. The comparator is an example of an op amp circuit where there is no negative feedback, and the circuit exhibits infinite gain. The result is that the op amp will ***saturate***. Saturation implies that the output remains at its most positive or most negative output value. Certain op amps are specifically designed to operate as comparators (e.g., LM339). The output of the comparator is defined by

$$V_{out} = \begin{cases} +V_{sat} & V_{in} > V_{ref} \\ -V_{sat} & V_{in} < V_{ref} \end{cases} \tag{5.41}$$

where $V_{sat}$ is the saturation voltage of the comparator and $V_{ref}$ is the reference voltage to which the input voltage $V_{in}$ is being compared. The positive saturation value is slightly less than the positive supply voltage, and the negative saturation value is greater than the negative supply voltage.

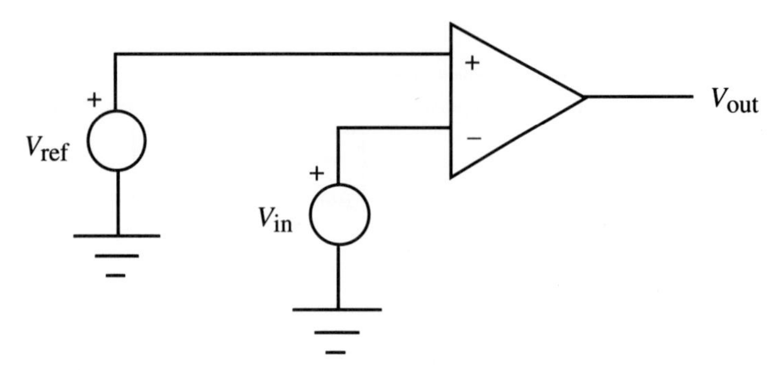


**FIGURE 5.22**
Comparator.

## 5.14 THE REAL OP AMP

An actual operational amplifier deviates somewhat in characteristics from an ideal op amp. The best way to familiarize yourself with an IC is to review its specifications in the data book provided by the manufacturer. Complete descriptions of op amps and many other analog ICs are found in manufacturers' LINEAR data books. Some of the more important parameters that can be found on op amp data sheets are described in the next section.

Corresponding to ideal operational amplifiers, real op amps have a very high input impedance, so very little current is drawn at the inputs. At the same time, there is very little voltage difference between the input terminals. However, the input impedance of a real op amp is not infinite, and its magnitude is an important terminal characteristic of the op amp.

Another important terminal characteristic of any real op amp is related to the maximum output voltage that can be obtained from the amplifier. Consider an op amp circuit with a gain of 100 set by the external resistors in a noninverting amplifier configuration. For a 1 V input you would expect a 100 V output. In reality, the maximum voltage output will be about 1.4 V less than the supply voltage $V_{cc}$ to the op amp for a large load impedance. So if a ±15 V supply is being used, the maximum voltage output would be approximately 13.6 V and the minimum would be −13.6 V.

Two other important characteristics of a real op amp are associated with its response to a square wave input. Ideally, when you apply a square wave input to an amplifier circuit you would expect a square wave output. However, the output cannot change infinitely fast; instead, it exhibits a ramp from one level to the next. The response of an amplifier circuit to a square wave input is illustrated in Figure 5.23.

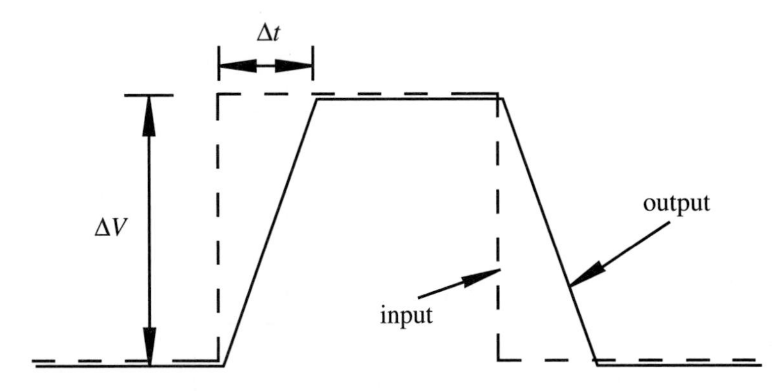


**FIGURE 5.23**
Effect of slew rate on a square wave.

In order to quantify the step response to a square wave, two operational amplifier parameters are defined:

- ***Slew Rate***—The maximum time rate of change possible for the output voltage:

$$SR = \frac{\Delta V}{\Delta t} \tag{5.42}$$

- ***Rise Time***—The time required for the output voltage to go from 10% to 90% of its final value. This parameter is specified by manufacturers for specific load and input parameters.

Another important characteristic of a real op amp is its frequency response. An ideal op amp would exhibit infinite bandwidth. In practice, however, a real op amp has a finite bandwidth, which is a function of the gain established by external components. To quantify this dependence of bandwidth on the gain, another definition is used: the ***gain bandwidth product (GBP).*** The GBP of an op amp is the product of the open loop gain and the bandwidth at that gain. This measure is constant over a wide range of frequencies since, as shown in Figure 5.24, typical op amps exhibit a linear log-log relationship between open loop gain and frequency. Note how the op amp's gain degrades with input signal frequency. Higher quality op amps have larger GBPs. The open loop gain is a characteristic of the op amp without feedback. The closed loop gain is the overall gain of an op amp circuit with feedback. The closed loop gain is always limited by the open loop gain of the op amp. For example, a noninverting amplifier with a closed loop gain of 100 would have a bandwidth of approximately 10,000 Hz as illustrated in Figure 5.24. As you increase the gain of a circuit, you limit its bandwidth. Likewise, if your application only requires a small bandwidth (e.g., in a low-frequency application), larger gains can be used without signal attenuation or distortion.

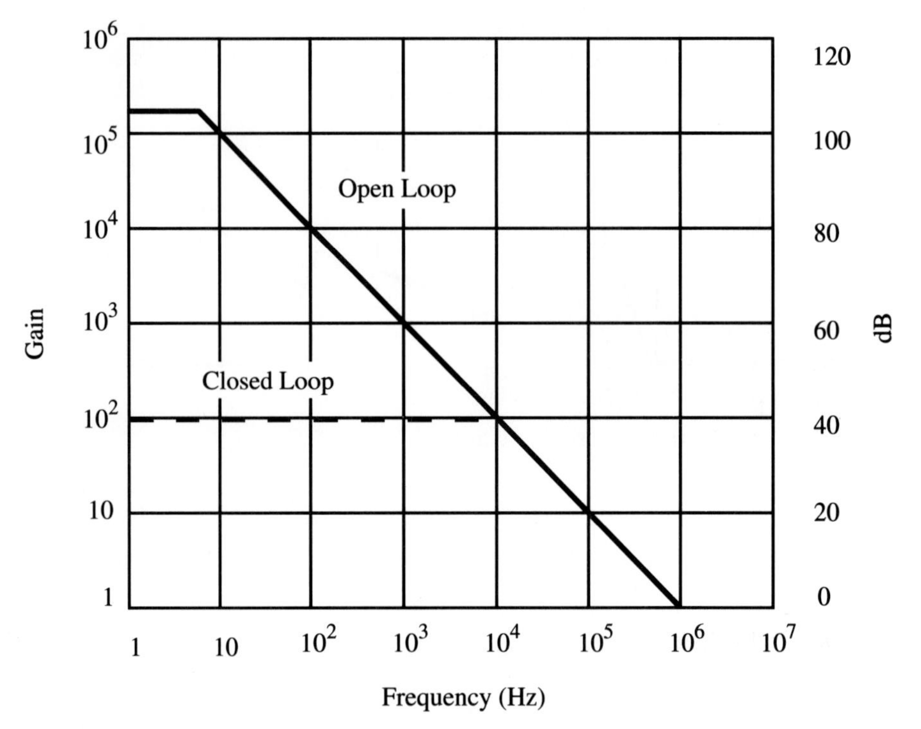


**FIGURE 5.24**
Typical op amp open and closed loop response.

### 5.14.1 Important Parameters from Op Amp Data Sheets

Most of the parameters used to describe the characteristics of actual op amps are listed and described below. These parameters are important when designing and using op amp circuits.

A. Input Parameters

- Input Voltage ($V_{icm}$): This is the maximum input voltage that can be applied between either input and ground. In general, this voltage is equal to the supply voltage.
- Input Offset Voltage ($V_{io}$): This is the voltage that must be applied to one of the input terminals, with the other input being at 0 V, to give a zero output voltage. Remember, for an ideal op amp, the output voltage offset is zero.
- Input Bias Current ($I_{ib}$): This is the average of the currents flowing into both inputs. Ideally, the two input bias currents are equal.
- Input Offset Current ($I_{io}$): This is the difference between the input bias currents when the output voltage is zero.

- Input Voltage Range ($V_{cm}$): This is the range of allowable common mode input voltage, where the same voltage is placed on both inputs.
- Input Resistance ($Z_i$): This is the resistance "looking into" either input with the other input grounded.

B. Output Parameters

- Output Resistance ($Z_{oi}$): This is the resistance seen "looking into" the op amp's output.
- Output Short Circuit Current ($I_{osc}$): This is the maximum output current that the op amp can deliver to a load.
- Output Voltage Swing ($\pm V_{omax}$): This is the maximum peak output voltage that the op amp can supply without saturating or clipping.

C. Dynamic Parameters

- Open Loop Voltage Gain ($A_{OL}$): This is the ratio of the output to the differential input voltage of the op amp without external feedback.
- Large Signal Voltage Gain: This is the ratio of the maximum voltage swing to the change in the input voltage required to drive the output from zero to a specified voltage.
- ***Slew Rate*** (***SR***): This is the time rate of change of the output voltage with the op amp circuit having a voltage gain of 1.

D. Other Parameters

- Maximum Supply Voltage ($\pm V_s$): This is the maximum positive and negative voltage permitted to power the op amp.
- Supply Current: This is the current that the op amp will draw from the power supply.
- ***Common-Mode Rejection Ratio*** (***CMRR***): This is a measure of the ability of the op amp to reject signals of equal value at the inputs. It is the ratio of the common mode input voltage to the generated output voltage, usually expressed in decibels (dB).
- Channel Separation: Whenever there is more than one op amp in a single package, such as a type 747 op amp, a certain amount of ***cross-talk*** will be present. That is, a signal applied to the input of one section of a dual op amp will produce a finite output signal in the remaining section, even though there is no input to that section.

Data for each of the parameters above are usually provided in IC manufacturers' LINEAR data books. Figure 5.25 is a reproduction of the LM741 data sheet from National Semiconductor. It is divided into a maximum ratings section and an electrical characteristics section. This data sheet is typical of those from other manufacturers. Figure 5.26 shows the frequency response characteristics of the TL071. These graphs are also provided on op amp data sheets.

## Absolute Maximum Ratings

**If Military/Aerospace specified devices are required, please contact the National Semiconductor Sales Office/ Distributors for availability and specifications.**
**(Note 5)**

| | LM741A | LM741E | LM741 | LM741C |
|---|---|---|---|---|
| Supply Voltage | ±22V | ±22V | ±22V | ±18V |
| Power Dissipation (Note 1) | 500 mW | 500 mW | 500 mW | 500 mW |
| Differential Input Voltage | ±30V | ±30V | ±30V | ±30V |
| Input Voltage (Note 2) | ±15V | ±15V | ±15V | ±15V |
| Output Short Circuit Duration | Continuous | Continuous | Continuous | Continuous |
| Operating Temperature Range | −55°C to +125°C | 0°C to +70°C | −55°C to +125°C | 0°C to +70°C |
| Storage Temperature Range | −65°C to +150°C | −65°C to +150°C | −65°C to +150°C | −65°C to +150°C |
| Junction Temperature | 150°C | 100°C | 150°C | 100°C |
| Soldering Information | | | | |
| N-Package (10 seconds) | 260°C | 260°C | 260°C | 260°C |
| J- or H-Package (10 seconds) | 300°C | 300°C | 300°C | 300°C |
| M-Package | | | | |
| Vapor Phase (60 seconds) | 215°C | 215°C | 215°C | 215°C |
| Infrared (15 seconds) | 215°C | 215°C | 215°C | 215°C |

See AN-450 "Surface Mounting Methods and Their Effect on Product Reliability" for other methods of soldering surface mount devices.

| | LM741A | LM741E | LM741 | LM741C |
|---|---|---|---|---|
| ESD Tolerance (Note 6) | 400V | 400V | 400V | 400V |

## Electrical Characteristics (Note 3)

| Parameter | Conditions | LM741A/LM741E | | | LM741 | | | LM741C | | | Units |
|---|---|---|---|---|---|---|---|---|---|---|---|
| | | Min | Typ | Max | Min | Typ | Max | Min | Typ | Max | |
| Input Offset Voltage | $T_A = 25°C$ | | | | | | | | | | |
| | $R_S \le 10\ k\Omega$ | | | | | 1.0 | 5.0 | | 2.0 | 6.0 | mV |
| | $R_S \le 50\Omega$ | | 0.8 | 3.0 | | | | | | | mV |
| | $T_{AMIN} \le T_A \le T_{AMAX}$ | | | | | | | | | | |
| | $R_S \le 50\Omega$ | | | 4.0 | | | | | | | mV |
| | $R_S \le 10\ k\Omega$ | | | | | | 6.0 | | | 7.5 | mV |
| Average Input Offset Voltage Drift | | | | 15 | | | | | | | μV/°C |
| Input Offset Voltage Adjustment Range | $T_A = 25°C, V_S = \pm 20V$ | ±10 | | | | ±15 | | | ±15 | | mV |
| Input Offset Current | $T_A = 25°C$ | | 3.0 | 30 | | 20 | 200 | | 20 | 200 | nA |
| | $T_{AMIN} \le T_A \le T_{AMAX}$ | | | 70 | | 85 | 500 | | | 300 | nA |
| Average Input Offset Current Drift | | | | 0.5 | | | | | | | nA/°C |
| Input Bias Current | $T_A = 25°C$ | | 30 | 80 | | 80 | 500 | | 80 | 500 | nA |
| | $T_{AMIN} \le T_A \le T_{AMAX}$ | | | 0.210 | | | 1.5 | | | 0.8 | μA |
| Input Resistance | $T_A = 25°C, V_S = \pm 20V$ | 1.0 | 6.0 | | 0.3 | 2.0 | | 0.3 | 2.0 | | MΩ |
| | $T_{AMIN} \le T_A \le T_{AMAX}$, $V_S = \pm 20V$ | 0.5 | | | | | | | | | MΩ |
| Input Voltage Range | $T_A = 25°C$ | | | | | | | ±12 | ±13 | | V |
| | $T_{AMIN} \le T_A \le T_{AMAX}$ | | | | ±12 | ±13 | | | | | V |
| Large Signal Voltage Gain | $T_A = 25°C, R_L \ge 2\ k\Omega$ | | | | | | | | | | |
| | $V_S = \pm 20V, V_O = \pm 15V$ | 50 | | | | | | | | | V/mV |
| | $V_S = \pm 15V, V_O = \pm 10V$ | | | | 50 | 200 | | 20 | 200 | | V/mV |
| | $T_{AMIN} \le T_A \le T_{AMAX}$, $R_L \ge 2\ k\Omega$, | | | | | | | | | | |
| | $V_S = \pm 20V, V_O = \pm 15V$ | 32 | | | | | | | | | V/mV |
| | $V_S = \pm 15V, V_O = \pm 10V$ | | | | 25 | | | 15 | | | V/mV |
| | $V_S = \pm 5V, V_O = \pm 2V$ | 10 | | | | | | | | | V/mV |

**FIGURE 5.25**
Example op amp data sheet. *(Courtesy of National Semiconductor, Santa Clara, CA)*

## Electrical Characteristics (Note 3) (Continued)

| Parameter | Conditions | LM741A/LM741E | | | LM741 | | | LM741C | | | Units |
|---|---|---|---|---|---|---|---|---|---|---|---|
| | | Min | Typ | Max | Min | Typ | Max | Min | Typ | Max | |
| Output Voltage Swing | $V_S = \pm 20V$ | | | | | | | | | | |
| | $R_L \geq 10\ k\Omega$ | ±16 | | | | | | | | | V |
| | $R_L \geq 2\ k\Omega$ | ±15 | | | | | | | | | V |
| | $V_S = \pm 15V$ | | | | | | | | | | |
| | $R_L \geq 10\ k\Omega$ | | | | ±12 | ±14 | | ±12 | ±14 | | V |
| | $R_L \geq 2\ k\Omega$ | | | | ±10 | ±13 | | ±10 | ±13 | | V |
| Output Short Circuit Current | $T_A = 25°C$ | 10 | 25 | 35 | | 25 | | | 25 | | mA |
| | $T_{AMIN} \leq T_A \leq T_{AMAX}$ | 10 | | 40 | | | | | | | mA |
| Common-Mode Rejection Ratio | $T_{AMIN} \leq T_A \leq T_{AMAX}$ | | | | | | | | | | |
| | $R_S \leq 10\ k\Omega$, $V_{CM} = \pm 12V$ | | | | 70 | 90 | | 70 | 90 | | dB |
| | $R_S \leq 50\Omega$, $V_{CM} = \pm 12V$ | 80 | 95 | | | | | | | | dB |
| Supply Voltage Rejection Ratio | $T_{AMIN} \leq T_A \leq T_{AMAX}$, $V_S = \pm 20V$ to $V_S = \pm 5V$ | | | | | | | | | | |
| | $R_S \leq 50\Omega$ | 86 | 96 | | | | | | | | dB |
| | $R_S \leq 10\ k\Omega$ | | | | 77 | 96 | | 77 | 96 | | dB |
| Transient Response | $T_A = 25°C$, Unity Gain | | | | | | | | | | |
| Rise Time | | | 0.25 | 0.8 | | 0.3 | | | 0.3 | | μs |
| Overshoot | | | 6.0 | 20 | | 5 | | | 5 | | % |
| Bandwidth (Note 4) | $T_A = 25°C$ | 0.437 | 1.5 | | | | | | | | MHz |
| Slew Rate | $T_A = 25°C$, Unity Gain | 0.3 | 0.7 | | | 0.5 | | | 0.5 | | V/μs |
| Supply Current | $T_A = 25°C$ | | | | | 1.7 | 2.8 | | 1.7 | 2.8 | mA |
| Power Consumption | $T_A = 25°C$ | | | | | | | | | | |
| | $V_S = \pm 20V$ | | 80 | 150 | | | | | | | mW |
| | $V_S = \pm 15V$ | | | | | 50 | 85 | | 50 | 85 | mW |
| LM741A | $V_S = \pm 20V$ | | | | | | | | | | |
| | $T_A = T_{AMIN}$ | | | 165 | | | | | | | mW |
| | $T_A = T_{AMAX}$ | | | 135 | | | | | | | mW |
| LM741E | $V_S = \pm 20V$ | | | | | | | | | | |
| | $T_A = T_{AMIN}$ | | | 150 | | | | | | | mW |
| | $T_A = T_{AMAX}$ | | | 150 | | | | | | | mW |
| LM741 | $V_S = \pm 15V$ | | | | | | | | | | |
| | $T_A = T_{AMIN}$ | | | | | 60 | 100 | | | | mW |
| | $T_A = T_{AMAX}$ | | | | | 45 | 75 | | | | mW |

**Note 1:** For operation at elevated temperatures, these devices must be derated based on thermal resistance, and $T_j$ max. (listed under "Absolute Maximum Ratings"). $T_j = T_A + (\theta_{jA} P_D)$.

| Thermal Resistance | Cerdip (J) | DIP (N) | HO8 (H) | SO-8 (M) |
|---|---|---|---|---|
| $\theta_{jA}$ (Junction to Ambient) | 100°C/W | 100°C/W | 170°C/W | 195°C/W |
| $\theta_{jC}$ (Junction to Case) | N/A | N/A | 25°C/W | N/A |

**Note 2:** For supply voltages less than ±15V, the absolute maximum input voltage is equal to the supply voltage.

**Note 3:** Unless otherwise specified, these specifications apply for $V_S = \pm 15V$, $-55°C \leq T_A \leq +125°C$ (LM741/LM741A). For the LM741C/LM741E, these specifications are limited to $0°C \leq T_A \leq +70°C$.

**Note 4:** Calculated value from: BW (MHz) = 0.35/Rise Time(μs).

**Note 5:** For military specifications see RETS741X for LM741 and RETS741AX for LM741A.

**Note 6:** Human body model, 1.5 kΩ in series with 100 pF.

**FIGURE 5.25**
*(continued)*

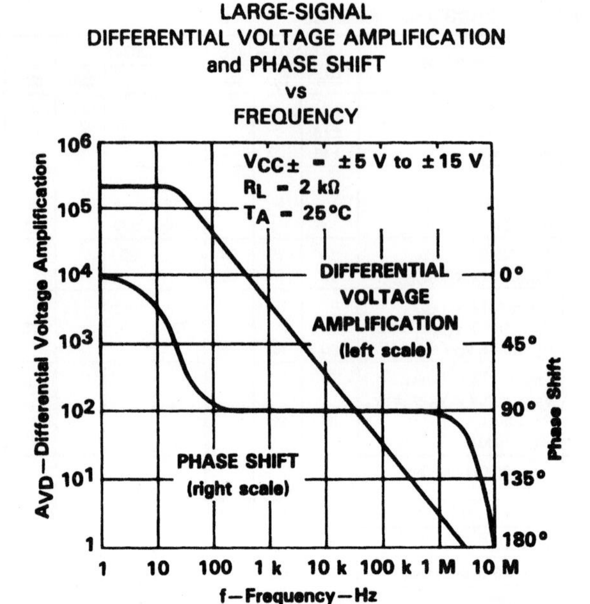


**FIGURE 5.26**
TL071 FET input op amp. *(Courtesy of Texas Instruments, Dallas, TX)*

As you gain practical experience with op amp circuits, you should develop an appreciation of the significance of the many parameters affecting op amp performance.

---

▼ ***EXAMPLE 5.1 Sizing Resistors in OP Amp Circuits.*** The ideal model of the op amp would imply that if you constructed the following two op amp circuits in the laboratory, they would have the same gain. Ideally, both circuits would have gain of –2. However, the top circuit would be a very poor design and would not function as expected. The reason for this can be found by considering the Output Short Circuit Current found on the specification sheet for the op amp. From Figure 5.25, the value for a LM741 is typically 25 mA. This is the largest current that the output can source. But looking at the circuit, the output current is $V_{out}/2\Omega$, and since $V_{out} = -2V_{in} = -$ 10V, the output current would be 5 A! This is far above the current sourcing capability of the op amp. To avoid this problem, larger resistances such as the ones shown in the bottom circuit are used. Here, the output current is 5 mA, which is well within the op amp specification.

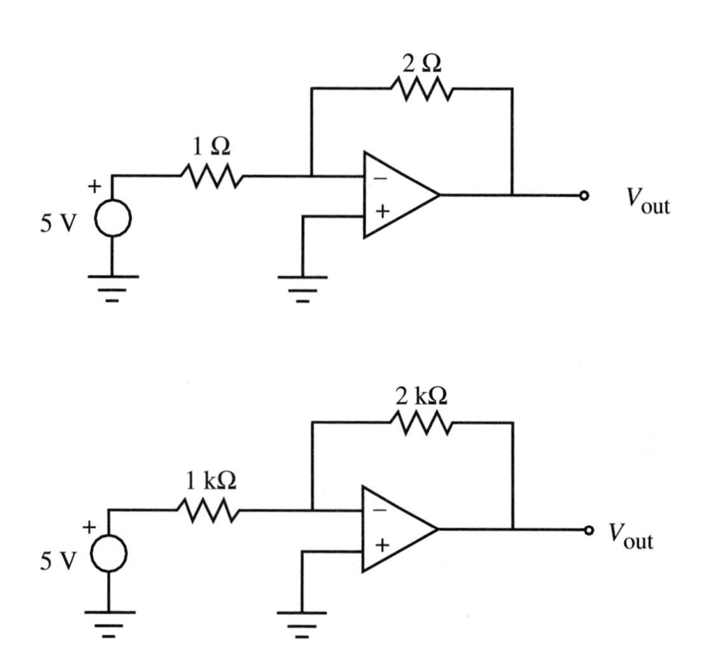


---

▼▼ ***DESIGN EXAMPLE 5.1 Myogenic Control of a Prosthetic Limb.*** Interfacing prosthetic devices to the human body presents one of the most interesting and challenging problems for engineers. The problem poses a number of medical and engineering challenges in the fields of materials, fluids, electronics, control, and mechanics. Think of the artificial heart, dialysis machine, prosthetic hip joint, osmotic skin patch, and artificial retina as examples. As we develop improved technological products, we find very important applications in bioengineering. Let us consider one important problem that uses our knowledge of operational amplifiers.

Suppose you would like to design a prosthetic limb, that is, an artificial arm or leg that could be controlled by the thoughts of the wearer. Early prosthetic limbs were either purely passive, like Peg-Leg Pete's wooden stump, or somewhat mechanically controlled by twitching physical locations on the body connected by cables to the limb prosthesis. A novel approach however, would be to provide some measure of thought control of the limb, where physical motion would not be required. Two possibilities present themselves: neural control or myogenic control. For neural control we would have to electrically tap the nervous system, which still is a problem that has not been completely solved technically. Myogenic control is easier to realize. When a muscle is caused to move or twitch, the tiny movement of electrolytes in the muscles below the skin cause an electric field that induces a small voltage on the surface of the skin. Certainly this voltage is quite small, otherwise we would shock each other every time we touched. In actuality it ranges from microvolts to millivolts and may be mixed with other biopotential signals. The problem then is to sense and isolate this small voltage and convert it into a signal that is capable of switching something like an electric motor that could be attached to a prosthetic device. So here is our problem: How can we design a mechatronic system that uses the surface skin potential from a muscle as an input to control an actuator such as an electric motor?

Let us begin by outlining in broad brush strokes what the approach will be. First we will have to tap the skin potential with a special surface electrode. Then we have to amplify the signal and filter it to eliminate unwanted noise components and to achieve the correct frequency response. Then we need to convert it to a form that allows setting different levels for a control strategy. Finally, we need to drive an electric motor that requires significant current. You currently have the design capability to do all or these things. We will start by looking at the transducer to sense the skin potential.

The electric fields that occur in living tissue are caused by charge separations in electrolytes and not by the movement of electrons. In order to sense the voltage at the skin we need a transducer, a sensor that converts subdermal (below the skin) electrolytic currents to electron currents for our electronic system. Silver–silver chloride electrodes have this property. So if we place a silver chloride electrode on the skin and couple it with a conducting gel, we can sense the body's voltage at that location. The magnitude of the voltage will be related to how much a subcutaneous muscle is contracted (it is the myoelectric signal we are interested in). The problem that remains is that the electrode produces a very small signal, at best a few millivolts. Also, there is a considerable amount of 60 Hz background noise and other signals obscuring the signal associated with the muscle. Moreover, the electrode has a high impedance.

This is an application where an instrumentation amplifier is necessary to provide the high input impedance, high common mode rejection ratio, and gain necessary to extract the biopotential signal produced by the contracting muscle. The figure on the next page displays the preamplifier stage of our Electro Myogenic (EMG) detector. For the components shown, it should be easy to create a CMRR in excess of 60 dB and a gain of 125 with an input impedance of 10 megaohms. Notice that there are two active (differential) electrodes 1 and 2 that will be mounted close together above the muscle. The third electrode is a ground reference. This circuit will be satisfactory in capturing the EMG signal.

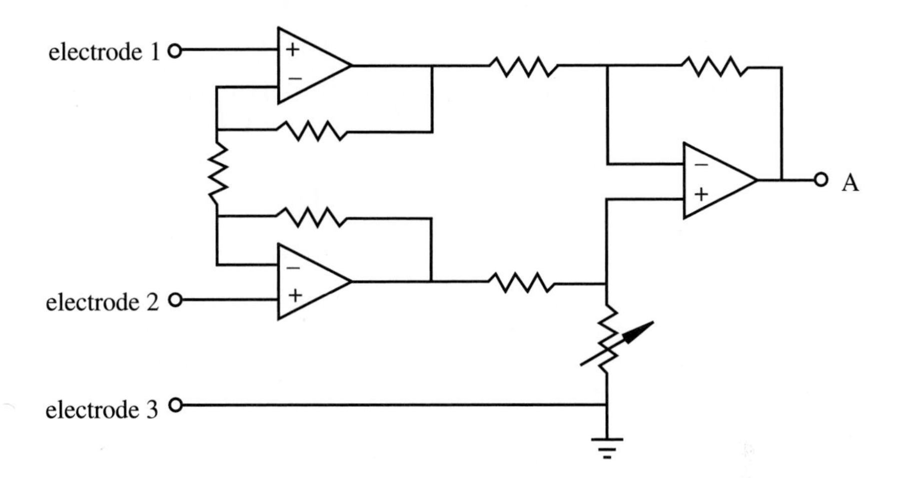


The instrumentation amplifier was chosen because it can extract a very small signal common to both electrodes even when the noise at that electrode is orders of magnitude larger! Although we now have eliminated a considerable amount of 60 Hz common mode noise and other signals not common to both electrodes with the input stage of the amplifier, there is still something called a *motion artifact* that can occur due to movement of the subject. This will produce voltages sufficient to saturate the second stage amplifier. The 2 Hz high-pass filter on the input of the second stage of the amplifier shown below reduces artifacts and adds further gain to the system.

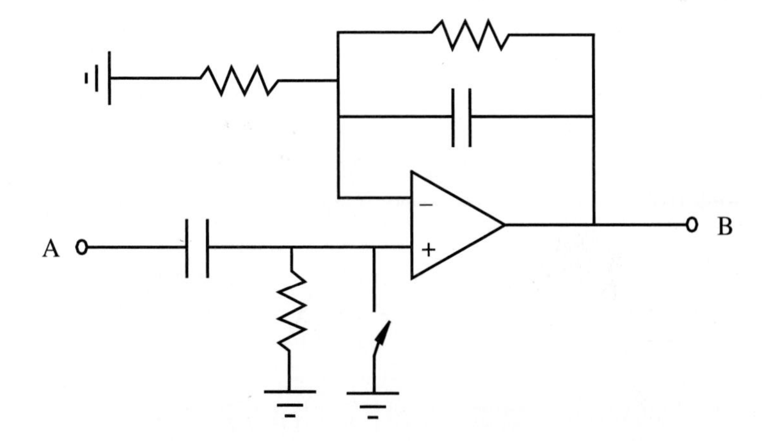


At this point the EMG signal presented by the oscilloscope would look like the following:

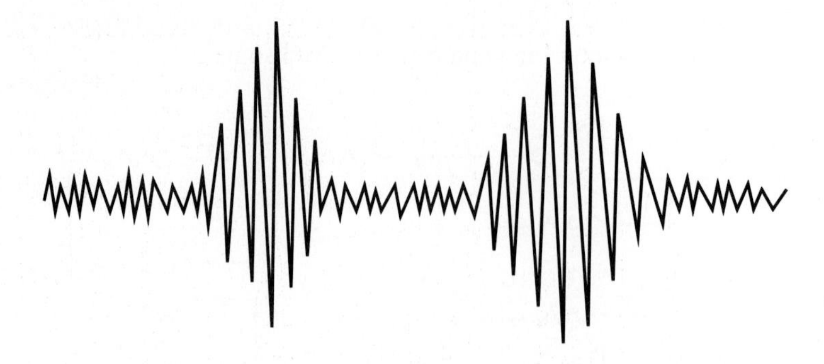

This is a rather high-frequency signal with components between a few Hz and 250 Hz. We need to do two things at this point: rectify the signal so that specific levels of the signal can be used for control, and low-pass filter it so the envelope of the high-frequency waveform results. The problem here is that the signal is still very small, and a normal diode rectifier will not be sensitive enough to work due to the 0.7 V turn-on voltage necessary. We require a precision rectifier, one that approximates the action of an ideal diode. Using the high gain of an op amp in a circuit known as a *precision rectifier* (see figure below) improves the performance so that the diode function is close to ideal.

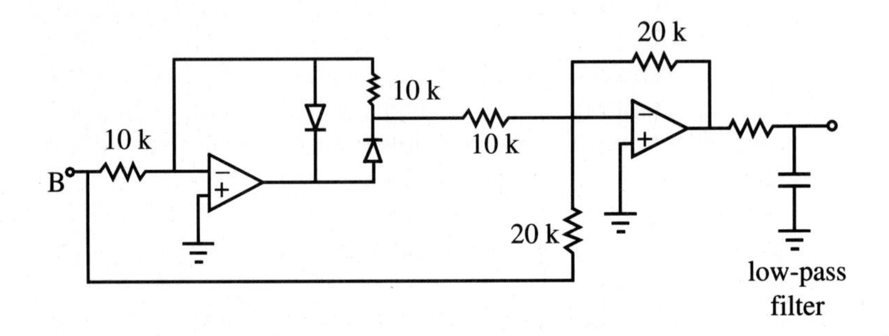


The precision-rectified EMG and the resulting low-pass-filtered signals look like:

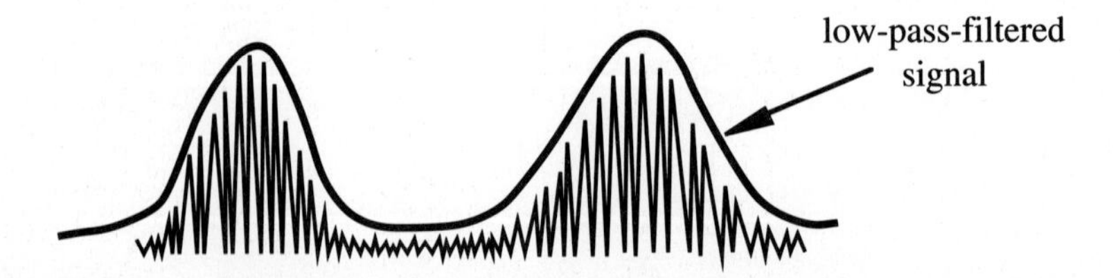


The low-pass-filtered signal is basically the envelope of the rectified signal. Now we have a signal that can be input to a comparator (see figure below) to provide a binary control signal, one that will be on when the muscle is contracted and off when relaxed. The designer would have to set the reference voltage for the particular application.

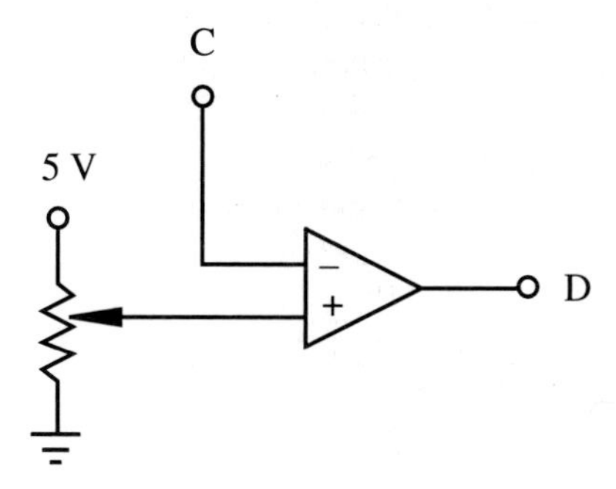


As shown below, the output of the rectifier can then be input to a power transistor circuit to control the current in a motor.

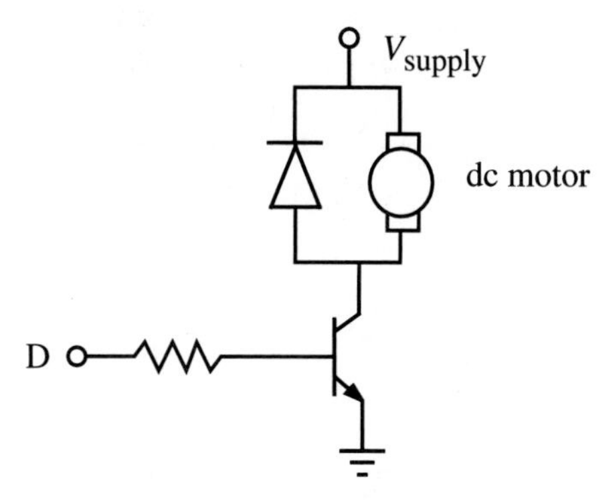


In summary, we have used a variety of op amp circuits to process an analog signal. It exemplifies the extraction of a very low level signal in the presence of noise, a variety of analog signal processing methods, and the interface to an actuator to control mechanical power. In this case we have converted the EMG signal into a binary control signal for a dc motor controlling, say, the elbow in a prosthetic arm.

---

▼▼▼ ***CLASS DISCUSSION ITEM 5.7. Bidirectional EMG Controller.*** In Design Example 5.1 we discussed turning on and off the motor via an EMG signal. Unfortunately, the controller can actuate the joint in one direction only. Discuss how you might change the design to allow bidirectional movement based on the principles we have discussed.

## QUESTIONS AND EXERCISES

**5.1.** Analyze the summer circuit in Figure 5.13 and determine an equation for the output voltage $V_{out}$ in terms of the input voltages $V_1$ and $V_2$ and the resistances $R_1$, $R_2$, and $R_F$. Use this result to verify that Equation 5.19 is correct. Show and explain all work.

**5.2.** Using the input waveform below, sketch the corresponding output waveform for each op amp circuit. Assume ideal op amp behavior.

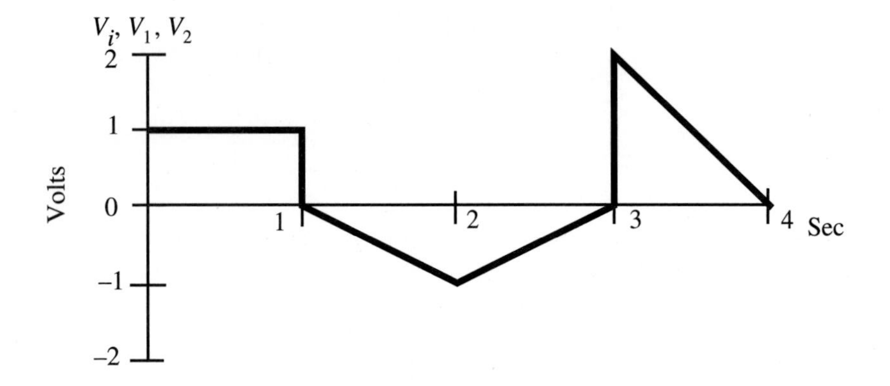


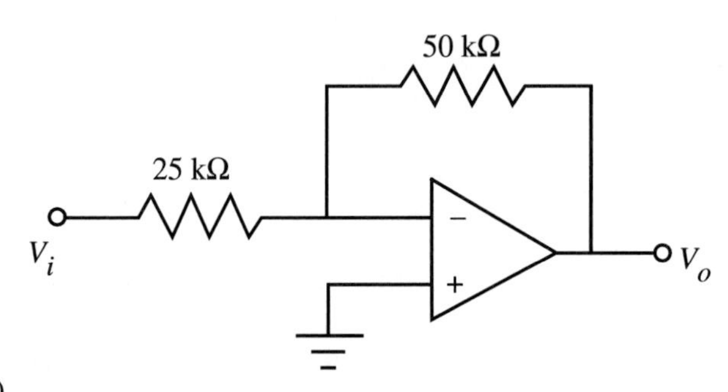


(*a*)

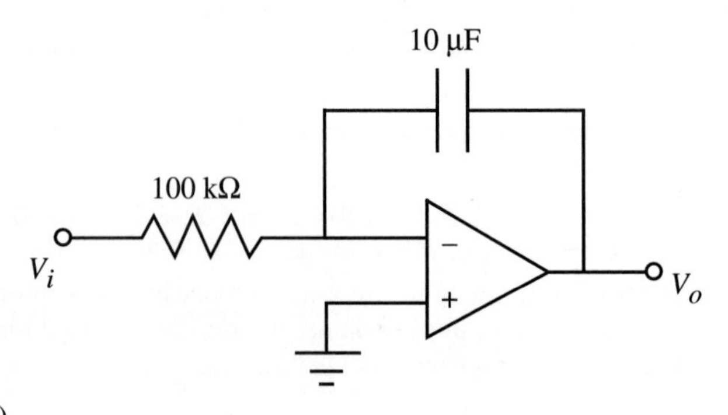


(*b*)

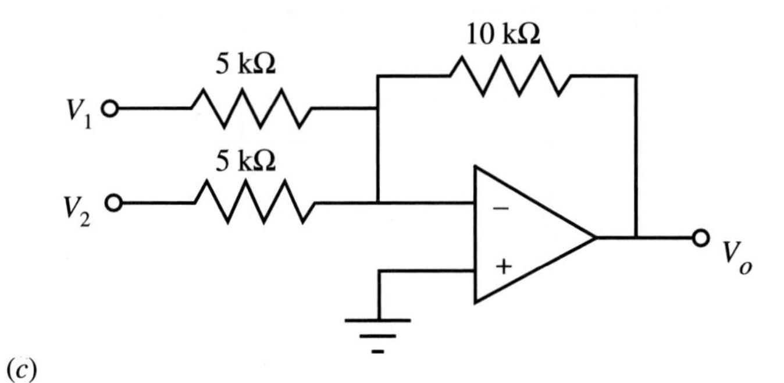


(c)

5 kΩ
5 kΩ
$V_1$
10 kΩ
$V_2$
−
+
$V_o$

(d)

**5.3.** Derive Equation 5.24 for the difference amplifier without using the principle of superposition.

**5.4.** Derive Equation 5.31 that expresses $V_{\text{out}}$ in terms of the $V_3$ and $V_4$ shown in Figure 5.17.

**5.5.** Determine $V_o$ as a function of $I$ for each of the op amp circuits shown below. Assume ideal op amp behavior.

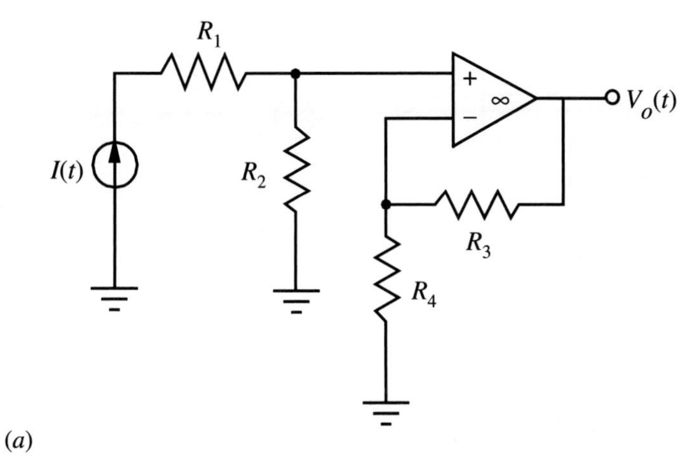


(a)

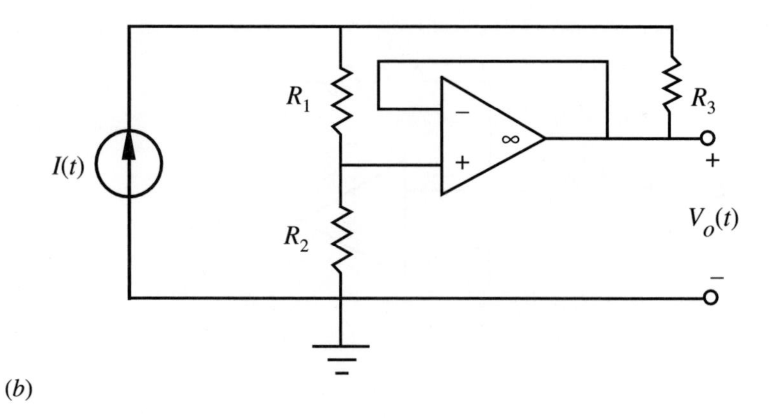


(*b*)

**5.6.** Determine $V_{out}$ in the circuits below with $R_1 = R_2 = R_3 = 1\ \text{k}\Omega$, $V_1 = 10$ V, and $V_2 = 5$ V. Assume ideal op amp behavior.

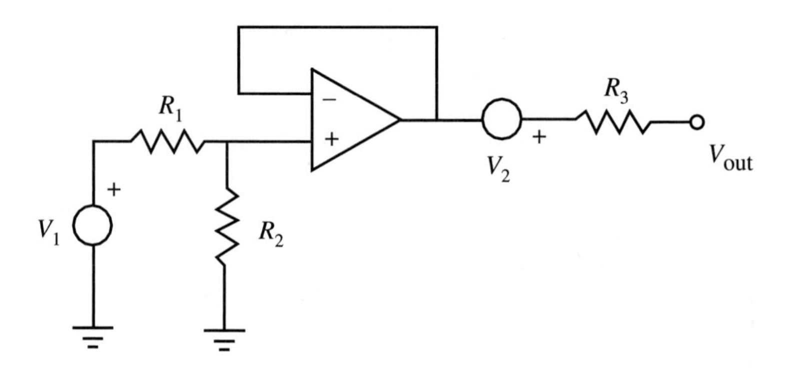


(*a*)

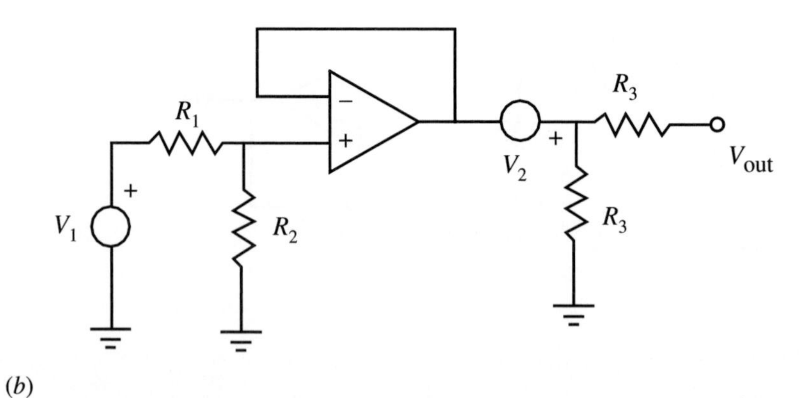


(*b*)

**5.7.** Determine $I_4$ in terms of $V_i$, $R_1$, $R_2$, $R_3$, and $R_4$ in the circuit below. Assume ideal op amp behavior.

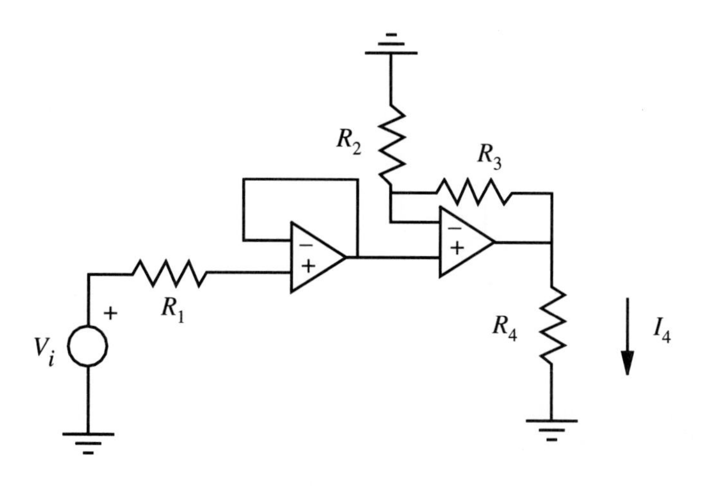


**5.8.** Find $V_o$ as a function of $V_i$ in the op amp circuit below.

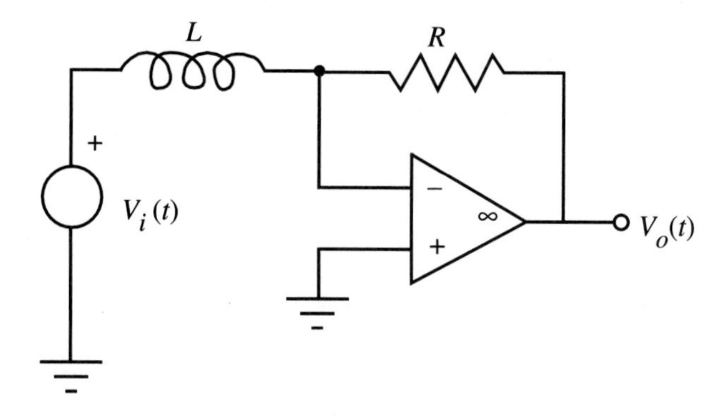


**5.9.** Explain why $V_o \neq V_i$ in the circuit below.

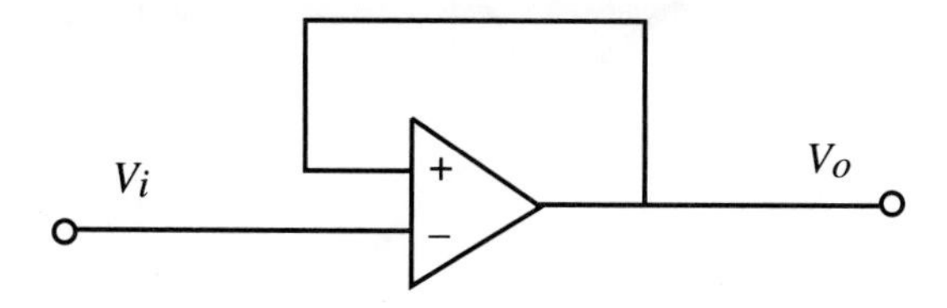


**5.10.** Use the principle of superposition to derive an expression for the output voltage in the circuit on the next page and explain why the circuit is called a ***level shifter***.

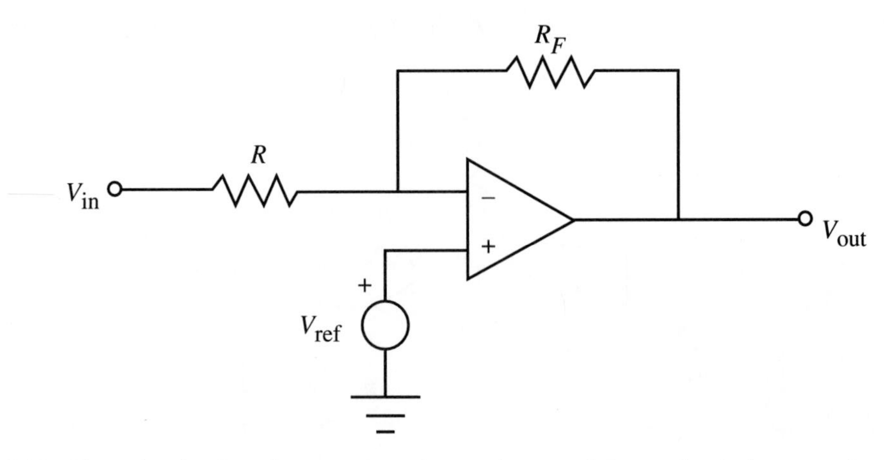


**5.11.** Given the circuit and op amp open loop gain curve below, what is the approximate bandwidth for the circuit when $R_F = 20\ \text{k}\Omega$ and $R = 2\ \text{k}\Omega$?

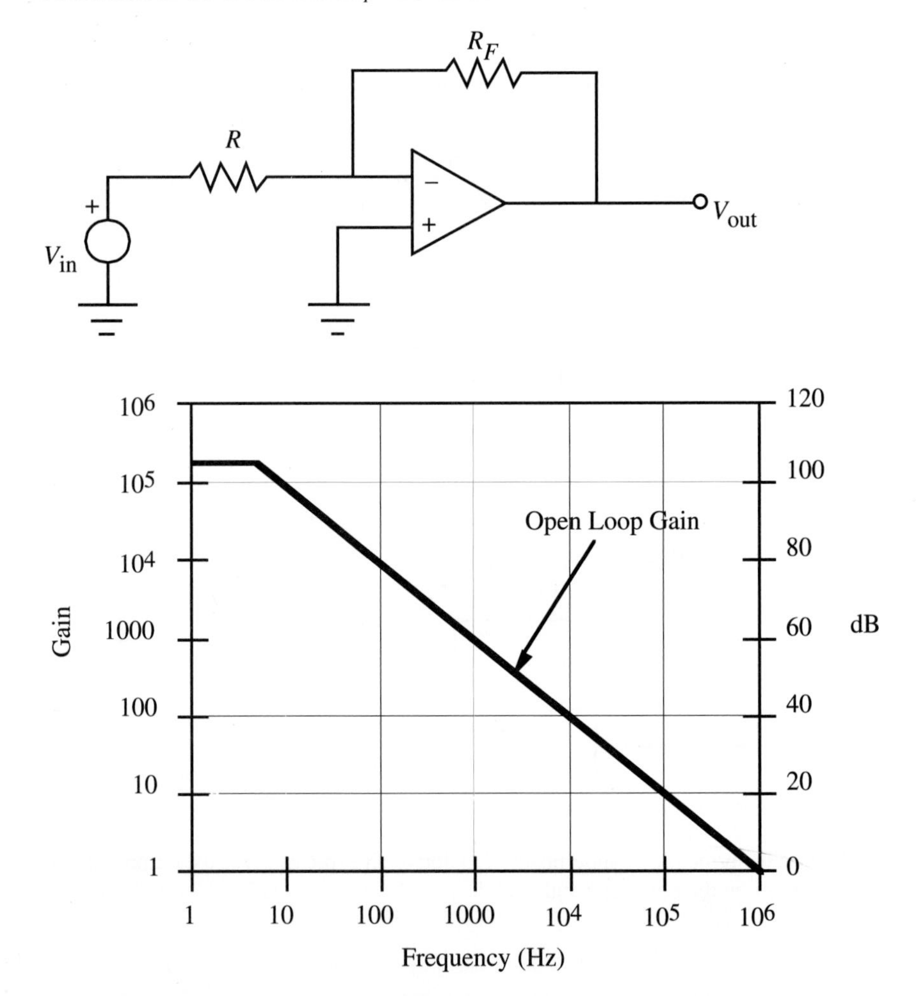

## BIBLIOGRAPHY

Coughlin, R. and Driscoll, F., *Operational Amplifiers & Linear Integrated Circuits,* 4th Edition, Prentice-Hall, Englewood Cliffs, NJ, 1991.

Horowitz, P. and Winfield, H., *The Art of Electronics,* 2nd Edition, Cambridge University Press, New York, 1989.

Johnson, D., Hilburn, J., and Johnson, J., *Basic Electric Circuit Analysis,* 2nd Edition, Prentice-Hall, Englewood Cliffs, NJ, 1984.

McWhorter, G. and Evans, A., *Basic Electronics,* Master Publishing, Inc., Richardson, TX, 1994.

Mims, F., *Engineer's Mini-Notebook: Op Amp IC Circuits,* Radio Shack Archer Catalog No. 276-5011, 1985.

Mims, F., *Getting Started in Electronics,* Radio Shack Archer Catalog No. 276-5003A, 1991.

Texas Instruments, *Linear Circuits Data Book, Volume 1–Operational Amplifiers,* Dallas, TX, 1992.

# 6

# DIGITAL CIRCUITS AND SYSTEMS

**OBJECTIVES:** After you read, discuss, study, and apply ideas in this chapter, you will be able to:

- Define a digital signal
- Understand how the binary number system is used in coding digital data
- Contrast combinational and sequential logic
- Draw a timing diagram for a digital circuit
- Use Boolean mathematics to analyze logic circuits
- Design logic networks
- Use a variety of flip-flops for storing information
- Use counters for different counting applications
- Display numerical data using light emitting diodes
- Understand the differences among microprocessors, microcomputers, and microcontrollers, including their component parts and functionality

## 6.1 INTRODUCTION

This chapter presents the fundamentals of digital devices and digital systems that are used to control modern mechatronic and measurement systems. We begin with terminology and the basics of digital logic, continue with sequential circuits and their applications, and end with an introduction to microprocessor-based systems.

In contrast to an analog signal, a digital signal exists only at specific levels or states and changes its level in discrete steps. An analog and a digital signal are illustrated in Figure 6.1. Most digital signals have only two states. The two-state system allows the application of Boolean logic and binary number representation, which form the foundation for the design of all digital devices.

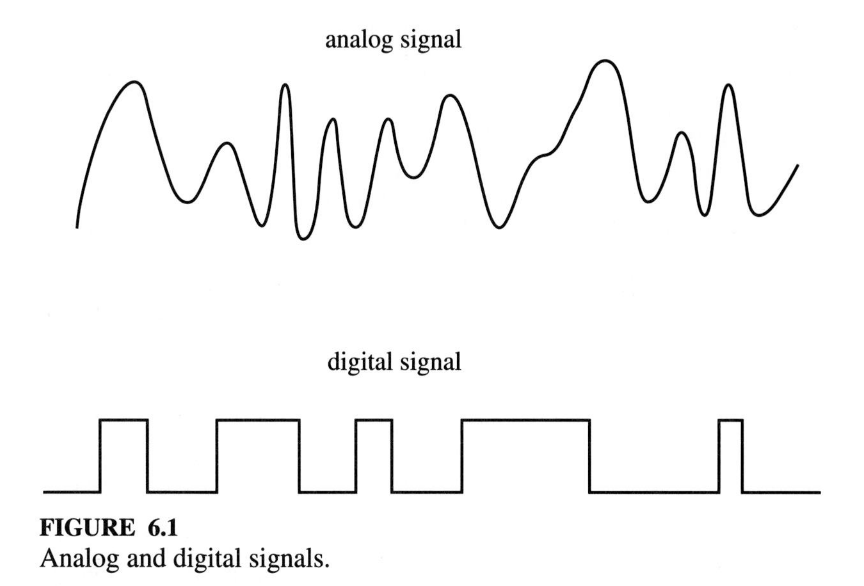


**FIGURE 6.1**
Analog and digital signals.

Digital devices are categorized according to their function as ***combinational logic*** or ***sequential logic*** devices. All digital devices convert digital inputs into one or more digital outputs. The difference between the two categories is based on signal timing. For sequential logic devices, the timing, or sequencing history, of the input signals plays a role in determining the output. This is not the case with combinational logic devices whose outputs only depend on the instantaneous values of the inputs.

Before we introduce the various digital devices, we will first review the binary number system and the application of binary numbers in digital calculations. Then we discuss Boolean algebra, which is the mathematical basis for digital computations and digital electronic devices. Finally, we discuss a number of specific combinational and sequential logic devices and their applications.

## 6.2 DIGITAL REPRESENTATIONS

We grow up becoming proficient using the base 10 decimal number system. The ***base*** of the number system indicates the number of different symbols required to represent whole numbers and fractions. In base 10, the symbols are: 0, 1, 2, 3, 4, 5, 6, 7, 8, and 9. Each digit in a decimal number is a placeholder for different powers of 10 according to

$$\begin{aligned} &d_n d_{n-1} \ldots d_3 d_2 d_1 d_0 \\ &= d_n \cdot 10^n + d_{n-1} \cdot 10^{n-1} + \cdots + d_2 \cdot 10^2 + d_1 \cdot 10^1 + d_0 \cdot 10^0 \end{aligned} \tag{6.1}$$

where each digit $d_i$ is one of the ten symbols. As an example, the decimal number 123 can be expanded as

$$123 = 1 \text{ x } 10^2 + 2 \text{ x } 10^1 + 3 \text{ x } 10^0 \quad (6.2)$$

Fractions may also be included if digits for negative powers of 10 are included ($d_{-1}$, $d_{-2}$, . . .).

In order to represent and manipulate numbers with digital devices such as computers, we use a base 2 number system called the ***binary number system***. The reason for this is that the operation of digital devices is based on transistors that switch between two states: the ON or saturated state and the OFF or cutoff state. These states are designated by the symbols 1 (ON) and 0 (OFF) in the base 2 system. The digits in a binary number, as with the base 10 system, correspond to different powers of the base. A binary number can be expanded as

$$(d_n d_{n-1} \ldots d_3 d_2 d_1 d_0)_2 = d_n \cdot 2^n + d_{n-1} \cdot 2^{n-1} + \cdots + d_2 \cdot 2^2 + d_1 \cdot 2^1 + d_0 \cdot 2^0 \quad (6.3)$$

where each digit $d_i$ is one of the two symbols 0 and 1. The trailing subscript 2 is used to indicate that the number is base 2 and not the normally assumed base 10. As an example of Equation 6.3, the binary number 1101 can be expanded as

$$1101_2 = 1 \cdot 2^3 + 1 \cdot 2^2 + 0 \cdot 2^1 + 1 \cdot 2^0 = 8_{10} + 4_{10} + 1_{10} = 13_{10} \quad (6.4)$$

The digits of a binary number are also called ***bits***, and the first and last bits have special names. The first or leftmost bit is known as the ***most significant bit (MSB)*** since it represents the largest power of 2. The last or rightmost bit is known as the ***least significant bit (LSB)*** since it represents the smallest power of 2.

In general, the value of a number represented in any base can be expanded and computed with

$$(d_n d_{n-1} \ldots d_3 d_2 d_1 d_0)_b = (d_n \cdot b^n + d_{n-1} \cdot b^{n-1} + \cdots + d_2 \cdot b^2 + d_1 \cdot b^1 + d_0 \cdot b^0 \quad (6.5)$$

where $b$ is the base. Often it is necessary to convert from one base system to another. Equation 6.5 provides a mechanism to convert from an arbitrary base to base 10. To convert a number from base 10 to some other base, the procedure successively divides the decimal number by the base and records the remainders after each division. The remainders, when written in reverse order from left to right, form the digits of the number represented in the new base. Table 6.1 illustrates this procedure by converting the decimal number 123 to its binary equivalent. You can use Equation 6.5 to verify the binary result by calculating its expansion.

**TABLE 6.1**
**Decimal binary conversion**

| Successive divisions | Remainder | |
|---|---|---|
| 123/2 | 1 | LSB |
| 61/2 | 1 | |
| 30/2 | 0 | |
| 15/2 | 1 | |
| 7/2 | 1 | |
| 3/2 | 1 | |
| 1/2 | 1 | MSB |
| result | 1111011 | |

Binary arithmetic is carried out in the same way as our familiar base 10 arithmetic. The following example illustrates the similarities.

---

▼ ***EXAMPLE 6.1. Binary Arithmetic.*** This example illustrates the analogy between decimal addition and multiplication and binary addition and multiplication.

$$\begin{array}{r} 9 \\ +\ 3 \\ \hline 12 \end{array} \qquad \begin{array}{r} 1001 \\ +\ 0011 \\ \hline 1100 \end{array} \qquad \begin{array}{r} 9 \\ \times\ 3 \\ \hline 27 \end{array} \qquad \begin{array}{r} 1001 \\ \times\ 0011 \\ \hline 1001 \\ +\ 1001\phantom{0} \\ +\ 0000\phantom{00} \\ +\ 0000\phantom{000} \\ \hline 11011 \end{array}$$

Note that when adding two 1 bits (1 + 1), the sum is 0 with a carry of 1 to the next higher order bit.

---

Since binary numbers can be long and cumbersome to write and display, often the ***hexadecimal*** (base 16) number system is used as an alternative representation. Table 6.2 lists the symbols for the hexadecimal system along with their binary and decimal equivalents. Note that the letters A through F are used to represent the digits larger than 9.

**TABLE 6.2**
**Hexadecimal symbols and equivalents**

| Binary | Hexadecimal | Decimal |
|---|---|---|
| 0 | 0 | 0 |
| 1 | 1 | 1 |
| 10 | 2 | 2 |
| 11 | 3 | 3 |
| 100 | 4 | 4 |
| 101 | 5 | 5 |
| 110 | 6 | 6 |
| 111 | 7 | 7 |
| 1000 | 8 | 8 |
| 1001 | 9 | 9 |
| 1010 | A | 10 |
| 1011 | B | 11 |
| 1100 | C | 12 |
| 1101 | D | 13 |
| 1110 | E | 14 |
| 1111 | F | 15 |

To convert a binary number to hexadecimal, divide the number into groups of four digits beginning with the least significant bit and replace each group with its hexadecimal equivalent. For example,

$$123_{10} = 111\ 1011_2 = 7B_{16} \tag{6.6}$$

In addition to numbers, alphanumeric characters can also be represented in digital (binary) form with ***ASCII*** codes. ASCII is short for the *A*merican *S*tandard *C*ode for *I*nformation *I*nterchange. ASCII codes are 7-bit codes used to denote all of the alphanumeric characters. There is a unique code for each alphanumeric character. Some example codes are:

$$\text{“A”: } 0100\ 0001 = 41_{16} = 65_{10}$$
$$\text{“B”: } 0100\ 0010 = 42_{16} = 66_{10}$$
$$\text{“0”: } 0011\ 0000 = 30_{16} = 48_{10}$$
$$\text{“1”: } 0011\ 0001 = 31_{16} = 49_{10}$$

***Binary coded decimal (BCD)*** is another type of digital representation that is commonly used for input and output of numerical data. With BCD, 4 bits are used to represent a single, base 10 digit. BCD is a convenient mechanism for representing decimal numbers in a binary number format, but it is inefficient since only 10 of the 16 possible states of the 4-bit number are used. To convert a decimal number to BCD, assemble the 4-bit codes for each decimal digit. For example,

$$123_{10} = 0001\ 0010\ 0011_{bcd} \tag{6.7}$$

Note that this is different from the binary representation:

$$123_{10} = 0111\ 1011_2 \tag{6.8}$$

▼▼▼ ***CLASS DISCUSSION ITEM 6.1. Computer Magic.*** How can a digital computer perform the complex operations it does given that its architecture and operation are based on simply manipulating bits (zeros and ones)?

## 6.3 COMBINATIONAL LOGIC

Combinational logic devices are digital devices that convert binary inputs into binary outputs based on the rules of mathematical logic. The basic operations, devices, and schematic symbols for ***combinational logic*** devices are shown in Table

6.3. These devices are also called ***gates*** because they control the flow of signals from the inputs to the single output. A small circle at the input or output of a digital device denotes signal inversion; that is, a 0 becomes a 1, or a 1 becomes a 0. The NAND and NOR gates are AND and OR gates respectively with the output inverted, hence the circle is shown on the output. The ***truth table*** for each device is shown on the right. The truth table is a compact means of displaying all combinations of inputs and their corresponding outputs realizable with the respective device. Usually the combination of inputs is written as the ascending list of binary numbers whose number of bits correspond to the number of inputs (e.g., 00, 01, 10, 11). The standard AND, NAND, OR, NOR, and XOR gates have only two inputs, but other forms are available with more than two inputs. In the case of a multiple input AND gate, the output is 1 if and only if all inputs are 1; otherwise, the output is 0. In the case of the OR gate, the output is 0 if and only if all inputs are 0; otherwise, it is 1. In the case of the XOR gate, the output is 0 if all of the inputs are 0 or if all of the inputs are 1; otherwise, it is 1. The algebraic symbols used to represent the logic functions are: plus (+) for logic OR, dot (·) for the logic AND, and an overbar ($\overline{X}$) for logic NOT, denoting inversion.

**TABLE 6.3**
**Combinational logic operations**

| Gate | Operation | Symbol | Expression | Truth table |
|---|---|---|---|---|
| buffer | increase output signal current | A — C | $C = A$ | A C<br>0 0<br>1 1 |
| inverter (INV, NOT) | invert signal (complement) | A — C | $C = \overline{A}$ | A C<br>0 1<br>1 0 |
| AND gate | AND logic | A, B — C | $C = A \cdot B$ | A B C<br>0 0 0<br>0 1 0<br>1 0 0<br>1 1 1 |
| NAND gate | inverted AND logic | A, B — C | $C = \overline{A \cdot B}$ | A B C<br>0 0 1<br>0 1 1<br>1 0 1<br>1 1 0 |
| OR gate | OR logic | A, B — C | $C = A + B$ | A B C<br>0 0 0<br>0 1 1<br>1 0 1<br>1 1 1 |

*(continued)*

| Gate | Operation | Symbol | Expression | Truth table |
|---|---|---|---|---|
| NOR gate | inverted OR logic | A, B → C | $C = \overline{A+B}$ | A B C<br>0 0 1<br>0 1 0<br>1 0 0<br>1 1 0 |
| XOR Gate | exclusive OR logic | A, B → C | $C = A \oplus B$ | A B C<br>0 0 0<br>0 1 1<br>1 0 1<br>1 1 0 |

The ***buffer*** is used to increase the current supplied at the output while not changing the digital state. This is important if you wish to drive multiple digital devices from a single output. A normal digital device has a limited ***fan-out,*** which defines the maximum number of similar digital devices that can be driven by the output. The buffer can help overcome fan-out limitations. A typical gate can only supply approximately 1 mA, but a buffer can increase the output to approximately 15 mA. Higher output currents require the use of a discrete transistor.

The most common digital protocol used to produce the two digital states is ***transistor-transistor logic***, or ***TTL***. For a digital input, ***logic zero*** (0) or ***low*** (*L*) is defined as a value less than 0.7 V, and ***logic one*** (1) or ***high*** (*H*) is defined as a value greater than 2.5 V. The digital output of a TTL device typically ranges between 0 and 0.4 V for low and between 2.5 V and 5 V for high. The input voltage range 0.7 V to 2.5 V between the logic 0 and logic 1 states is a dead zone where the input state is undefined.

An increasingly common class of digital devices is the ***complementary metal oxide semiconductor,*** or ***CMOS***. The logic levels for CMOS devices depend on the supply voltage. CMOS devices consume very little power; therefore, they are very valuable in battery operated applications where power is limited. A drawback of CMOS devices is their susceptibility to damage from static electricity.

Figure 6.2 illustrates an equivalent circuit for the NAND gate. Similar to other gates, the internal design is composed of transistors, resistors, and diodes, which are easily manufactured on a silicon chip. Usually the chip is assembled as a ***dual in-line package (DIP)*** with the pin connections illustrated in Figure 6.3. This particular DIP has four NAND gates on a single silicon chip, hence the name QUAD NAND gate. Figure 6.4 shows a typical manufacturer's data sheet, which lists the maximum ratings, operating conditions, electrical characteristics, and switching characteristics. Manufacturers provide similar data sheets for all digital devices contained in ***TTL data books***. The labeling system used in TTL data books is usually in the form AAxxyzz, where AA is the manufacturer's TTL IC prefix (SN: TI and others; DM: National Semiconductor); xx distinguishes between military (xx = 54) and industrial (xx = 74) quality; y distinguishes between different internal designs (no letter: standard TTL; L: low power dissipation; H: high power dissipation; S: Schottky type; AS: advanced Schottky, LS: low power Schottky; ALS:

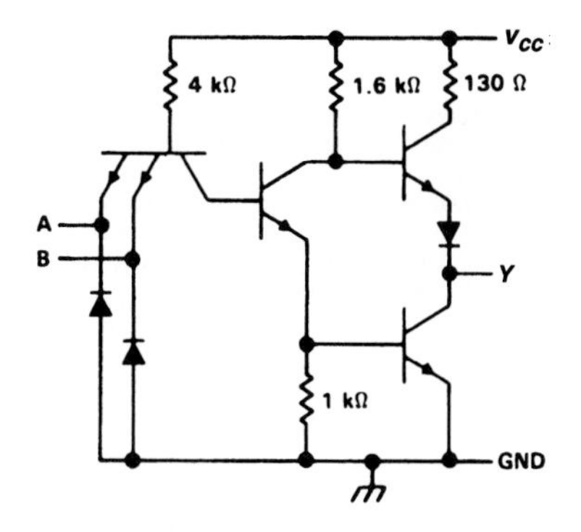


**FIGURE 6.2**
NAND gate internal design. *(Courtesy of Texas Instruments, Dallas, TX)*

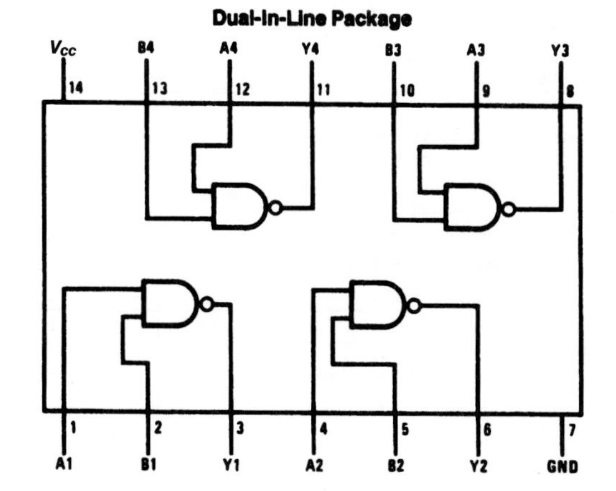


**FIGURE 6.3**
QUAD NAND gate IC pin-out. *(Courtesy of National Semiconductor, Santa Clara, CA)*

advanced low power Schottky); and zz is the sequence number in the data book. Schottky devices have faster switching speeds and require less power. The label for the QUAD NAND IC in Figure 6.3 is DM74LS00.

It is important to be aware of the type of output circuit being used in a digital device. Certain TTL devices (e.g., 7401, 7403, 7405, 7406) have ***open collector outputs,*** which means that the output signal appears on the collector of an output transistor that is not internally connected. It is necessary to complete the transistor circuit in order for the device to function properly. Normally this is done by including a ***pull-up resistor*** as shown in Figure 6.5.

▼▼▼ ***CLASS DISCUSSION ITEM 6.2. NAND Magic.*** Look at the NAND gate circuit shown in Figure 6.2 and convince yourself that it results in NAND logic.

## Absolute Maximum Ratings (Note)

**If Military/Aerospace specified devices are required, please contact the National Semiconductor Sales Office/Distributors for availability and specifications.**

| | |
|---|---|
| Supply Voltage | 7V |
| Input Voltage | 7V |
| Operating Free Air Temperature Range | |
| DM54LS and 54LS | −55°C to +125°C |
| DM74LS | 0°C to +70°C |
| Storage Temperature Range | −65°C to +150°C |

Note: *The "Absolute Maximum Ratings" are those values beyond which the safety of the device cannot be guaranteed. The device should not be operated at these limits. The parametric values defined in the "Electrical Characteristics" table are not guaranteed at the absolute maximum ratings. The "Recommended Operating Conditions" table will define the conditions for actual device operation.*

## Recommended Operating Conditions

| Symbol | Parameter | DM54LS00 | | | DM74LS00 | | | Units |
|---|---|---|---|---|---|---|---|---|
| | | Min | Nom | Max | Min | Nom | Max | |
| $V_{CC}$ | Supply Voltage | 4.5 | 5 | 5.5 | 4.75 | 5 | 5.25 | V |
| $V_{IH}$ | High Level Input Voltage | 2 | | | 2 | | | V |
| $V_{IL}$ | Low Level Input Voltage | | | 0.7 | | | 0.8 | V |
| $I_{OH}$ | High Level Output Current | | | −0.4 | | | −0.4 | mA |
| $I_{OL}$ | Low Level Output Current | | | 4 | | | 8 | mA |
| $T_A$ | Free Air Operating Temperature | −55 | | 125 | 0 | | 70 | °C |

## Electrical Characteristics over recommended operating free air temperature range (unless otherwise noted)

| Symbol | Parameter | Conditions | | Min | Typ (Note 1) | Max | Units |
|---|---|---|---|---|---|---|---|
| $V_I$ | Input Clamp Voltage | $V_{CC}$ = Min, $I_I$ = −18 mA | | | | −1.5 | V |
| $V_{OH}$ | High Level Output Voltage | $V_{CC}$ = Min, $I_{OH}$ = Max, $V_{IL}$ = Max | DM54 | 2.5 | 3.4 | | V |
| | | | DM74 | 2.7 | 3.4 | | |
| $V_{OL}$ | Low Level Output Voltage | $V_{CC}$ = Min, $I_{OL}$ = Max, $V_{IH}$ = Min | DM54 | | 0.25 | 0.4 | V |
| | | | DM74 | | 0.35 | 0.5 | |
| | | $I_{OL}$ = 4 mA, $V_{CC}$ = Min | DM74 | | 0.25 | 0.4 | |
| $I_I$ | Input Current @ Max Input Voltage | $V_{CC}$ = Max, $V_I$ = 7V | | | | 0.1 | mA |
| $I_{IH}$ | High Level Input Current | $V_{CC}$ = Max, $V_I$ = 2.7V | | | | 20 | μA |
| $I_{IL}$ | Low Level Input Current | $V_{CC}$ = Max, $V_I$ = 0.4V | | | | −0.36 | mA |
| $I_{OS}$ | Short Circuit Output Current | $V_{CC}$ = Max (Note 2) | DM54 | −20 | | −100 | mA |
| | | | DM74 | −20 | | −100 | |
| $I_{CCH}$ | Supply Current with Outputs High | $V_{CC}$ = Max | | | 0.8 | 1.6 | mA |
| $I_{CCL}$ | Supply Current with Outputs Low | $V_{CC}$ = Max | | | 2.4 | 4.4 | mA |

## Switching Characteristics at $V_{CC}$ = 5V and $T_A$ = 25°C (See Section 1 for Test Waveforms and Output Load)

| Symbol | Parameter | $R_L$ = 2 kΩ | | | | Units |
|---|---|---|---|---|---|---|
| | | $C_L$ = 15 pF | | $C_L$ = 50 pF | | |
| | | Min | Max | Min | Max | |
| $t_{PLH}$ | Propagation Delay Time Low to High Level Output | 3 | 10 | 4 | 15 | ns |
| $t_{PHL}$ | Propagation Delay Time High to Low Level Output | 3 | 10 | 4 | 15 | ns |

**FIGURE 6.4**
DM74LS00 NAND gate IC data sheet. *(Courtesy of National Semiconductor, Santa Clara, CA)*

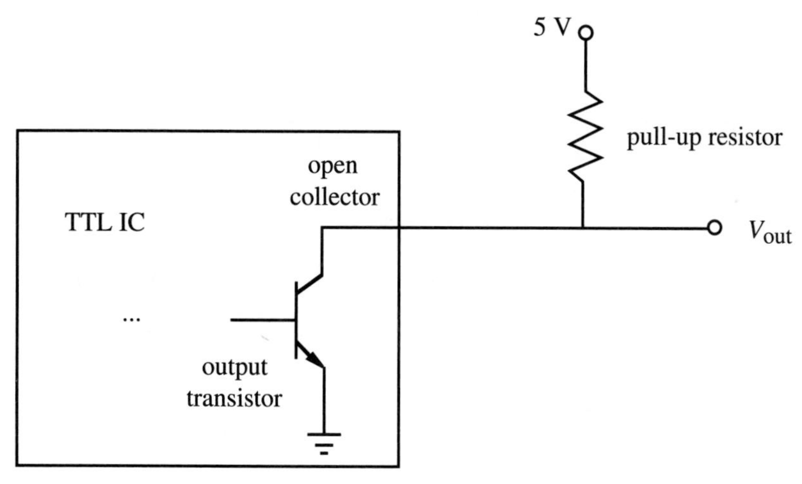


**FIGURE 6.5**
Open collector output with pull-up resistor.

## 6.4
## TIMING DIAGRAMS

In order to analyze complex logic circuits it often helps to sketch ***timing diagrams,*** which show the simultaneous levels of the inputs and outputs in a circuit. The timing diagram is used to illustrate every possible combination of input values and corresponding outputs, providing a graphical summary of the input-output relationships. Timing diagrams for the AND and OR gates are shown in Figures 6.6 and 6.7 as examples. Many digital oscilloscopes and logic analyzers have the capability to display timing diagrams for digital circuits.

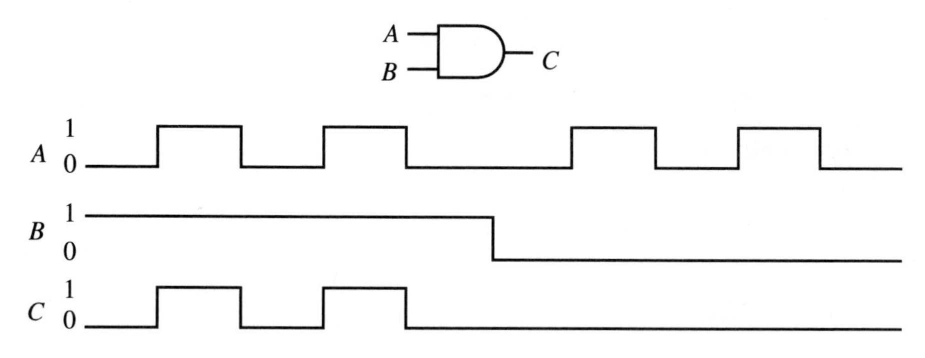


**FIGURE 6.6**
AND gate timing diagram.

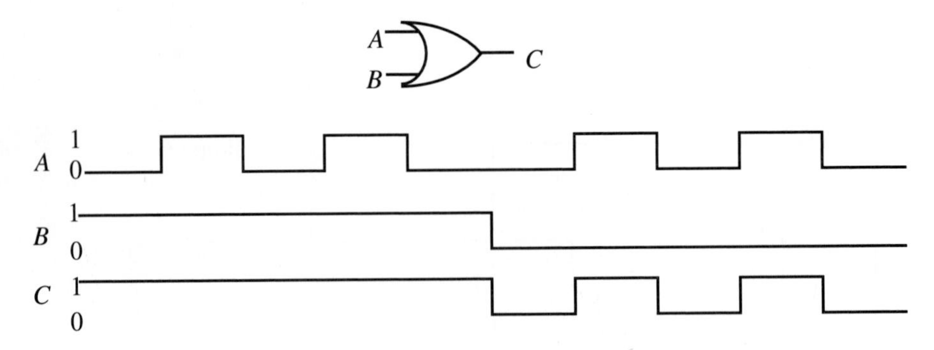


**FIGURE 6.7**
OR gate timing diagram.

## 6.5 BOOLEAN IDENTITIES

In formulating mathematical expressions for logic circuits, it is important to have knowledge of ***Boolean algebra***, which defines the rules for expressing and simplifying binary logic statements. The basic Boolean laws and identities are listed below. A bar over a symbol indicates the Boolean operation NOT, which corresponds to inversion of a signal.

### Boolean Algebra Laws and Identities

#### Fundamental Laws

| OR | AND | NOT |
|---|---|---|
| $A + 0 = A$ | $A \cdot 0 = 0$ | |
| $A + 1 = 1$ | $A \cdot 1 = A$ | |
| $A + A = A$ | $A \cdot A = A$ | $\overline{\overline{A}} = A$ |
| $A + \overline{A} = 1$ | $A \cdot \overline{A} = 0$ | (double inversion) |

(6.9)

#### Commutative Laws

$$A + B = B + A \tag{6.10}$$

$$A \cdot B = B \cdot A \tag{6.11}$$

#### Associative Laws

$$(A + B) + C = A + (B + C) \tag{6.12}$$

$$(A \cdot B) \cdot C = A \cdot (B \cdot C) \tag{6.13}$$

**Distributive Laws**

$$A \cdot (B + C) = (A \cdot B) + (A \cdot C) \tag{6.14}$$

$$A + (B \cdot C) = (A + B) \cdot (A + C) \tag{6.15}$$

**Other Useful Identities**

$$A + (A \cdot B) = A \tag{6.16}$$

$$A \cdot (A + B) = A \tag{6.17}$$

$$A + (\overline{A} \cdot B) = A + B \tag{6.18}$$

$$(A + B) \cdot (A + \overline{B}) = A \tag{6.19}$$

$$(A + B) \cdot (A + C) = A + (B \cdot C) \tag{6.20}$$

$$A + B + (A \cdot \overline{B}) = A + B \tag{6.21}$$

$$(A \cdot B) + (B \cdot C) + (\overline{B} \cdot C) = (A \cdot B) + C \tag{6.22}$$

$$(A \cdot B) + (A \cdot C) + (\overline{B} \cdot C) = (A \cdot B) + (\overline{B} \cdot C) \tag{6.23}$$

***DeMorgan's Laws*** are also useful in rearranging or simplifying longer Boolean expressions or in converting between AND and OR gates:

$$\overline{A + B + C + \cdots} = \overline{A} \cdot \overline{B} \cdot \overline{C} \cdot \cdots \tag{6.24}$$

$$\overline{A \cdot B \cdot C \cdot \cdots} = \overline{A} + \overline{B} + \overline{C} + \cdots \tag{6.25}$$

If we invert both sides of these equations and apply the double NOT law from Equation set 6.9 we can write DeMorgan's Laws in the following form:

$$A + B + C + \cdots = \overline{\overline{A} \cdot \overline{B} \cdot \overline{C} \cdot \cdots} \tag{6.26}$$

$$A \cdot B \cdot C \cdot \cdots = \overline{\overline{A} + \overline{B} + \overline{C} + \cdots} \tag{6.27}$$

Truth tables can be very helpful in proving an identity. For example, to show that Equation 6.18 is valid, we can construct the following truth table where each term in the identity is evaluated for all input combinations:

| A | $\overline{A}$ | B | $\overline{A} \cdot B$ | $A + (\overline{A} \cdot B)$ | A + B |
|---|---|---|---|---|---|
| 1 | 0 | 0 | 0 | 1 | 1 |
| 1 | 0 | 1 | 0 | 1 | 1 |
| 0 | 1 | 0 | 0 | 0 | 0 |
| 0 | 1 | 1 | 1 | 1 | 1 |

Since both sides of the identity are equal for every input combination, the identity is valid.

▼ ***EXAMPLE 6.2. Simplifying a Boolean Expression.*** Simplify the following expression using Boolean laws and identities.

$$X = (A \cdot B \cdot C) + (B \cdot C) + (\bar{A} \cdot B)$$

First, we can rewrite this equation using the associative law and the fundamental law that $Z \cdot 1 = Z$:

$$X = A \cdot (B \cdot C) + 1 \cdot (B \cdot C) + (\bar{A} \cdot B)$$

In this form, it is clear that we can use the distributive law to factor out the $(B \cdot C)$ term:

$$X = (A + 1) \cdot (B \cdot C) + (\bar{A} \cdot B)$$

Since $A + 1 = 1$,

$$X = (B \cdot C) + (\bar{A} \cdot B)$$

Furthermore, we can factor out $B$:

$$X = B \cdot (C + \bar{A})$$

We have reduced the number of operators from seven in the original expression to three in the final expression. This is important because it reduces the number of gates required to build the circuit.

## 6.6 DESIGN OF LOGIC NETWORKS

As an example illustrating the application of combinational logic to a real engineering problem, suppose you are asked to design a circuit for a simple security protection system for a home. The home-owner wants the alarm to sound if someone breaks into the house through a door or window or if something is moving around in the house while the occupants are away. Under certain conditions, the users may also want to disable portions of the alarm system. We will assume that there are sensors to detect if windows or doors are disturbed and to detect motion. To accomplish the goals of this security system, we will design a combinational logic controller using two switches that can be set by the owner.

The following steps facilitate the design of a digital circuit to solve this and similar problems:

1. Define the problem in words.
2. Write quasi-logic statements in English that can be translated into Boolean expressions.
3. Write the Boolean expressions.
4. Simplify and optimize the Boolean expressions if possible.
5. Write an all-AND or all-NAND or all-OR or all-NOR realization of the circuit to minimize the number of required logic gate IC components.

6. Draw the logic schematic for the electronic realization of the circuit.

Each of these steps is carried out for the security lock example in the following sections.

### 6.6.1 Define the Problem in Words

We begin our logic design by translating the problem into a series of word statements that reflect what should be happening in the system. We want the alarm system to create a high signal sounding the alarm for certain combinations of the house sensors. Also, we want the user to be able to select one of the three operating states:

1. Active state where the alarm will sound only if the windows or doors are disturbed. This state is useful when the occupants are sleeping.
2. Active state where the alarm will sound if the windows or doors are disturbed or if there is motion in the house. This state is useful when the occupants are away.
3. Disabled state where the alarm will not sound. This state is useful during normal household activity.

At this time, we must define Boolean variables that will represent the inputs and outputs of the circuit. The following Boolean variables will be used to design the security system logic:

- *A:* state of the door and window sensors
- *B:* state of the motion detector
- *Y:* output used to sound the alarm
- *C D:* 2-bit code set by the user to select the operating state defined by

$$C\,D = \begin{cases} 0\,1 & \text{operating state 1} \\ 1\,0 & \text{operating state 2} \\ 0\,0 & \text{operating state 3} \end{cases}$$

The inputs to the system are $A$, $B$, $C$, and $D$, and the output is $Y$. We will assume ***positive logic*** for $A$, $B$, and $Y$, where a 1 implies active or ON and a 0 implies inactive or OFF.

### 6.6.2 Write Quasi-Logic Statements

We further translate the word statements into logic-like statements. The quasi-logic statements for the security system are:

> Activate the alarm ($Y = 1$) if $A$ is high *and* the code "$C\,D$" is "0 1" *or* activate the alarm if $A$ *or* $B$ is high *and* the code is "1 0".

Note the italicized quasi-Boolean operators, which should aid in writing the Boolean expression.

### 6.6.3 Write the Boolean Expression

Now we write the Boolean expressions based on the quasi-logic statement. To create a product of 1 for the active control code "0 1", we need to form the expression $\overline{C} \cdot D$; alternatively, to create a product of 1 for the other active control code "1 0", we need to form the expression $C \cdot \overline{D}$. Based on this, the complete Boolean expression for the security system is

$$Y = A \cdot (\overline{C} \cdot D) + (A + B) \cdot (C \cdot \overline{D}) \tag{6.28}$$

The alarm will sound ($Y = 1$) if the expression $A \cdot (\overline{C} \cdot D)$ is 1 or if the expression $(A + B) \cdot (C \cdot \overline{D})$ is 1; otherwise the alarm will not sound ($Y = 0$). The first expression will be 1 if and only if $A$ is 1 and $C$ is 0 and $D$ is 1; the second expression will be 1 if and only if $C$ is 1 and $D$ is 0 and $A$ or $B$ is 1.

For this particular problem, we can simplify Equation 6.28 by looking at a truth table for the $C$ and $D$ terms for the different control code combinations.

| $C$ | $D$ | $(\overline{C} \cdot D)$ | $(C \cdot \overline{D})$ |
|---|---|---|---|
| 0 | 0 | 0 | 0 |
| 1 | 0 | 0 | 1 |
| 0 | 1 | 1 | 0 |

Note that $(\overline{C} \cdot D) = D$ and $(C \cdot \overline{D}) = C$ for the control code combinations. Since we disallow the state $C\,D = 1\ 1$, Equation 6.28 can be simplified as

$$Y = (A \cdot D) + (A + B) \cdot C \tag{6.29}$$

### 6.6.4 AND Realization

Once a Boolean expression is simplified, it may be desirable to manipulate the result further in order to convert all operations to a preferred type of gate (e.g., AND or OR). The reason for this is that logic gates come packaged on integrated circuit chips (logic ICs) in groups of four, six, or eight. Therefore, we may be able to reduce the total number of chips required by using all of one type of gate. Converting from one gate type to another is easily accomplished with a repeated application of DeMorgan's Laws. For the security system example, an all-AND representation is achieved by applying DeMorgan's Law (Equation 6.26):

$$Y = (A \cdot D) + (A + B) \cdot C \tag{6.30}$$

$$Y = (A \cdot D) + (\overline{\overline{A} \cdot \overline{B}}) \cdot C \tag{6.31}$$

$$Y = \overline{\overline{A \cdot D} \cdot \overline{(\overline{\overline{A} \cdot \overline{B}}) \cdot C}} \tag{6.32}$$

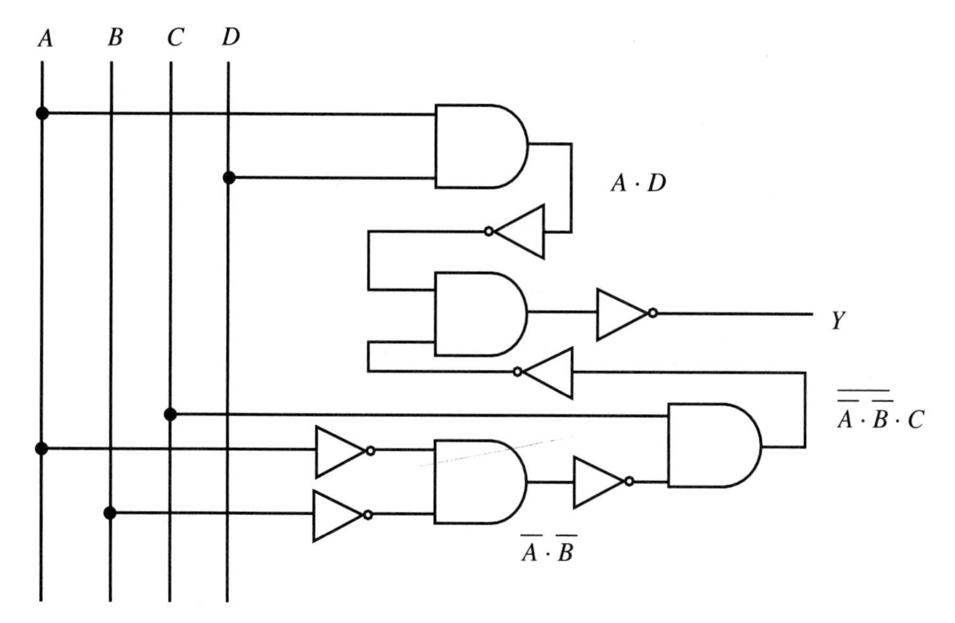


**FIGURE 6.8**
AND realization schematic of the security system.

### 6.6.5 Draw the Circuit Diagram

Now it is relatively straightforward to draw the circuit using only AND gates and inverters from inspection of the final Boolean expression in Equation 6.32, by building the subexpressions one at a time and connecting them as shown in Figure 6.8. Since there are a total of four AND gates and six inverters, the circuit can be constructed with two ICs: one QUAD AND gate IC and one HEX INVERTER IC. Equation 6.30 could also be implemented using two ICs since only two OR gates and two AND gates are required. Therefore, for the security system, the all-AND realization did not reduce the number of ICs. However, for more complex Boolean expressions, an all-AND realization will usually reduce the number of ICs.

▼▼▼ ***CLASS DISCUSSION ITEM 6.3. Everyday Logic.*** Make a list of devices that you interact with on a daily basis that use logic for control purposes. For each, describe what logic is being performed.

## 6.7
## FINDING A BOOLEAN EXPRESSION GIVEN A TRUTH TABLE

As an alternative to Sections 6.6.1 and 6.6.2 where we defined a logic problem in words and then wrote quasi-logic statements, sometimes it is convenient to express the complete input and output combinations with a truth table. Sometimes, a truth table may already be specified. In these situations, there are two methods for directly obtaining the Boolean expression that performs the logic specified in the truth table. Both methods are described below, and an example is given to demonstrate their application.

The first method is known as the ***sum-of-products method***. It is based on the fact that we can represent an output as a sum of products containing combinations of the inputs. For example, if we have three inputs $A$, $B$, and $C$ and an output $X$, the sum of products would be a Boolean expression containing input terms AND-ed together to form product terms that are OR-ed together to define the output $X$ as a Boolean sum. The following equation is an example of what a sum-of-products expression looks like:

$$X = (\bar{A} \cdot B \cdot C) + (\bar{A} \cdot \bar{B} \cdot C) + (A \cdot B \cdot \bar{C}) \tag{6.33}$$

If we form a product for every row in the truth table that results in an output of 1 and take the sum of the products, we can represent the complete logic of the table. For rows whose output values are 1, we must ensure that the product representing that row is 1. In order to do this, any input whose value is 0 in the row must be inverted in the product. By expressing a product for every input combination (row) whose value is 1, we have completely modeled the logic of the truth table since every other combination will result in a 0.

The second method is known as the ***product-of-sums method***. It is based on the fact that we can represent an output as a product of sums containing combinations of the inputs. For example, if we have three inputs $A$, $B$, and $C$ and an output $X$, the product of sums would be a Boolean expression containing input terms OR-ed together to form sum terms that are AND-ed together to define the output $X$ as a Boolean product. The following equation is an example of what a product-of-sums expression looks like:

$$X = (\bar{A} + B + C) \cdot (\bar{A} + \bar{B} + C) \cdot (A + B + \bar{C}) \tag{6.34}$$

If we form a sum for every row in the truth table that results in an output of 0 and take the product of the sums, we can represent the complete logic of the table. For rows whose output values are 0, we must ensure that the sum representing that row is 0. In order to do this, any input whose value is 1 in the row must be inverted in the sum. By expressing a sum for every input combination (row) whose value is 0, we have completely modeled the logic of the truth table since every other combination will result in a 1.

---

▼ ***EXAMPLE 6.3. Sum of Products and Product of Sums.*** In performing binary arithmetic, the simplest operation is summing the two least significant bits resulting in a sum bit and a carry bit. The truth table for this operation is

| $A$ | $B$ | $S$ | $C$ |
|---|---|---|---|
| 0 | 0 | 0 | 0 |
| 0 | 1 | 1 | 0 |
| 1 | 0 | 1 | 0 |
| 1 | 1 | 0 | 1 |

where $A$ and $B$ are the input bits, $S$ is the sum bit, and $C$ is the carry bit. We will apply the sum-of-products and product-of-sums methods to both of the these outputs to illustrate how the methods differ. The sum-of-products method applied to output $S$ yields:

$$S = (\bar{A} \cdot B) + (A \cdot \bar{B})$$

The product-of-sums method applied to output $S$ yields:

$$S = (A + B) \cdot (\bar{A} + \bar{B})$$

The sum-of-products method applied to output $C$ yields:

$$C = (A \cdot B)$$

The product-of-sums method applied to output $C$ yields:

$$C = (A + B) \cdot (A + \bar{B}) \cdot (\bar{A} + B)$$

Note that the sum-of-products method was easier to apply to output $C$ since only a single row has an output of 1.

If we use the product-of-sums result for $S$ and the sum-of-products result for $C$, we obtain a circuit using the minimum number of gates:

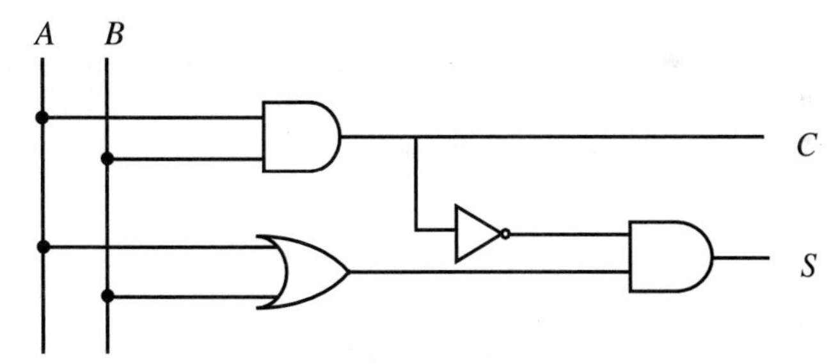


This circuit is known as a ***half-adder*** since it only applies to the two least significant bits of a sum. Higher order bits require a lower order carry as an input (see Question 6.12).

▼▼▼ *CLASS DISCUSSION ITEM 6.4. **Equivalence of Sum of Products and Product of Sums.*** Draw the logic circuits for $S$ and $C$ in Example 6.3 using only the product-of-sums results and then do the same for the sum-of-products results. Compare your circuits to the one shown above. Also, show that the sum-of-products and product-of-sums results are equivalent.

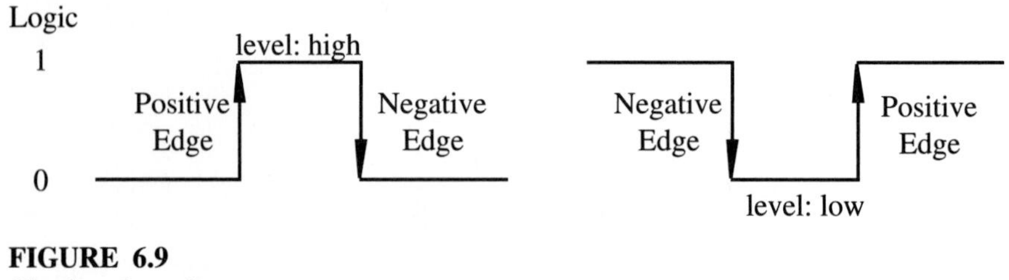


**FIGURE 6.9**
Clock pulse edges.

## 6.8 SEQUENTIAL LOGIC

In addition to combinational logic devices that generate an output based on the input values independent of the input timing, there is another class of devices called ***sequential logic*** devices for which timing or sequencing of the input signals is important. Devices in this class include flip-flops, counters, monostables, latches, and more complex devices such as microprocessors. Specific devices trigger events in different ways, usually by detecting the transitions of a trigger signal from one level to another. The trigger signal is usually referred to as the ***clock*** (*CK*) signal. The clock signal can be a periodic square wave or an aperiodic collection of pulses. A typical clock signal in a personal computer microprocessor such as the INTEL Pentium may be a 120 MHz square wave.

Output states of a sequential logic device will change at specific transitions or edges of a digital signal. Figure 6.9 illustrates edge terminology in relation to a clock pulse, where an arrow is used to indicate edges where state transitions occur. ***Positive edge-triggered*** devices respond to a low-to-high (0 to 1) transition, and ***negative edge-triggered*** devices respond to a high-to-low (1 to 0) transition. This topic is addressed again in Section 6.10 where it is applied to flip-flops.

## 6.9 FLIP-FLOPS

Since digital data is stored in the form of bits, digital memory devices such as computer random access memory require a means for storing and switching between the two binary states. A ***flip-flop*** is a sequential logic device that can perform this function. The flip-flop is called a ***bistable*** device since it has two and only two possible stable states: 1 (high) and 0 (low). It has the capability of remaining in a particular state (i.e., storing a bit) until an external signal causes it to change state. This is the basis of all semiconductor information storage and processing in digital computers; in fact, flip-flops perform many of the basic functions critical to the operation of almost all digital devices. The flip-flop is constructed from logic gates, but it is different from combinational logic circuits since it includes internal feedback from the output to the inputs.

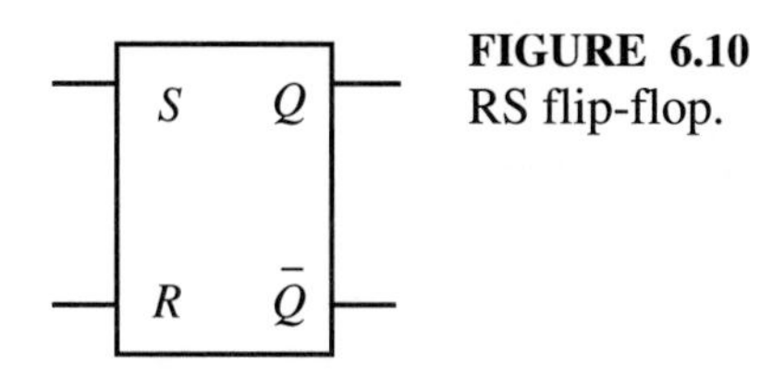


**FIGURE 6.10**
RS flip-flop.

The simplest and most fundamental flip-flop, an ***RS flip-flop***, is schematically shown in Figure 6.10. $S$ is the ***set*** input, $R$ is the ***reset*** input, and $Q$ and $\overline{Q}$ are the complementary outputs of the flip-flop. Most flip-flops provide these complementary outputs where one is the inverse (NOT) of the other. The RS flip-flop obeys the following rules:

1. As long as the inputs $S$ and $R$ are both 0, the outputs of the flip-flop remain unchanged.
2. When $S$ is 1 and $R$ is 0, the flip-flop is set to $Q = 1$ and $\overline{Q} = 0$.
3. When $S$ is 0 and $R$ is 1, the flip-flop is reset to $Q = 0$ and $\overline{Q} = 1$.
4. It is "not allowed" (NA) to place a 1 on $S$ and $R$ simultaneously since the output will be unpredictable.

A truth table is a valuable tool for describing the functionality of a flip-flop. The truth table for a basic RS flip-flop is given in Table 6.4. $Q_0$ is the value of the output $Q$ before the indicated input conditions were established. 1 is logic high and 0 is logic low. The **NA** in the last row indicates that the input condition for that row is not allowed. Because we are precluded from applying the $S = 1$, $R = 1$ input condition, the RS flip-flop is seldom used in actual designs. Other more versatile flip-flops that avoid the NA limitation are presented in subsequent sections.

To understand how flip-flops and other sequential logic circuits function, we will look at the internal design of an RS flip-flop illustrated in Figure 6.11a. It consists of combinational logic gates with feedback from the outputs to the inputs of the NAND gates. Figure 6.11b illustrates the timing of the various signals, which are affected by very short propagation delays through the NAND gates. Immediately after signal $R$ transitions from 0 to 1, the inputs to the lower NAND gate are 0 and $Q$, which is still 1. This changes $\overline{Q}$ to 1 after a slight propagation delay $\Delta t_1$. Feedback of $\overline{Q}$ to the top NAND gate drives $Q$ to 0 after a slight delay $\Delta t_2$. Now

**TABLE 6.4**
**Truth table for the RS flip-flop**

| Inputs | | Outputs | |
|---|---|---|---|
| $S$ | $R$ | $Q$ | $\overline{Q}$ |
| 0 | 0 | $Q_0$ | $\overline{Q_0}$ |
| 1 | 0 | 1 | 0 |
| 0 | 1 | 0 | 1 |
| 1 | 1 | NA | |

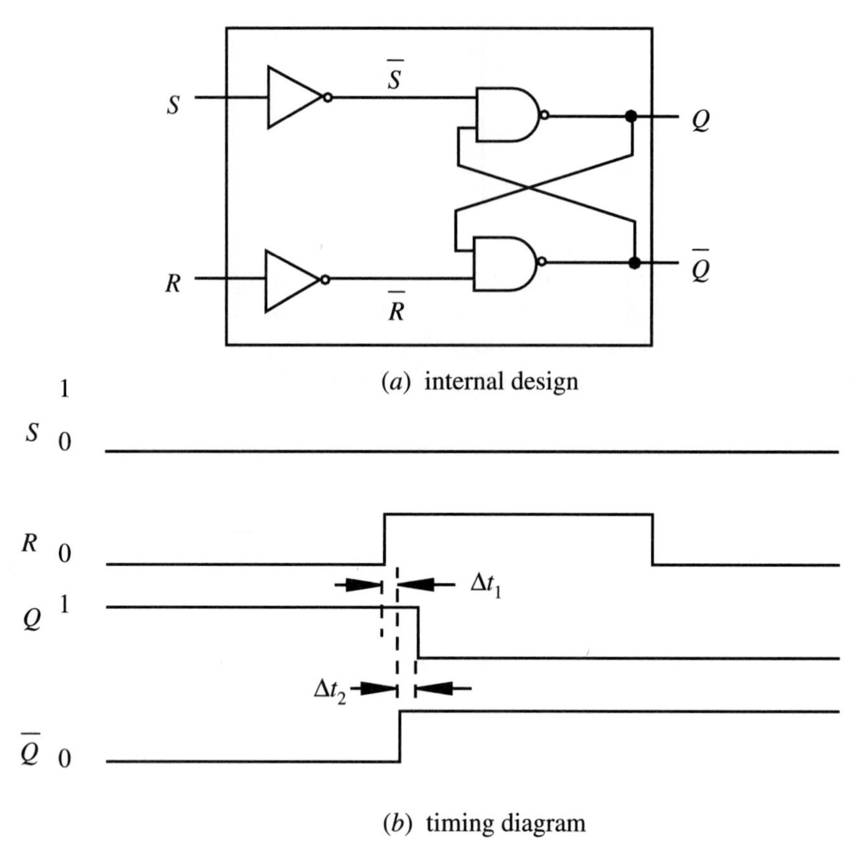


**FIGURE 6.11**
RS flip-flop internal design and timing.

the flip-flop is reset, and it remains in this state even after $R$ returns to 0. The set operation functions in a similar manner. The propagation delays $\Delta t_1$ and $\Delta t_2$ are usually in the nanosecond range. All sequential logic devices depend on feedback and propagation delays for their operation.

## 6.10 TRIGGERING OF FLIP-FLOPS

Many flip-flops are clocked; that is, there is a master signal in the circuit that coordinates or synchronizes the changes of the output states of the device. Hence complex circuits such as a microprocessor can be designed where all system changes are triggered by a common clock signal. This is called ***synchronous*** operation since changes in state are coordinated by the clock pulses. Digital circuits that do not switch in a prescribed clocked pattern are said to exhibit ***asynchronous*** opera-

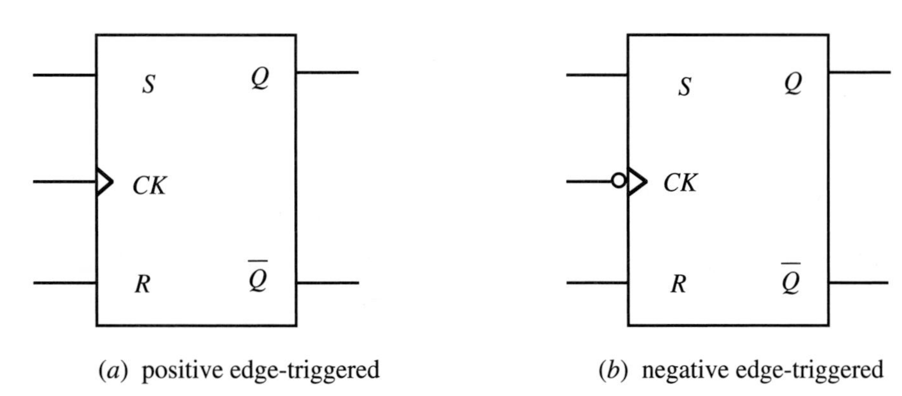


(*a*) positive edge-triggered (*b*) negative edge-triggered

**FIGURE 6.12**
Edge-triggered RS flip-flops.

tion. Different types of flip-flops are designed to respond either to a positive edge or to a negative edge of a clock pulse. These flip-flops are termed ***edge-triggered*** flip-flops. Positive edge triggering is indicated schematically by a small angle bracket on the clock input to the flip-flop (see Figure 6.12a). Negative edge triggering is indicated schematically by a small circle and angle bracket on the clock input (see Figure 6.12b).

The function of the edge-triggered RS flip-flop is defined by the following rules:

1. If $S$ and $R$ are both 0 when the clock edge is encountered, the output state remains unchanged.
2. If $S$ is 1 and $R$ is 0 when the clock edge is encountered, the flip-flop output is set to 1. If the output was at 1 already, there is no change.
3. If $S$ is 0 and $R$ is 1 when the clock edge is encountered, the flip-flop output is reset to 0. If the output was at 0 already, there is no change.
4. $S$ and $R$ should never both be 1 when the clock edge is encountered.

The truth table for a positive edge-triggered RS flip-flop is given in Table 6.5. The up-arrow ↑ in the clock (*CK*) column represents the positive edge transition from 0 to 1. The NA in the last row indicates that the input condition for that row is not allowed. A timing diagram is shown in Figure 6.13.

**TABLE 6.5**
**Positive edge-triggered RS flip-flop truth table**

| $S$ | $R$ | $CK$ | $Q$ | $\overline{Q}$ |
|---|---|---|---|---|
| 0 | 0 | ↑ | $Q_0$ | $\overline{Q_0}$ |
| 1 | 0 | ↑ | 1 | 0 |
| 0 | 1 | ↑ | 0 | 1 |
| 1 | 1 | ↑ | NA | |

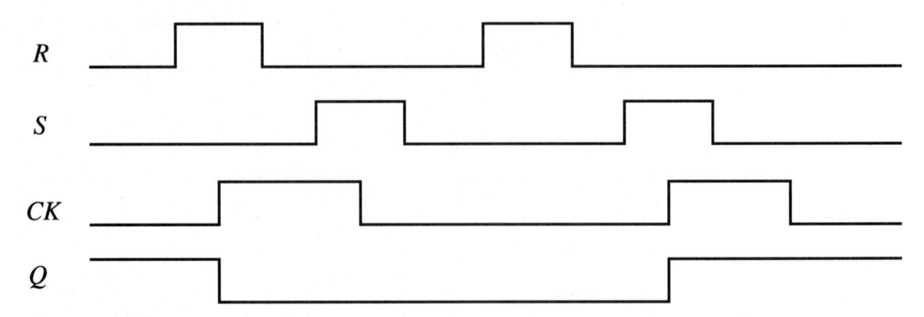


**FIGURE 6.13**
Positive edge-triggered RS flip-flop timing diagram.

Devices that are not edge triggered are called ***level-triggered*** devices. They respond to their inputs while the clock signal is at a high level and retain their output values after the level changes. If there were an inversion circle at the clock input, the device would respond during a low level instead. Level-triggered operation is described as asynchronous since the output is not synchronized by a distinct trigger. Rather, while the clock level is high the output can change whenever the input changes. An important example of a level-triggered flip-flop is the ***latch*** (e.g., 7475). Its schematic symbol is shown in Figure 6.14. The output $Q$ tracks the input $D$ while $CK$ is high. At a negative edge (i.e., when $CK$ goes low), the flip-flop output will hold or "latch" the value that $D$ had at the edge transition. The truth table for a latch is given in Table 6.6, and a timing diagram example is shown in Figure 6.15. The **x** in the last row of the table indicates that the value of $D$ has no effect on the output as long as $CK$ is low.

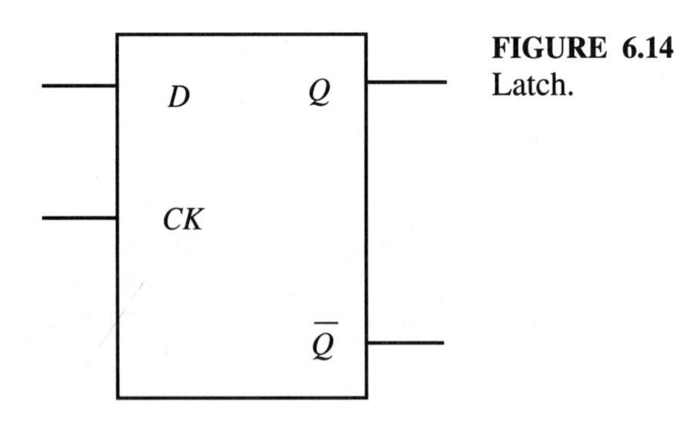


**FIGURE 6.14**
Latch.

**TABLE 6.6**
**Latch truth table**

| $D$ | $CK$ | $Q$ | $\overline{Q}$ |
|---|---|---|---|
| 0 | 1 | 0 | 1 |
| 1 | 1 | 1 | 0 |
| x | 0 | $Q_0$ | $\overline{Q_0}$ |

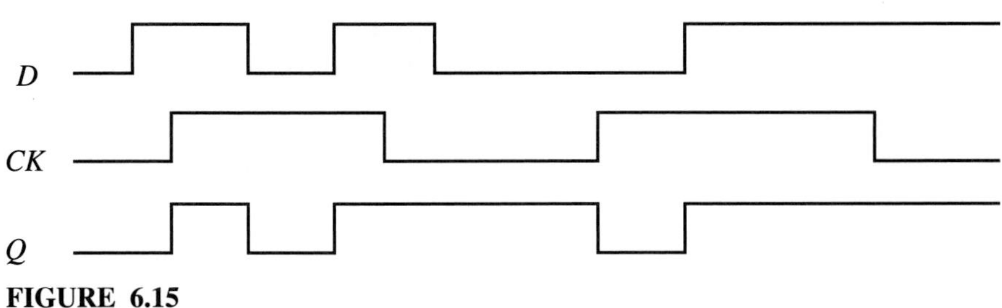


**FIGURE 6.15**
Latch timing diagram.

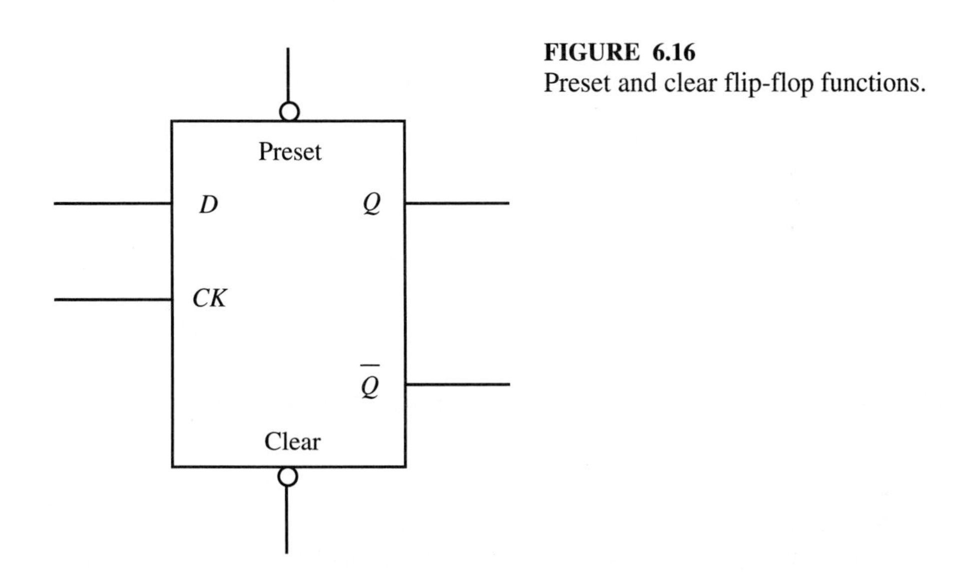


**FIGURE 6.16**
Preset and clear flip-flop functions.

Often flip-flops have preset and clear functions such as those illustrated in Figure 6.16. The ***preset*** input is used to set or initialize the output $Q$ of the flip-flop to 1 or high. The ***clear*** input is used to clear or reset the output $Q$ of the flip-flop to 0 or low. Either of these inputs can be used to define the state of a flip-flop after power-up. Otherwise, at power-up the output of a flip-flop is arbitrary. The circles at the inputs as shown in Figure 6.16 denote that they are active low, which means a negative logic state or 0 is required to execute the function. If the circles were not present, the functions would be activated by a 1 or high state. Preset and clear are referred to as asynchronous functions because they do not depend on the clock signal.

## 6.11 D FLIP-FLOP

The ***D flip-flop*** (e.g., 7474, 74171, or 74174), also called a data flip-flop, has a single input $D$ whose value is stored and presented at the output $Q$ at the edge of a

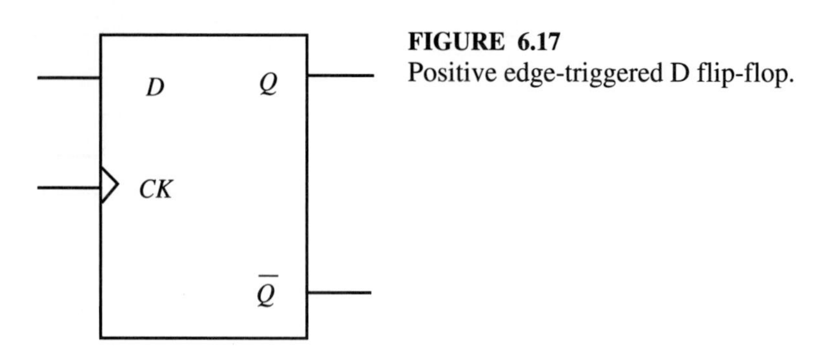


**FIGURE 6.17**
Positive edge-triggered D flip-flop.

**TABLE 6.7**
**Positive edge-triggered D flip-flop truth table**

| $D$ | $CK$ | $Q$ | $\overline{Q}$ |
|---|---|---|---|
| 0 | ↑ | 0 | 1 |
| 1 | ↑ | 1 | 0 |
| x | 0 | $Q_0$ | $\overline{Q_0}$ |
| x | 1 | $Q_0$ | $\overline{Q_0}$ |

clock pulse. A positive edge-triggered D flip-flop is illustrated in Figure 6.17, and its truth table is given in Table 6.7.

▼▼▼ ***CLASS DISCUSSION ITEM 6.5. Computer Memory.*** With your knowledge of flip-flops, discuss how you think computer random access memory (RAM) works.

## 6.12 JK FLIP-FLOP

The ***JK flip-flop*** (e.g., 7473, 7476, or 74107) is similar to the RS flip-flop where the $J$ is analogous to the $S$ (set) input and the $K$ is analogous to the $R$ (reset) input. The major difference is that the $J$ and $K$ inputs may both be high simultaneously. This causes the output to **toggle**, which means the output changes to the opposite state. The symbol and truth table for a negative edge-triggered JK flip-flop are shown in Figure 6.18 and Table 6.8, respectively. The first two rows describe the preset or clear functions that can be used to initialize the output of the flip-flop. The third row indicates the impossibility of setting and clearing simultaneously. The symbol ↓ represents the negative edge of the clock signal, which causes the change in the output. The last row describes the memory feature of the flip-flop in the absence of a negative edge.

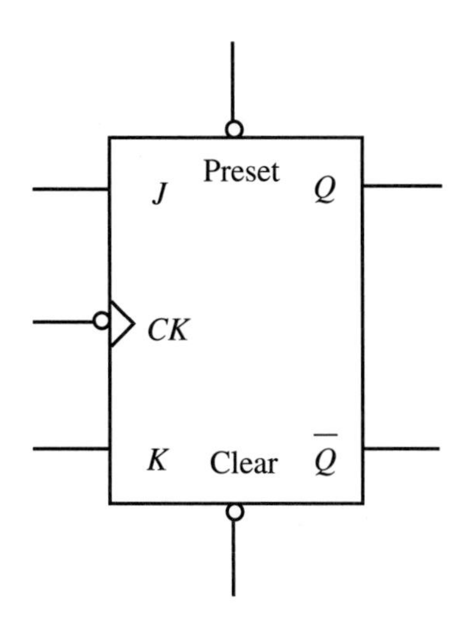


**FIGURE 6.18**
Negative edge-triggered JK flip-flop.

**TABLE 6.8**
**Truth table for a negative edge-triggered JK flip-flop**

| Preset | Clear | *CK* | *J* | *K* | *Q* | $\overline{Q}$ |
|---|---|---|---|---|---|---|
| 0 | 1 | x | x | x | 1 | 0 |
| 1 | 0 | x | x | x | 0 | 1 |
| 0 | 0 | | | NA | | |
| 1 | 1 | ↓ | 0 | 0 | $Q_0$ | $\overline{Q_0}$ |
| 1 | 1 | ↓ | 1 | 0 | 1 | 0 |
| 1 | 1 | ↓ | 0 | 1 | 0 | 1 |
| 1 | 1 | ↓ | 1 | 1 | toggle | |
| 1 | 1 | 0, 1 | x | x | $Q_0$ | $\overline{Q_0}$ |

▼▼▼ ***CLASS DISCUSSION ITEM 6.6. JK Flip-Flop Timing Diagram.*** Construct a timing diagram for the JK flip-flop illustrating its complete functionality.

The JK flip-flop has a wide range of applications, and all flip-flops can easily be constructed from it with proper external wiring. The ***T (toggle) flip-flop*** serves as a good example of this. The symbol for a positive edge-triggered T flip-flop and the equivalent JK flip-flop configuration are shown in Figure 6.19. The T flip-flop simply toggles the output every time it is triggered. The preset and

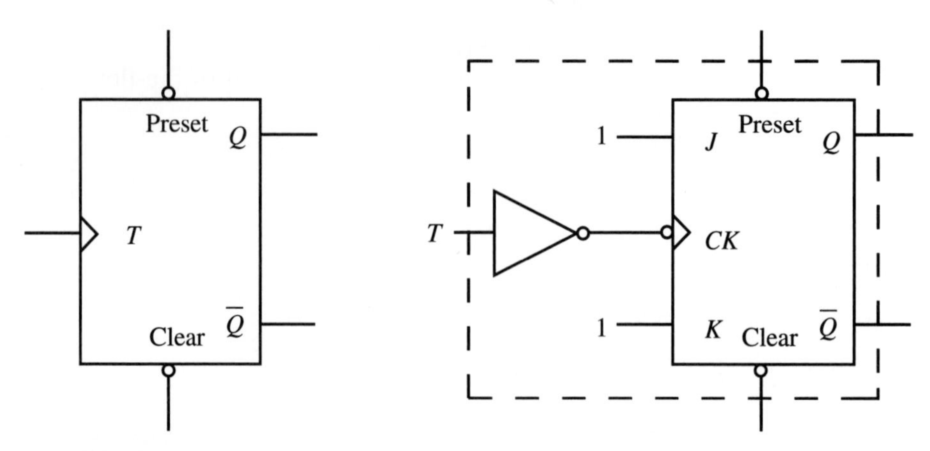


**FIGURE 6.19**
Positive edge-triggered T flip-flop.

**TABLE 6.9**
**Positive edge-triggered T flip-flop truth table**

| T | Preset | Clear | $Q$ | $\bar{Q}$ |
|---|---|---|---|---|
| ↑ | 1 | 1 | $\overline{Q_0}$ | $Q_0$ |
| 0 | 1 | 1 | $Q_0$ | $\overline{Q_0}$ |
| 1 | 1 | 1 | $Q_0$ | $\overline{Q_0}$ |
| x | 0 | 1 | 1 | 0 |
| x | 1 | 0 | 0 | 1 |

clear functions are necessary to provide direct control over the output since the $T$ input alone provides no mechanism for initializing the output value. The truth table is given in Table 6.9.

---

▼ ***EXAMPLE 6.4. Flip-Flop Circuit Timing Diagram.*** Given the digital circuit on the next page that includes RS, T, and JK flip-flops with the inputs as indicated in the timing diagram, the digital outputs $D$, $E$, and $F$ will be as shown. The signals $D$, $E$, and $F$ are assumed to be low at the beginning of the timing diagram. Observe how signal $D$ updates during a high level of $C$, signal $E$ updates at positive edges of $D$, and signal $F$ updates at negative edges of $C$.

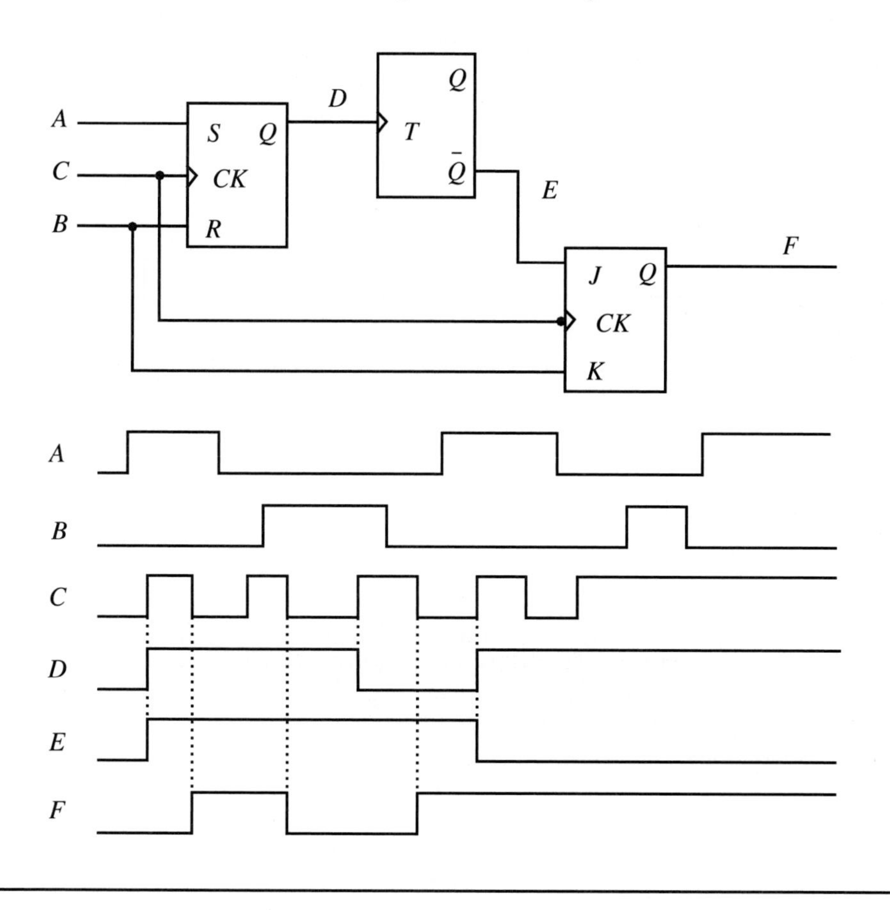


## 6.13 MASTER-SLAVE FLIP-FLOPS

***Pulse-triggered*** devices, also called ***master-slave*** devices, consist of several gates and flip-flops integrated together. They require an entire clock pulse to transfer data from the input to the output. The ***master-slave JK flip-flop*** is the most common example. It behaves like a positive edge-triggered JK flip-flop in that the $J$ and $K$ values are sampled at the leading edge of a clock pulse; however, there are differences. Any high level of the $J$ or $K$ inputs that occur during the high level of the clock pulse $CK$ is retained, and the output only changes on the trailing edge of the clock pulse in response to the stored $J$ and $K$ values. The timing diagram shown in Figure 6.20 illustrates this functionality. Clock pulses 1 through 5 and 7 exhibit a change in $Q$ on the trailing edge of $CK$ based on the leading edge values of $J$ and $K$. Clock pulses 6, 8, and 9 illustrate the feature where high inputs occurring during the high level of the clock are retained and applied at the trailing edge of the clock pulse.

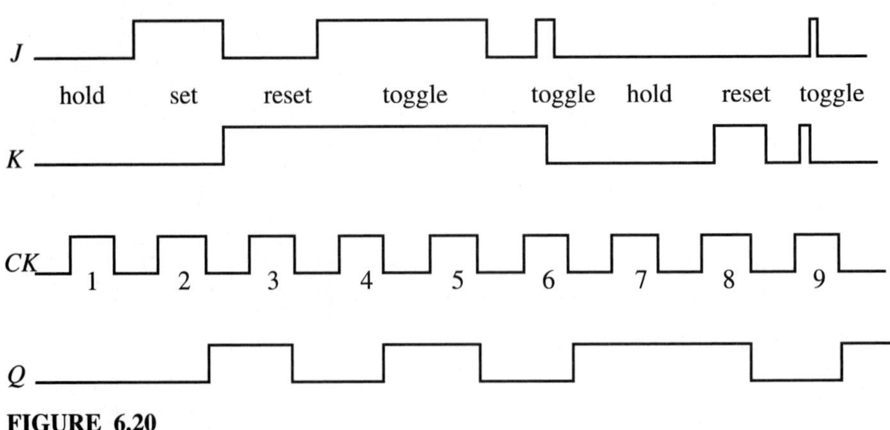


**FIGURE 6.20**
Master-slave JK flip-flop.

▼▼▼ ***CLASS DISCUSSION ITEM 6.7. Everyday Use of Logic Devices.*** Discuss the use of combinational and sequential logic devices in items you use every day. For each, identify the purpose of the logic device and what type you think it is (combinational or sequential).

## 6.14 APPLICATIONS OF FLIP-FLOPS

We have just seen that there are a variety of flip-flops. In the subsections below, we illustrate a number of applications that use flip-flops as their functional units.

### 6.14.1 Switch Debouncing

When mechanical switches are opened or closed, there are brief current oscillations due to mechanical bouncing or electrical arcing. This phenomenon is called ***switch bounce***. As illustrated in Figure 6.21, the mechanical contact associated with a switch closing results in multiple voltage transitions over a short period of time.

The sequential logic circuit shown in Figure 6.22 can provide an output that is free from multiple transitions associated with switch bounce. As the switch breaks contact with $B$, signal bounce occurs on the $B$ line. There is a small delay as the switch moves from contacts $B$ to $A$, and then signal bounce occurs on the $A$ line as contact is established with $A$. The output of the debouncer $Q$ is a single transition

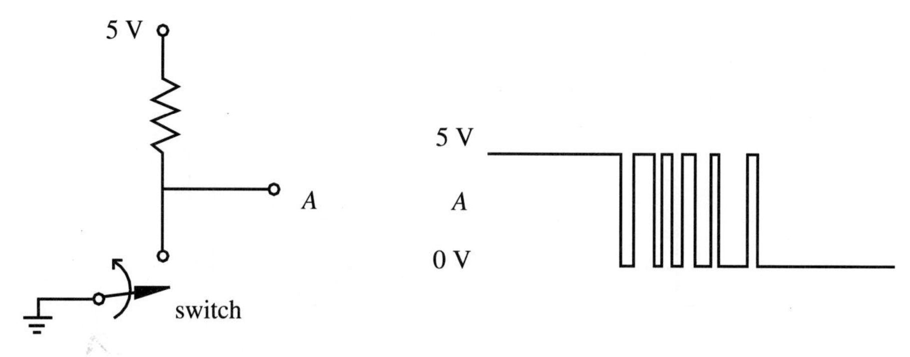


**FIGURE 6.21**
Switch bounce.

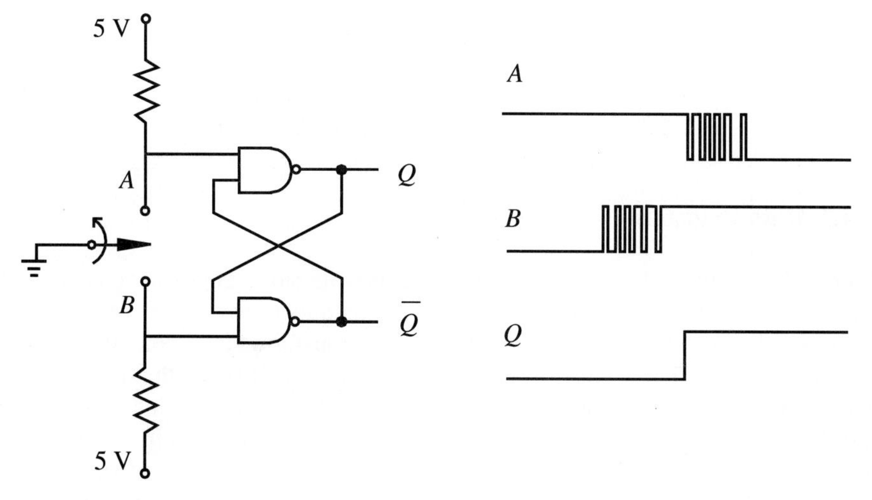


**FIGURE 6.22**
Switch debouncer circuit.

from 0 V to 5 V. Thus, the switch debouncer maintains its output state and doesn't change its output state until the inputs change in specific ways. Thus, the circuit functions very much like a single flip-flop (see Class Discussion Item 6.8).

▼▼▼ ***CLASS DISCUSSION ITEM 6.8. Switch Debouncer Function.*** Track the inputs and outputs of the two NAND gates in Figure 6.22 as the switch moves from contact *B* to contact *A* and create a timing diagram. Also, draw an equivalent circuit for the debouncer using an RS flip-flop.

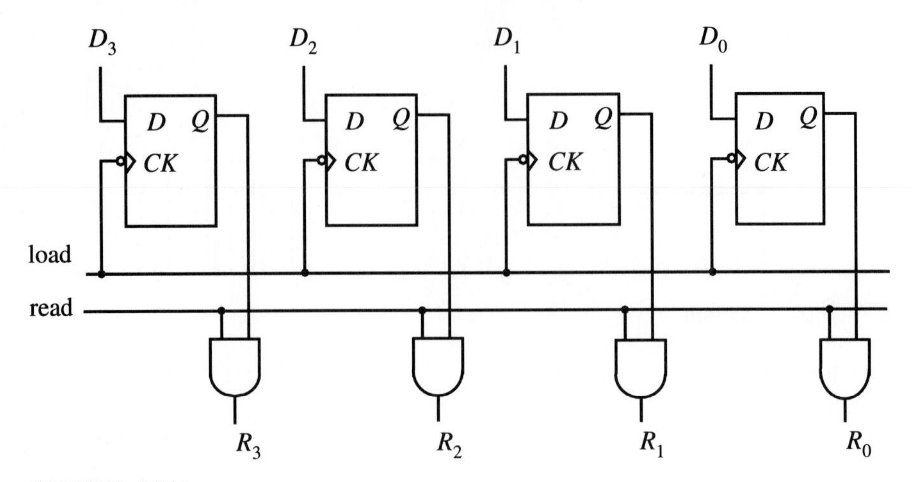


**FIGURE 6.23**
4-Bit data register.

### 6.14.2 Data Register

Figure 6.23 shows a 4-bit ***data register*** that uses negative edge-triggered D flip-flops to transfer data from four data lines to the outputs of four AND gates. It does this in two distinct steps. First, data values $D_i$ are transferred to the outputs $Q$ of the flip-flops on the negative edge of the load signal. Then a pulse on the read line presents the data $D_i$ at the register outputs $R_i$ of the AND gates. In this way, data is loaded and then transferred to the register. Data registers are used in microprocessors to hold data for arithmetic calculations. Data registers can be cascaded to store a larger number of bits.

### 6.14.3 Binary Counter

Figure 6.24 shows a 4-bit ***binary counter*** consisting of four negative edge-triggered toggle flip-flops connected in sequence. The timing diagram is also shown for the first ten input pulses. The four outputs $B_i$ change according to the binary number coding sequence. This circuit may also be used as a ***frequency divider***. Output $B_0$ is a divide-by-two output since its frequency is 1/2 of the input pulse train frequency. $B_1$, $B_2$, and $B_3$ are divide-by-four, -eight, and -sixteen outputs, respectively.

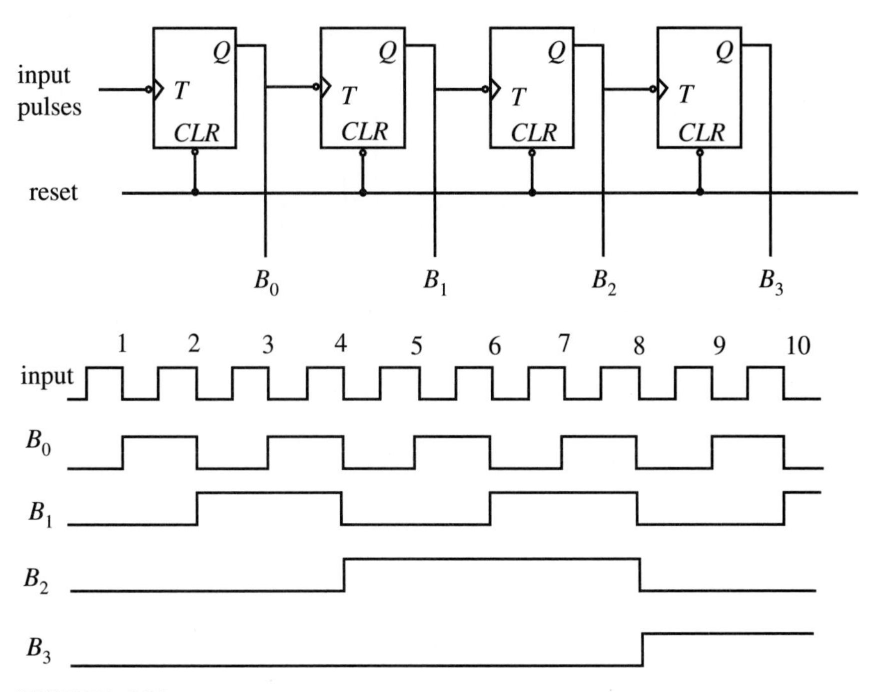


**FIGURE 6.24**
4-Bit binary counter.

### 6.14.4 Serial and Parallel Interfaces

Figures 6.25 and 6.26 show flip-flop circuits that convert between serial and parallel data. ***Serial data*** is a sequence of bits, or train of pulses, that occur on a single data line. ***Parallel data*** consists of a set of bits that occur simultaneously in parallel on a set of data lines. The serial-to-parallel converter utilizes negative edge-triggered D flip-flops, and the parallel-to-serial converter utilizes negative edge-triggered JK flip-flops. In both circuits, the serial input or output is synchronized by a clock signal. A reset line is used to clear the flip-flops prior to loading a set of bits. The reset line is active low, meaning that when the line goes low, the flip-flops are cleared, causing the outputs $Q$ to go low (0). The load line for the parallel-to-serial converter passes the data through the NAND gates, storing the data line values in the flip-flops using the active low flip-flop preset. When the load line goes high and a parallel bit ($P_i$) is high (1), then the respective flip-flop is preset resulting in a high (1) output $Q$. The figures show 4-bit converters, but the flip-flops can be cascaded for a larger number of bits.

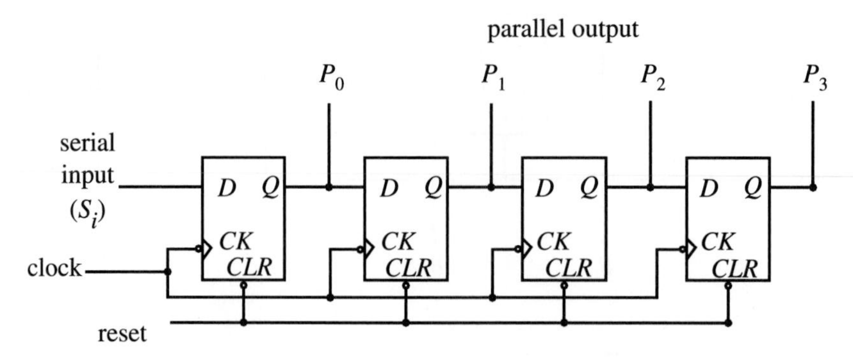


**FIGURE 6.25**
Serial-to-parallel converter.

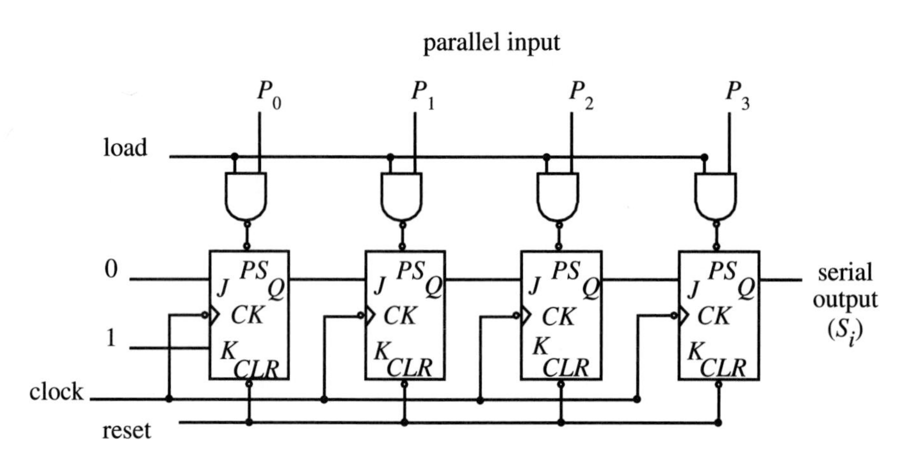


**FIGURE 6.26**
Parallel-to-serial converter.

---

▼ ***EXAMPLE 6.5 Converting between Serial and Parallel Data.*** Looking at the figures above, explain in detail the functioning of the circuits that convert between serial and parallel data. Also, how is the ***baud rate*** (bits per second) for serial transmission related to the converter's clock speed?

---

## 6.15 DECADE COUNTER

Section 6.14.3 presented a flip-flop circuit that can be used to perform binary counting. Another common counter called a ***decade counter*** can be constructed using a 7490 IC (see Question 6.18). It is a negative edge-triggered counter, and the output is in ***binary coded decimal (BCD)*** consisting of four bits. Table 6.10 shows the output sequence for the four bits as the counter increments from 0 to 9, making it useful for decimal counting applications. The timing diagram for the input and four outputs of the decade counter are shown in Figure 6.27. Note that the counter output cycles back to 0000 ($0_{10}$) after 1001 ($9_{10}$).

**TABLE 6.10**
**7490 Decade counter BCD coding**

| Decimal count | BCD output *D* | *C* | *B* | *A* |
|---|---|---|---|---|
| 0 | 0 | 0 | 0 | 0 |
| 1 | 0 | 0 | 0 | 1 |
| 2 | 0 | 0 | 1 | 0 |
| 3 | 0 | 0 | 1 | 1 |
| 4 | 0 | 1 | 0 | 0 |
| 5 | 0 | 1 | 0 | 1 |
| 6 | 0 | 1 | 1 | 0 |
| 7 | 0 | 1 | 1 | 1 |
| 8 | 1 | 0 | 0 | 0 |
| 9 | 1 | 0 | 0 | 1 |
| 0 | 0 | 0 | 0 | 0 |

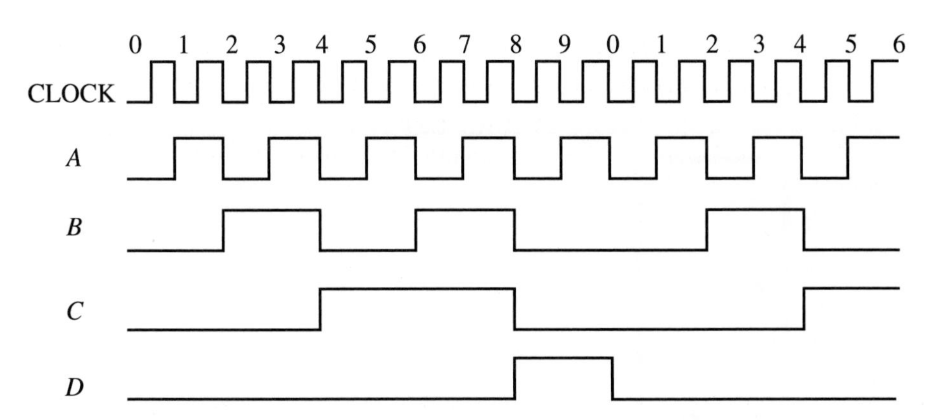


**FIGURE 6.27**
Decade counter timing.

As illustrated in Figure 6.28, BCD counters can be cascaded in order to count in powers of 10. Output *D* can be used as the clock input for a second 7490, thus cascading the two together in order to raise the range for counting from 0 to 99. Further cascading of the counters allows higher power of 10 ranges. The 7490 is a versatile IC and can be wired in a variety of useful ways. Examples include binary counters, divide-by-two counters, and divide-by-four counters.

A convenient device for viewing a BCD output is a 7-segment LED display (see Figure 6.29) driven by a 7447 BCD-to-seven-segment decoder. The 7447 converts the BCD bits at its inputs into a 7-bit code to properly drive the LED segments. The function table describing the input (BCD) to output (negative logic 7-segment LED code) relationship for the 7447 is shown in Table 6.11. Verify some of the rows in this table by drawing the digits with the appropriate segments. Figure 6.30 illustrates the internal design of the 7447, which consists of combinational logic circuits.

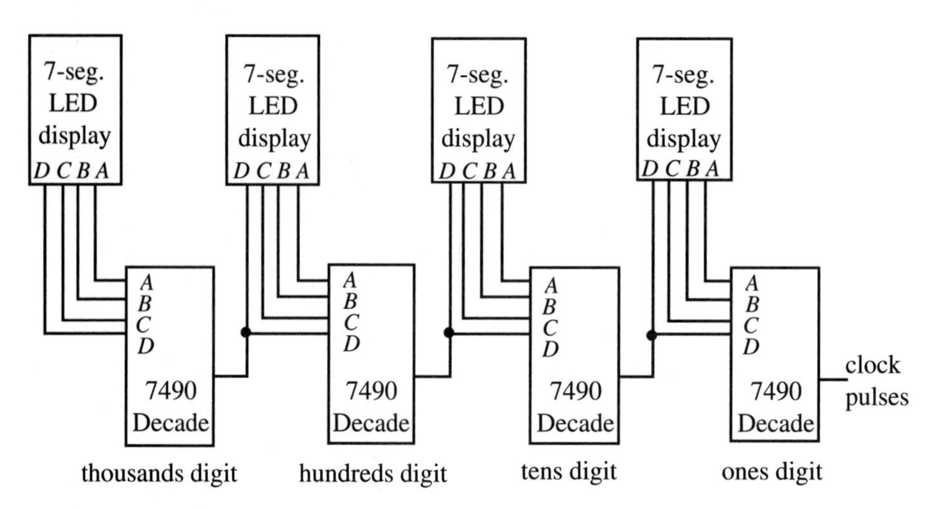


**FIGURE 6.28**
Cascaded decade counters.

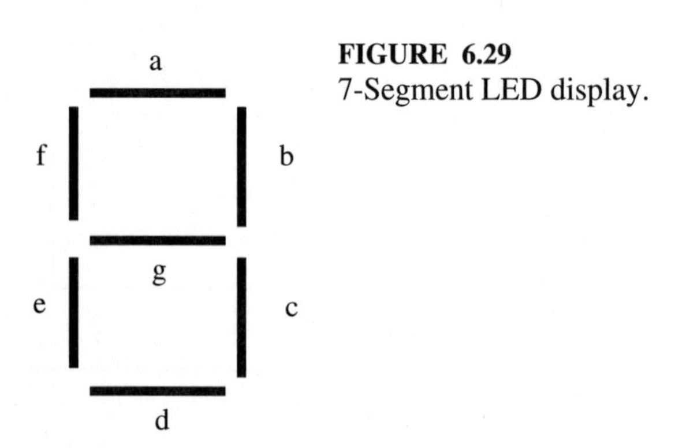


**FIGURE 6.29**
7-Segment LED display.

**TABLE 6.11**
**7447 BCD to 7-segment decoder**

| Decimal digit | Input | | | | Output | | | | | | |
|---|---|---|---|---|---|---|---|---|---|---|---|
| | $Q_D$ | $Q_C$ | $Q_B$ | $Q_A$ | *a* | *b* | *c* | *d* | *e* | *f* | *g* |
| 0 | 0 | 0 | 0 | 0 | 0 | 0 | 0 | 0 | 0 | 0 | 1 |
| 1 | 0 | 0 | 0 | 1 | 1 | 0 | 0 | 1 | 1 | 1 | 1 |
| 2 | 0 | 0 | 1 | 0 | 0 | 0 | 1 | 0 | 0 | 1 | 0 |
| 3 | 0 | 0 | 1 | 1 | 0 | 0 | 0 | 0 | 1 | 1 | 0 |
| 4 | 0 | 1 | 0 | 0 | 1 | 0 | 0 | 1 | 1 | 0 | 0 |
| 5 | 0 | 1 | 0 | 1 | 0 | 1 | 0 | 0 | 1 | 0 | 0 |
| 6 | 0 | 1 | 1 | 0 | 1 | 1 | 0 | 0 | 0 | 0 | 0 |
| 7 | 0 | 1 | 1 | 1 | 0 | 0 | 0 | 1 | 1 | 1 | 1 |
| 8 | 1 | 0 | 0 | 0 | 0 | 0 | 0 | 0 | 0 | 0 | 0 |
| 9 | 1 | 0 | 0 | 1 | 0 | 0 | 0 | 0 | 1 | 0 | 0 |

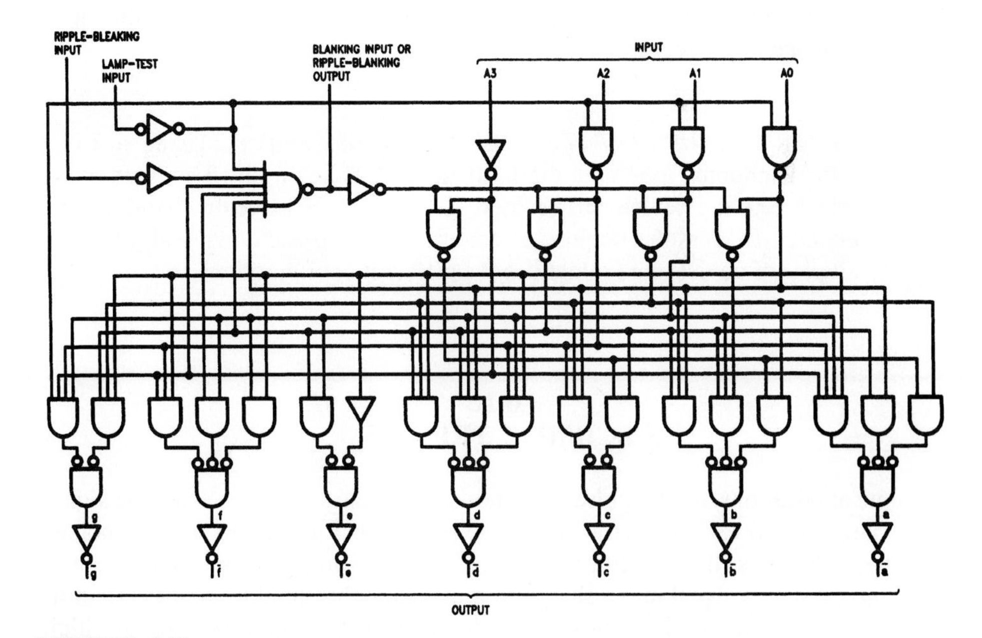


**FIGURE 6.30**
7447 internal design. *(Courtesy of National Semiconductor, Santa Clara, CA)*

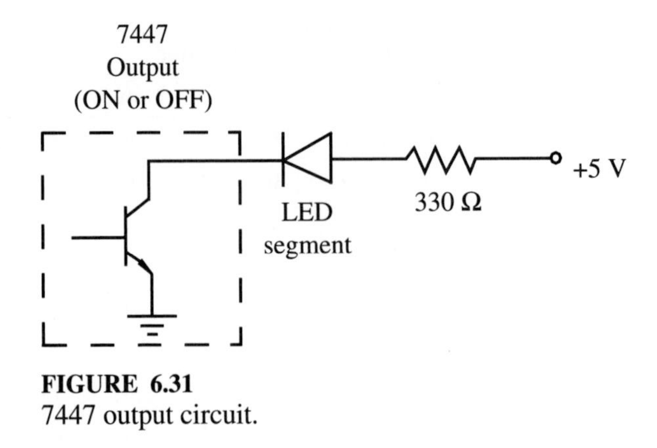


**FIGURE 6.31**
7447 output circuit.

If the 7447 decoder driver is properly connected to a 7-segment LED display, the output from a counter may be displayed in an easily recognizable numerical form. Note that the decoder driver does not actually drive the segment LEDs by supplying current to them; rather, it sinks current from them as illustrated in Figure 6.31. The outputs of the 7447 are called open collector outputs since each output transistor has a floating collector. The resistor and LED segment complete the circuit for each 7447 output. Therefore, the LED is ON (i.e., illuminated) when the 7447 output is low (0), allowing current to flow to ground. 330 Ω resistors are used in series with the segments to limit the current into the decoder driver, thus preventing damage to the segments.

▼▼▼ ***CLASS DISCUSSION ITEM 6.9. Up-Down Counters.*** Using a TTL logic handbook, look up a digital IC known as an up-down counter. You will notice the device can be used as a binary, hexadecimal, or decimal counter. Discuss the purpose of counting both up and down and give examples of devices that might use the different configurations.

## 6.16 SCHMITT TRIGGER

In some applications, digital pulses may not exhibit sharp edges; instead, the signal may ramp from 0 to 5 V over a finite time period, and it may do so in a "noisy" (jumpy) fashion as illustrated in Figure 6.32. The ***Schmitt trigger*** is a device that can convert such a signal into a sharp pulse using the threshold hysteresis effect illustrated in the figure. The output goes high when the input exceeds the high threshold and remains high until the input falls below the low threshold. The hysteresis between the low and high thresholds results in the distinct edges in the output. Six Schmitt triggers are usually packaged on a single IC (e.g., LM7414).

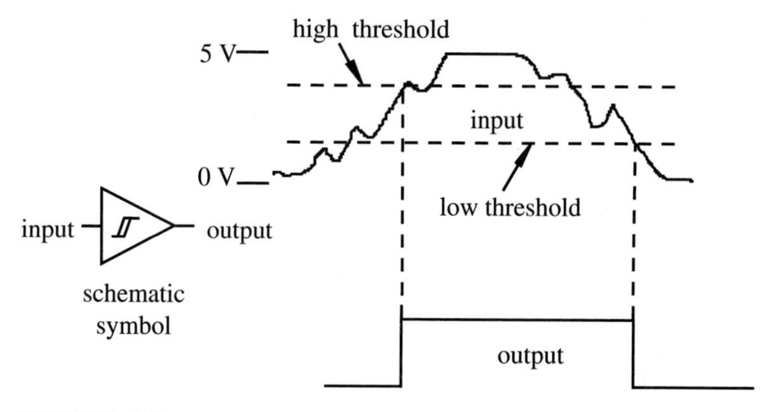


**FIGURE 6.32**
Input and output of a Schmitt trigger.

## 6.17
## THE 555 TIMER

The 555 integrated circuit is known as the "time machine" since it performs a wide variety of timing tasks. It is a combination of both digital and analog circuits. The block diagram for the 555, including pin terminology and numbering, is shown in Figure 6.33. Two packages and pin configurations for the 555 are shown in

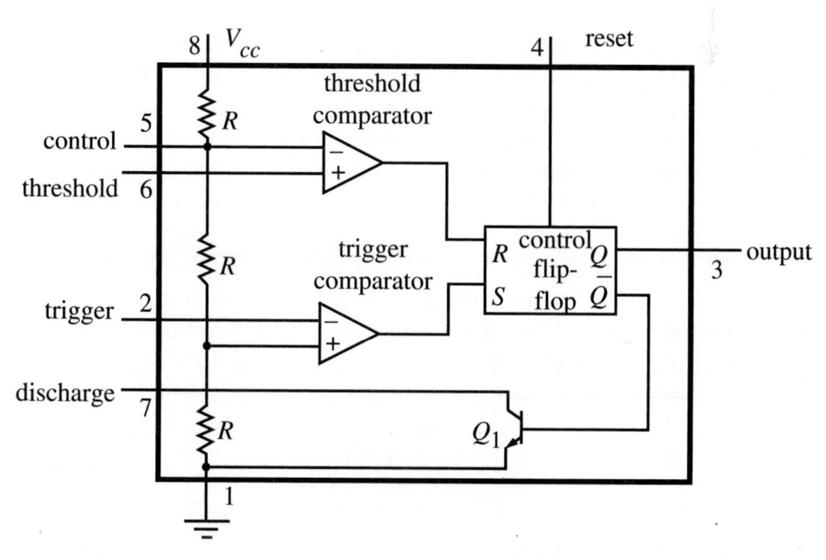


**FIGURE 6.33**
Block diagram of the 555 IC.

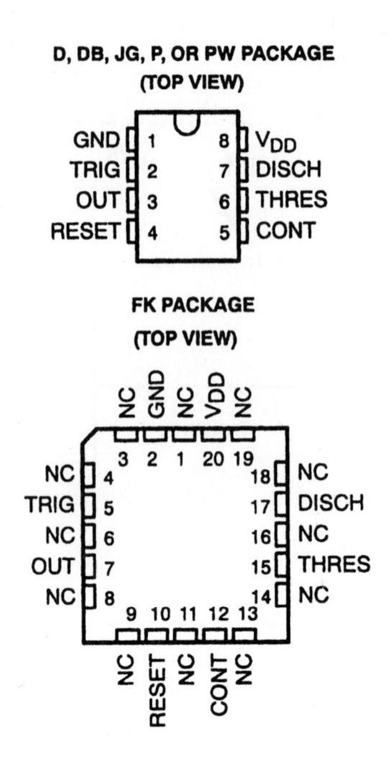


**FIGURE 6.34**
555 pin-out. *(Courtesy of Texas Instruments, Dallas, TX)*

Figure 6.34. Applications for the 555 include bounce-free switches, cascaded timers, frequency dividers, voltage-controlled oscillators, pulse generators, LED flashers, and many other useful circuits.

A circuit easily built with a 555 IC is the ***monostable multivibrator*** shown in Figure 6.35. It is constructed by adding an external capacitor and resistor to a 555. The circuit generates a single pulse of desired duration when it receives a trigger signal, hence it is also called a ***one-shot***. The time constant of the resistor-capacitor combination determines the length of the pulse. Referring to Figure 6.36, the sequence of operation is as follows: When the circuit is powered up or after the reset is activated, the output will be low ($Q = 0$), transistor $Q_1$ is saturated (ON) shorting capacitor $C$, and the outputs of both comparators are low. When the trigger pulse goes below 1/3 $V_{cc}$, the trigger comparator goes high, setting the flip-flop. Now the output is high ($Q = 1$), transistor $Q_1$ cuts off, and the capacitor begins to charge with time constant $\tau = R_a C$. The time constant is the time required for the capacitor voltage to reach 63.2% of its full charge value $V_{cc}$. When the capacitor voltage reaches 2/3 $V_{cc}$, the threshold comparator resets the flip-flop, which sets the output low ($Q = 0$) and discharges the capacitor again. If the trigger is pulsed while the output is high, it has no effect. The length of the pulse is given approximately (see Question 6.19) by

$$\Delta T \approx 1.1 R_a C \tag{6.35}$$

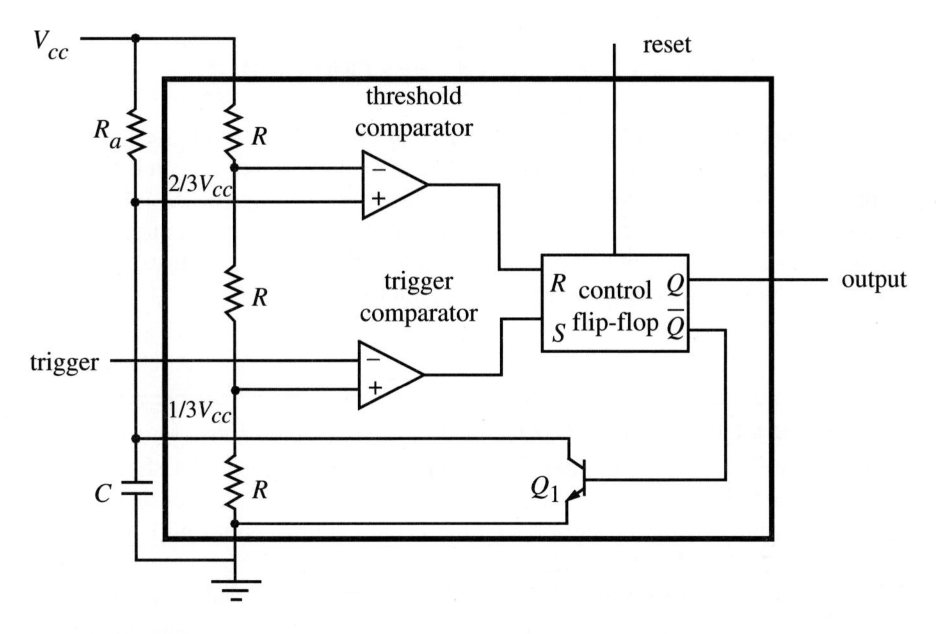


**FIGURE 6.35**
Monostable multivibrator (one-shot).

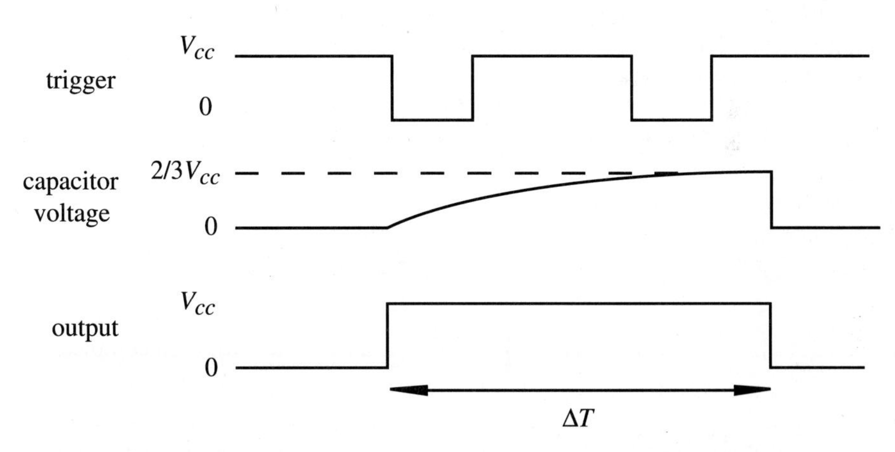


**FIGURE 6.36**
One-shot timing.

The 7400 series of the TTL logic family also includes devices that can be used as one-shots. Examples are the 74121 and the 74123. They also require an external resistor and capacitor to control the pulse width.

## 6.18 INTEGRATED CIRCUIT SYSTEM DESIGN

For most digital design applications, integrated circuits (ICs) can be used as building blocks to create required functionality. There are a myriad of ICs on the market that provide almost every conceivable digital function. Listings of ICs along with data sheets describing their functionality can be found in manufacturer TTL or logic data books. The digital tachometer example presented below illustrates a digital design solution employing several commercially available ICs.

---

▼▼ ***DESIGN EXAMPLE 6.1. Digital Tachometer.*** Our objective is to design a system to measure and display the rotational speed of a shaft. A simple method to measure rotational speed is to count the number of shaft rotations during a given period of time. The resulting count will be directly proportional to the shaft speed.

There are a number of different sensors that can detect shaft rotation. A proximity sensor that uses magnetic, optical, or mechanical principles to detect some feature on the shaft is one example. We can use an LED/phototransistor pair as an optical sensor and place a small piece of reflective tape of the shaft. Each time the tape passes by our sensor pair, our photo-optic pair and a Schmitt trigger will output a single pulse to a counting circuit. The figure below illustrates the sensor and the signal conditioning circuit.

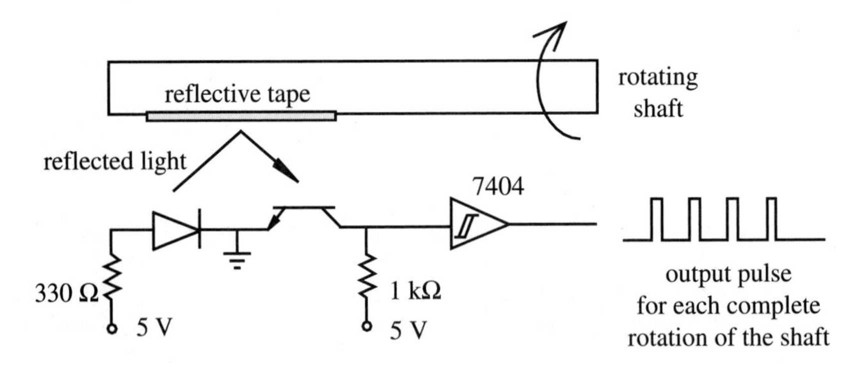


Now we need a circuit to count and display the pulses over a given interval of time $T$. The figure on the next page illustrates all of the required components. The 7490 decade counter counts the pulses and is reset by a negative edge on signal $R$ after the time period $T$. The period $T$ is set by a resistor-capacitor combination using a 555 oscillator circuit. If the count can exceed nine during the period $T$, additional 7490s must be cascaded to provide the full count. Just prior to counter reset, the output is stored by 7475 data latches that are enabled by a brief pulse on signal $L$. The latches are necessary to hold the previous count for display while the counter begins a new count cycle. One of the two 74123 one-shots is positive edge triggered by the clock signal $CK$ to generate a latch pulse $L$ of length $\Delta t$. Note that the latch and reset pulse widths must be small ($\Delta t \ll T$) to maintain count accuracy (see Class Discussion Item 6.10). The trailing edge of the latch pulse triggers the second one-shot, which is negative edge triggered, to produce a delayed reset pulse $R$ for the counter. The 7447 LED decoder and driver converts the latched BCD count in order to drive a 7-segment LED display. The display reports the number of pulses that have occurred during the counting period $T$.

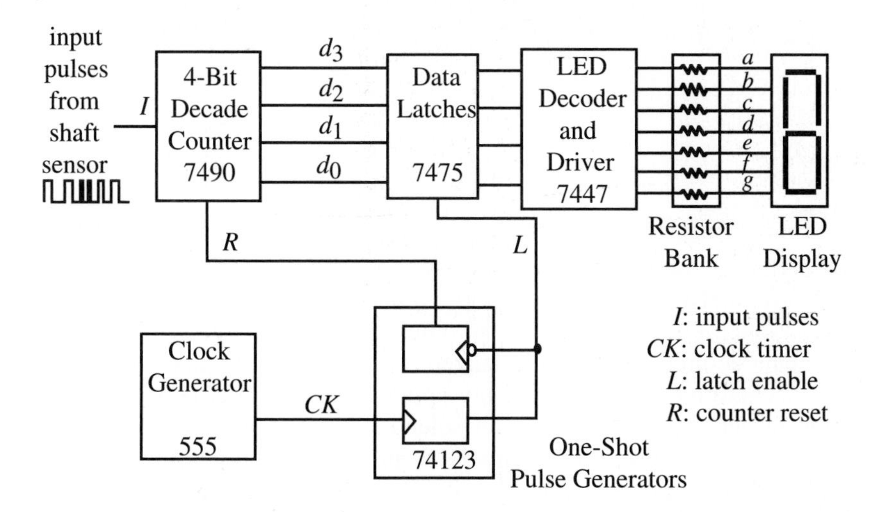


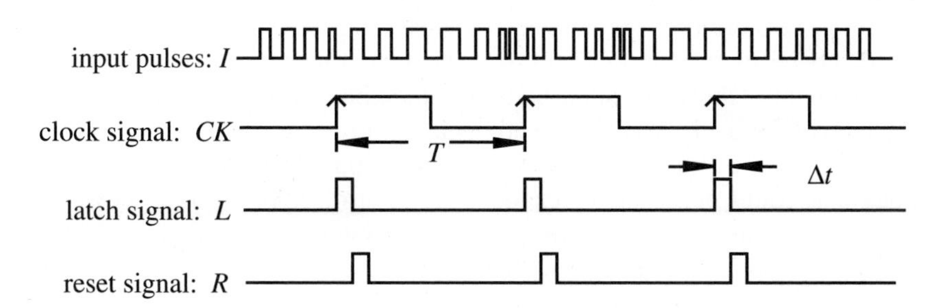


The shaft speed in revolutions per minute is related to the displayed pulse count by

$$rpm = \frac{pulse\ count/ppr}{T} 60$$

where $ppr$ is the number of pulses per revolution generated by the sensor.

▼▼▼ ***CLASS DISCUSSION ITEM 6.10. Digital Tachometer Accuracy.*** What effect does the choice of $\Delta t$ have on the accuracy of the counts displayed by the digital tachometer presented in the previous example?

▼▼▼ ***CLASS DISCUSSION ITEM 6.11. Digital Tachometer Latch Timing.*** One potential problem with the digital tachometer circuit could occur if latching happens at the exact instant when the counter is incrementing in response to an input pulse. Explain why. This problem can be resolved by blocking the input pulses during the latch with a logic gate. What would you need to add to the circuit schematic to accomplish this?

---

▼▼ ***DESIGN EXAMPLE 6.2. Digital Control of Power to a Load Using Specialized ICs.*** In Design Example 3.4 we introduced power drivers for peripheral devices in a mechatronic system. The large semiconductor manufacturers provide specialized integrated circuits that vastly aid the mechatronic designer in many applications, such as relay drivers, lamp drivers, motor drivers, and solenoid drivers. In this chapter we learned how to design logic circuits to provide digital control outputs. These outputs can then be interfaced to peripheral power devices to control mechanical power. Peripheral power requirements are quite varied, requiring interface integrated circuits that have a large degree of adaptability. Furthermore, specialized interface integrated circuits offer some added benefits that are difficult for the designer to include using discrete components: short circuit protection at outputs, glitch-free power-up, inductive fly-back protection, and negative transient protection.

Consider a situation where the digital output is provided via an 8-bit bus that could be used, for example, to control eight separate devices. This can occur with digital/microprocessor systems. A good choice for a peripheral driver is the National Semiconductor DP7311 octal latched peripheral driver. This device can provide up to 100 mA dc at each output at a maximum operating voltage of 30 V to drive LEDs, motors, sensors, solenoids, and relays. It has open collector outputs that require an external pull-up resistor when interfaced to high-current external discrete drivers capable of controlling much higher currents than the IC itself. Let us consider two possible applications. The first application shows a typical output (one of eight) for a high-side (load between supply and collector), large-current bipolar power transistor driver.

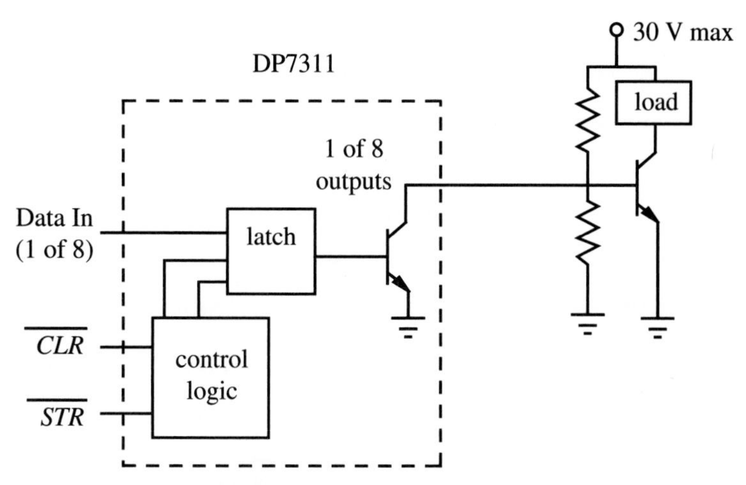


Data In is active high, and the Clear (*CLR*) and Strobe (*STR*) are active low. The latches are cleared (the outputs are turned off) when *CLR* is low, and the Data In values are latched when *STR* is low.

An alternative high-current n-channel MOSFET design is shown next:

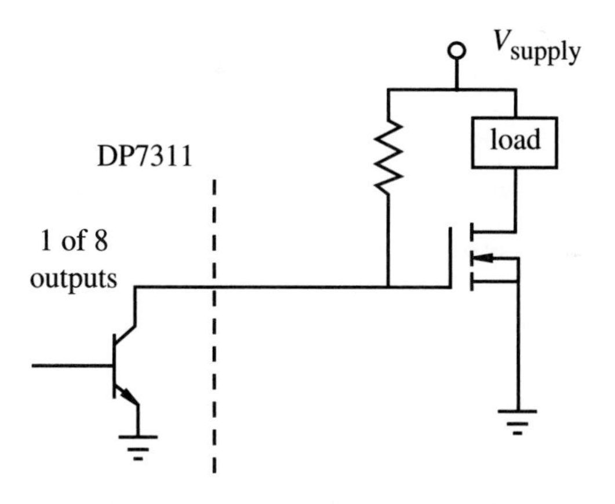


▼▼▼ ***CLASS DISCUSSION ITEM 6.12. Using Bypass Capacitors with ICs.*** It is standard practice to include one or more large tantalum or electrolytic capacitors across the power supply lines within a digital system and to include a small ceramic capacitor between the $V_{cc}$ and ground leads supplying power to each IC. Why?

The IEEE standard symbols for illustrating digital input and output conditions are shown in Figure 6.37. When drawing or reading digital schematics, it is important to recognize these conventions. You will find these symbols used in some manufacturers' TTL data books.

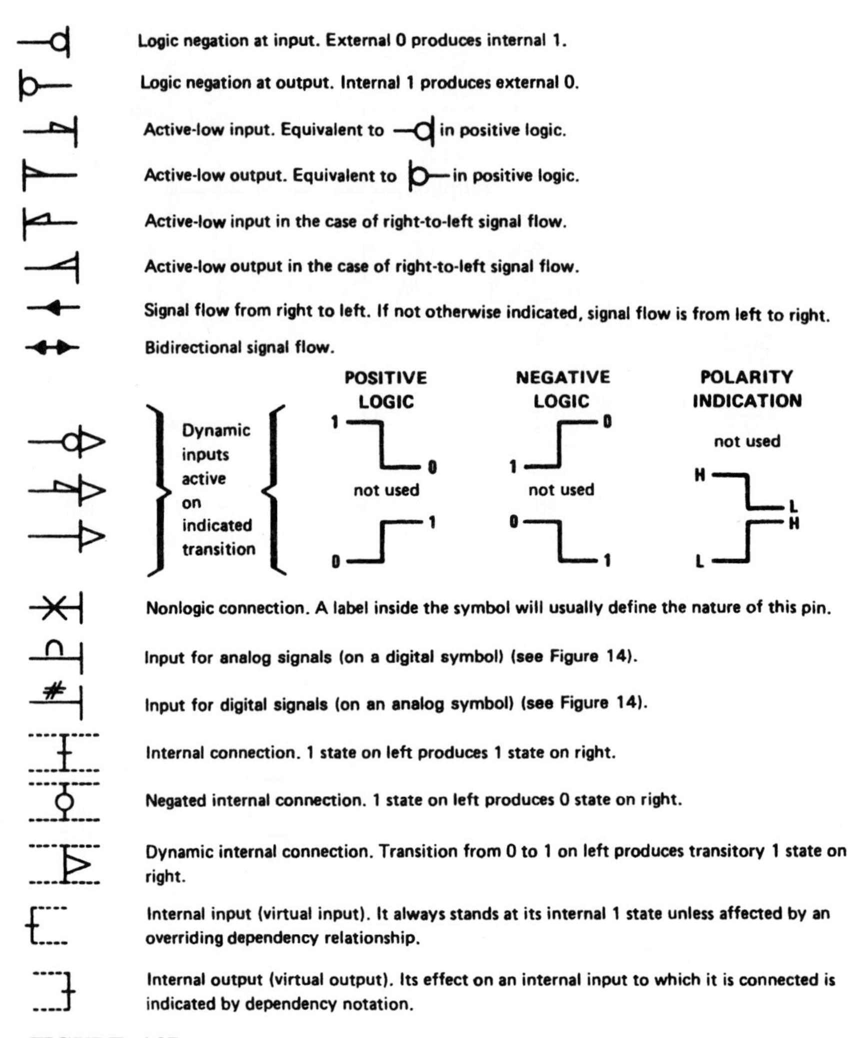


**FIGURE 6.37**
IEEE standard symbols for digital ICs. *(Courtesy of Texas Instruments, Dallas, TX)*

## 6.19
## MICROPROCESSORS AND MICROCOMPUTERS

The digital circuits presented so far in this chapter allow the implementation of logical and sequential operations by interconnecting ICs containing gates and flip-flops. This is considered a ***hardware*** solution since it consists of hard-wired ICs, which, once assembled, carry out a preset function. In order to make a change in the function, the hardware circuitry must be redesigned. This is a satisfactory approach for simple design tasks (e.g., the security system presented in Section 6.6 and the digital tachometer presented in Design Example 6.1). However, in many mechatronic systems, the control tasks are often much more complex, making a strictly hardware solution impractical. A better approach in these cases is using a microprocessor to implement a ***software*** solution, which is a procedural program consisting of a set of instructions. An advantage of a software solution is that the program can be easily modified to alter the mechatronic system's functionality without changing the hardware.

A ***microprocessor*** is a single, large scale integration (LSI) chip that contains many digital circuits that perform arithmetic, communication, and control functions. When a microprocessor is packaged on a printed circuit board with other components such as interface (I/O) and memory chips, the resulting package is referred to as a ***microcomputer*** or ***single board computer***. The overall architecture of a typical microprocessor and microcomputer is illustrated in Figure 6.38.

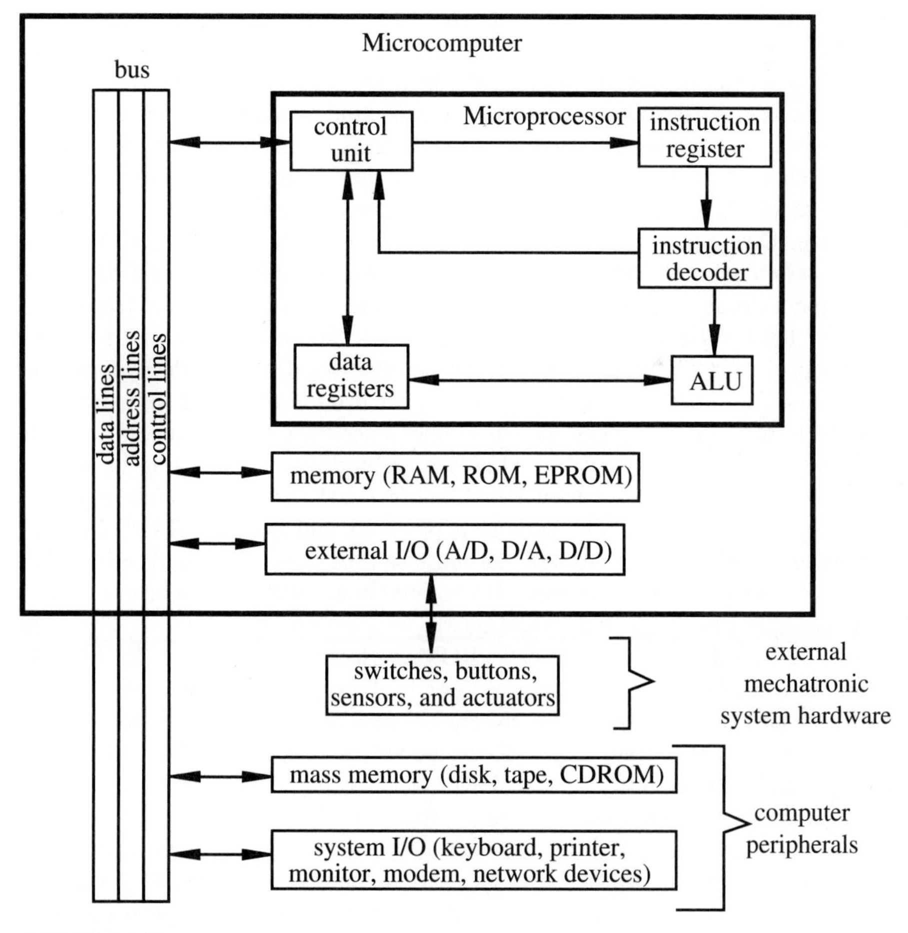


**FIGURE 6.38**
Microcomputer architecture.

The microprocessor, also called the ***central processing unit (CPU)*** or ***microprocessor unit (MPU)***, is where the primary computation and system control occur. The ***arithmetic logic unit (ALU)*** performs mathematical functions on data structured as binary words. A ***word*** is an ordered set of bits, usually 8, 16, 32, or 64 bits long. The instruction decoder interprets instructions that have been fetched sequentially from memory by the control unit and stored in the instruction register. Each instruction is a set of coded bits that command the ALU to perform bit manipulation, such as binary addition and logic functions, on words stored in the CPU data registers. The ALU results are also stored in data registers and then transferred to memory by the control unit.

The ***bus*** is a set of shared communication lines that serves as the central nervous system of the computer. Data, address, and control signals are shared by all system components via the bus. Each component connected to the bus communicates information to and from the bus via its own bus controller. The bus consists of data lines, address lines, and control lines that allow a specific component to access data addressed to that component. The data lines contain words that are usually transferred to and from data registers in system components. The address lines are used to select specific devices on the bus. Devices usually have a combinational logic address decoder circuit that identifies the address code for the device. The control lines contain read and write signals, the system clock, and other control signals such as system interrupts, which are described in the next section.

Key to a CPU's operation is the ability to store and retrieve data from a memory device. Different types of memory include ***read-only memory (ROM)***, ***random-access memory (RAM)***, and ***erasable-programmable ROM (EPROM)***. ROM is permanent storage memory that the CPU can read when needing to retrieve data. However, the CPU cannot write (store) data in ROM. ROM does not require a power supply to retain its data. RAM, on the other hand, is volatile memory that can be read from or written to numerous times but requires power to maintain its data. The term "random-access" is a misnomer since the CPU can access any individual data cell in ROM or RAM equally well by directly specifying its address (location). There are two types of RAM: ***static RAM (SRAM)***, which retains its data in flip-flops as long as the memory is powered, and ***dynamic RAM (DRAM)***, which consists of capacitor storage that must be refreshed (rewritten) periodically because of charge leakage over time. Data stored in an EPROM can be erased with ultraviolet light through a transparent quartz window on top of the IC. Then new data can be stored. Another type of EPROM is electrically erasable (***EEPROM***). Data in an EEPROM can be erased electrically through its leads without the need for ultraviolet light. Once old data are erased, new data can then be stored in the EEPROM by the CPU. Since data in RAM are volatile, ROM, EPROM, EEPROM, and peripheral mass memory storage devices such as magnetic disks and tapes and optical CDROM are sometimes needed to provide permanent storage.

Communication to and from the microprocessor occurs through input/output (I/O) devices connected to the bus. External computer peripheral I/O devices include keyboards, printers, displays, modems, and network devices. For mechatronic applications, analog to digital (A/D), digital to analog (D/A), and digital I/O (D/D) devices provide interfaces to hardware switches, buttons, sensors, and actuators.

The instructions executed by the CPU are in a code called ***machine code,*** which is processor dependent and usually listed as hexadecimal strings. Each code string causes the microprocessor to perform some low-level function (e.g., fetch a value from memory or add two integers). Microprocessors can be programmed using ***assembly language,*** which has an intuitive mnemonic command (e.g., MOV, ADD, PUSH, POP) for each corresponding microprocessor machine code function. Different microprocessor families have different assembly language sets that are not portable between families. When the set of instructions is small, the microprocessor is known as a ***RISC*** (reduced instruction set computer) microprocessor.

Microprocessors can be programmed at a higher level than machine code or assembly language by using a portable programming language such as C, C++, BASIC, or FORTRAN. Language compilers can convert the high-level instructions (e.g., assignment statements, IF-THEN-ELSE statements, loops) directly to machine code, precluding the need for assembly language programming.

### 6.19.1 Microcontrollers

There have been two trends in the development of the microprocessor. One trend supports the personal computer and workstation industry where the main constraints are high speed and large word size (16, 32, and 64 bits). The other trend includes development of a ***microcontroller***, which is a single IC containing a microprocessor, memory, I/O capabilities, and other on-chip resources. It is basically a microcomputer on a single IC. Examples of microcontrollers are Motorola's 68HC11, Intel's 8096, and National Semiconductor's HPC16040. Constraints that have driven development of the microcontroller are low cost, versatility, ease of programming, and small size. Microcontrollers are attractive in mechatronic system design since their small size and complete set of functionality allow them to be physically embedded in the system and perform all of the necessary control functions.

The block diagram for a Motorola 68HC11, a widely used microcontroller, is illustrated in Figure 6.39. As shown in the figure, it contains a microprocessor, control circuitry, ROM, RAM, I/O ports, an A/D converter, a programmable timer, a serial port, and a UART. The control circuitry includes a watchdog timer that can be used to automatically reset the microcontroller when a program stops executing correctly and an interrupt circuit that is used to suspend the execution of program instructions in response to some event (e.g., a timer event or some external input). The ROM is used for nonvolatile program storage. A designer can have a program permanently stored in ROM by the manufacturer, or the ROM could be in the form of EPROM that can be reprogrammed by the user. The EEPROM can also be used to store a program, but it is usually used to store variables and parameters between power-down and power-up cycles. The RAM is used to store and retrieve values for variables used in an executing program. The I/O ports, also called parallel ports, allow binary data to be transferred in and out of the microcontroller through external pins. The serial port and ***UART*** (*u*niversal *a*synchronous *r*eceiver *t*ransmitter) port allow the microcontroller to transmit or receive blocks of data or programs to or from external devices (e.g., a personal computer or another micro-controller). The A/D converter is an 8-bit

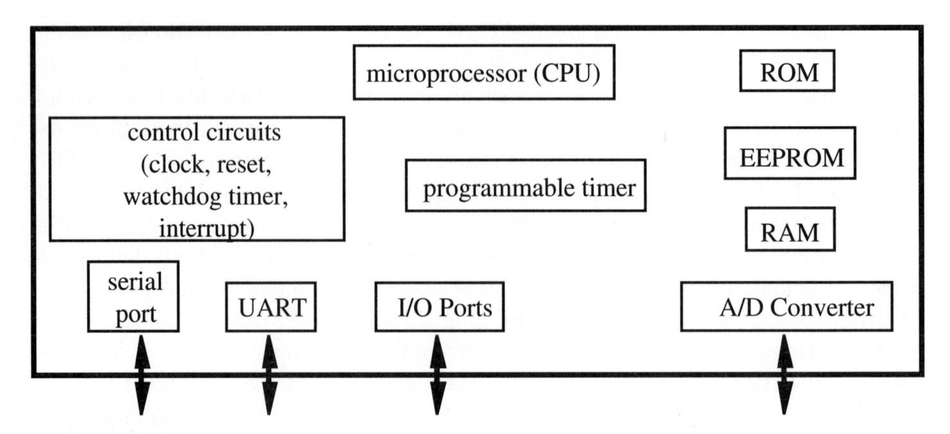


**FIGURE 6.39**
Motorola 68HC11 microcontroller.

successive approximation analog to digital converter (described in the next chapter) that allows the microcontroller to convert an external analog voltage (e.g., from a sensor) to a digital value that can be manipulated or stored by the microprocessor. The programmable timer can be used to perform delays or to generate events (e.g., read an input or output a pulse) at precisely defined intervals.

The Motorola 68HC11 comes in various packages and configurations to support different amounts and types of memory and different I/O capabilities (e.g., the MC68HC11EA9 has 12 kilobytes of EPROM, 512 bytes of EEPROM, and 512 bytes of RAM). However, due to size constraints, microcontrollers typically have relatively small memory capacities compared with microcomputers. A concern when using a microcontroller is that the on-chip resources will not have enough capacity (e.g., not enough memory) to meet the needs of the application. Fortunately, a microcontroller like the 68HC11 can be operated in what is called "expanded mode," which allows it to work in conjunction with other microcontrollers and other external resources such as memory.

To use a microcontroller in a mechatronic system, software must be written and tested and then stored in the ROM of the microcontroller. Usually, the software is developed and tested on a personal computer (PC) first and then downloaded to the microprocessor ROM as machine code. If the program is written in assembly language, the PC must have software called a ***cross-assembler*** that generates machine code for the microcontroller's microprocessor. An ***assembler*** is software that generates machine code for the resident microprocessor (on the PC), whereas a cross-assembler generates machine code for a different microprocessor (the one on the microcontroller). Programs can also be written in a higher-level language such as C, provided that a compiler is available that can generate machine code for the specific microcontroller being used. The advantages of using a high-level language such as C are that it is easier to learn and use, programs are easier to debug (the process of finding and removing errors), and programs are easy for you and others to comprehend. A disadvantage is that the resulting machine code will be less efficient (i.e., slower and require more memory) than a well-written assembly program. This issue of efficiency can be an important concern in some applications given the limited on-chip resources of a microcontroller.

There are various software development tools that can assist in testing and debugging programs written for a microcontroller. One such tool is a ***simulator,*** which is software that runs on a PC allowing the microcontroller code to be simulated (run) on the PC in response to inputs defined by the user. Another tool is an ***emulator,*** which is hardware that connects a PC to the microcontroller in a prototype mechatronic system. It usually consists of a printed circuit board that is connected to the mechatronic system through ribbon cables. The emulator can be used to load and run a program on the actual microcontroller attached to the mechatronic system hardware (sensors, actuators, and control circuits). The emulator allows the PC to monitor and control the operation of the microcontroller.

In a mechatronic system, the program (software) you write for the microcontroller typically has to perform some logic or control algorithm in response to external digital or analog sensor inputs. As a result, the software usually changes external digital outputs that control actuators or display/output devices. Example 6.6. illustrates how a program might be structured.

---

▼ ***EXAMPLE 6.6. C Program For Security System Example.*** Provided below is an example C program that could control the security system described in Section 6.6.

```
/* Example C program to perform the control functions of the secu-
   rity system presented in Section 6.6 */

/* Constants defining the I/O port addresses and loop parameter */
#define  INPUT_PORT  0x0002  /* hex address for the input port */
#define  OUTPUT_PORT  0x0004 /* hex address for the output port */
#define  ALWAYS  1           /* TRUE */

Main()
{

/* Define the layout of bits in the input byte */
struct
{
    unsigned int  a : 1;      /* sensor A's bit (MSB) */
    unsigned int  b : 1;      /* sensor B's bit */
    unsigned int  c : 1;      /* switch C's bit */
    unsigned int  d : 1;      /* switch D's bit */
    unsigned int  x : 4;      /* the remaining bits (not used) */
} input_byte;

/* Define the layout of bits in the output byte */
struct
{
    unsigned int  y : 1;       /* alarm bit (MSB) */
    unsigned int  x : 7;       /* the remaining bits (not used) */
}  output_byte;
```

```
output_byte = 0; /* initialize all bits in the output byte to 0 */

/* Main polling loop */
while (ALWAYS)/* infinite loop */
{
   /* Read the sensor states using the compiler function "inportb"
      which reads all 8 bits (1 byte) from an input port */
   input_byte = inportb (INPUT_PORT);

   /* Perform the security system logic */
   if ( (input_byte.c == 0) && (input_byte.d == 1) )
   {
      /* The system is in operating state 1 (check sensor A only) */
      if (input_byte.a == 1)
         output_byte.y = 1;
      else
         output_byte.y = 0;
   }
   else if ( (input_byte.c == 1) && (input_byte.d == 0) )
   {
      /* The system is in operating state 2 (check sensors A and B)
  */
      if ( (input_byte.a == 1) | | (input_byte.b == 1) )
            output_byte.y = 1;
      else
            output_byte.y = 0;
   }
   else if ( (input_byte.c == 0) && (input_byte.d == 0) )
   {
      /* The system is in operating state 3 (disabled) */
      output_byte.y = 0;
   }

   /* Sound or turn off the alarm using the compiler function
  "outportb" which writes a byte to an output port */
   outportb (OUTPUT_PORT, output_byte);
}  /* end of while */
}  /* end of main */
```

---

The program in Example 6.6 uses a technique called ***polling,*** where the program continually checks the sensor inputs and adjusts the output accordingly. The loop cycle is repeated over and over again as long as the microcontroller is powered. For more complex applications, polling may not be suitable since the loop may take too long to execute, especially if the program has tasks to perform other than continually checking inputs (e.g., performing complex calculations or writing information to memory). With a long loop cycle, the inputs may not be checked frequently enough. An alternative program structure for these cases is an ***interrupt*** driven program. With this method, the sensor inputs are tied to special input lines. When one of these lines changes level, the microcontroller temporarily suspends normal program execution until the change is acted upon by a subprogram or function called a service routine. When the service routine is done executing, control is returned to the main program and normal execution continues.

The topics of microcontrollers and single board computers are revisited briefly in Chapter 10, where various mechatronic system control architectures are described.

## QUESTIONS AND EXERCISES

**6.1.** Construct a truth table for a negative edge-triggered T flip-flop.

**6.2.** Design and draw the schematic for a 555 oscillator that generates a 1 Hz symmetrical pulse clock signal.

**6.3.** Draw a logic circuit for the following Boolean expression using 2-input NAND gates and INVERTERS only.

$$X = \overline{A} \cdot B \cdot C + (A + B) \cdot \overline{C}$$

**6.4.** Write out a simplified Boolean expression and construct a truth table for each of the following circuits. Assume 0 = low = 0 V and 1 = high = 5 V.

(*a*)

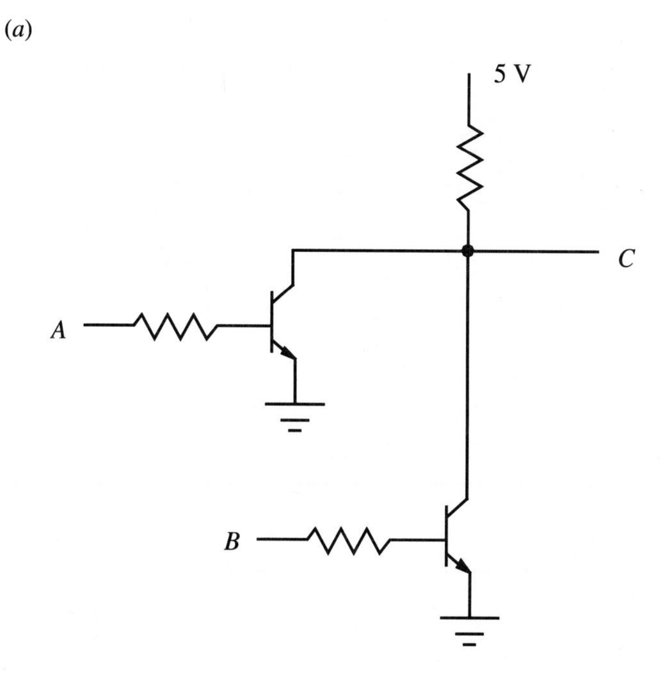


(*b*)

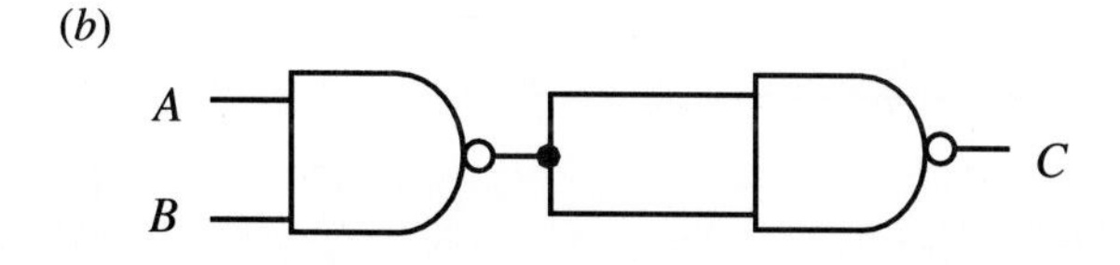

(*c*)

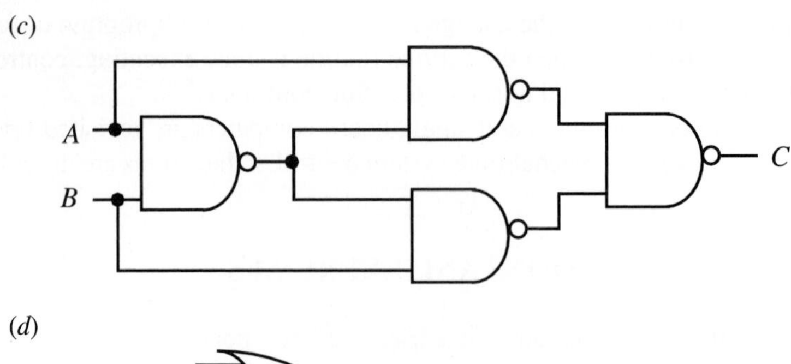


(*d*)

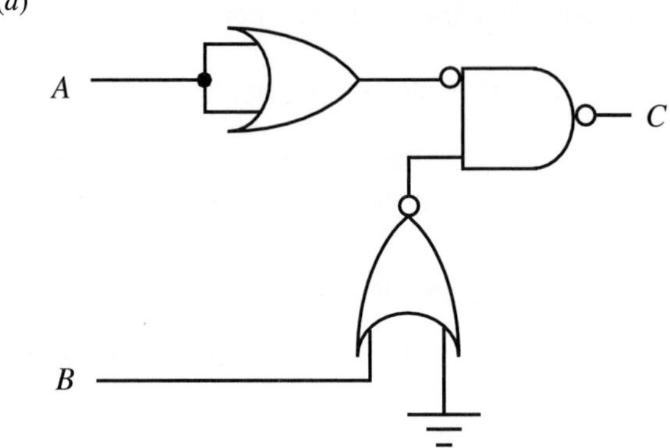


**6.5.** Complete the timing diagram shown below for the following combinational logic circuit.

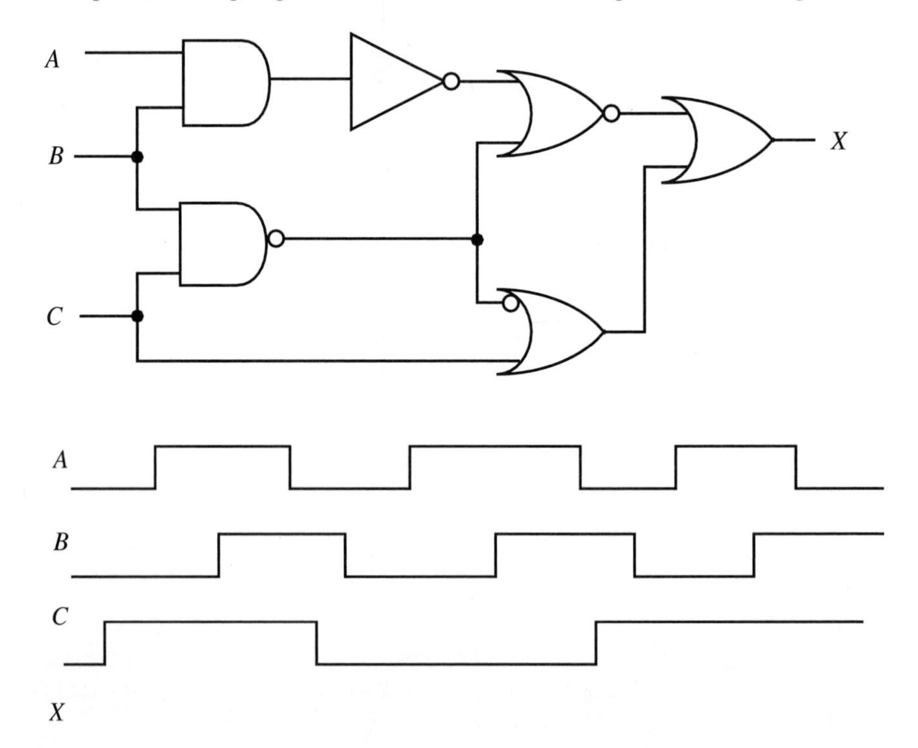

**6.6.** Design and draw a logic circuit that will drive segment $c$ of a 7-segment LED display (see below), given 4-bit BCD input DCBA representing decimal numbers from 0 to 9. Note that the logic circuit will be used in the driver circuit below.

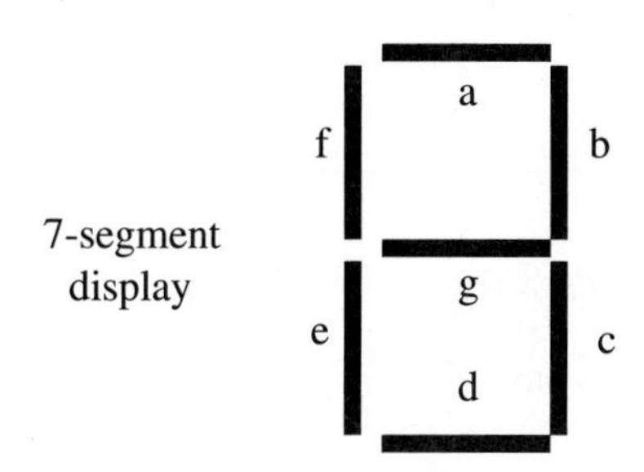


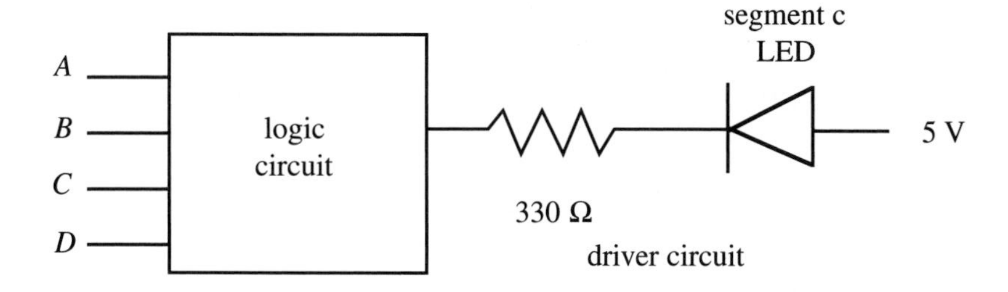


**6.7.** Prove Equation 6.19 using the Boolean algebra laws and identities.

**6.8.** Verify that Equation 6.21 is correct using a truth table.

**6.9.** Prove that the following Boolean identity is correct:

$$AB + AC + \overline{B}C = AB + \overline{B}C$$

**6.10.** Prove whether the following Boolean equations are valid or not:

(*a*) $(A \cdot B) + (B \cdot C) + (\overline{B} \cdot C) = (A \cdot B) + \overline{C}$

(*b*) $A \cdot B \cdot C = \overline{A + B + C}$

(*c*) $(A \cdot B) + (B \cdot C) + (\overline{B} \cdot C) = (A \cdot B) + C$

**6.11.** Determine a simplified Boolean expression for the circuit on the next page and draw an equivalent circuit using 2-input NOR gates and INVERTERS only.

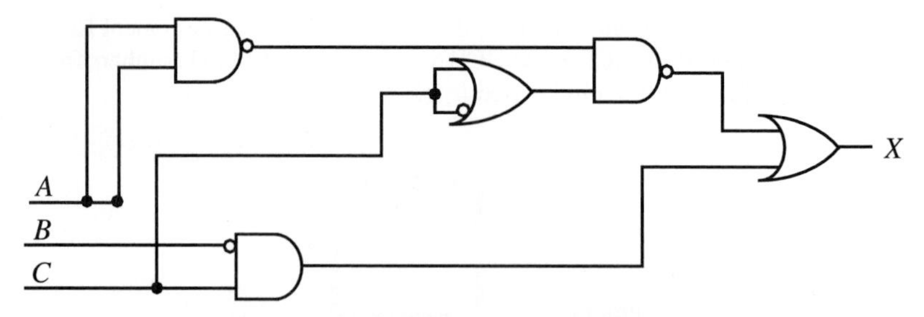


**6.12.** Design and draw a ***full-adder*** circuit that has two sum bits $A_i$ and $B_i$ and a lower order carry bit $C_{i-1}$ as inputs (see Example 6.3). Include a complete truth table and include simplified Boolean expressions for the output sum bit $S_i$ and carry bit $C_i$.

**6.13.** The circuit below is called a ***multiplexer***. Construct a truth table and derive a simplified Boolean expression. Also, explain why the circuit is termed a multiplexer.

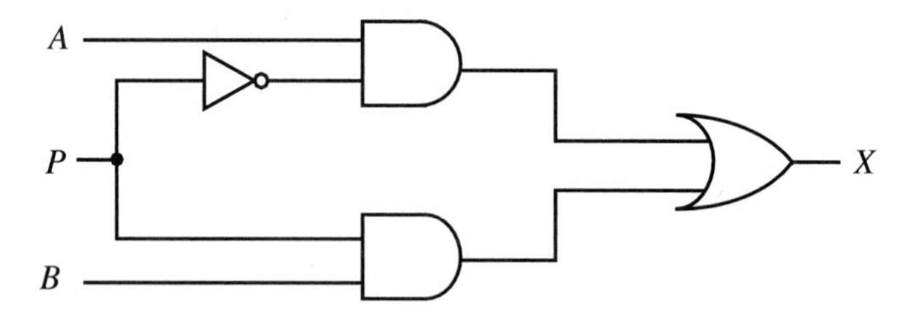


**6.14.** Derive a Boolean expression for the output columns *c* and *e* of Table 6.11 using either the product-of-sums or sum-of-products discussed in Section 6.7.

**6.15.** Complete the timing diagram for the circuit below.

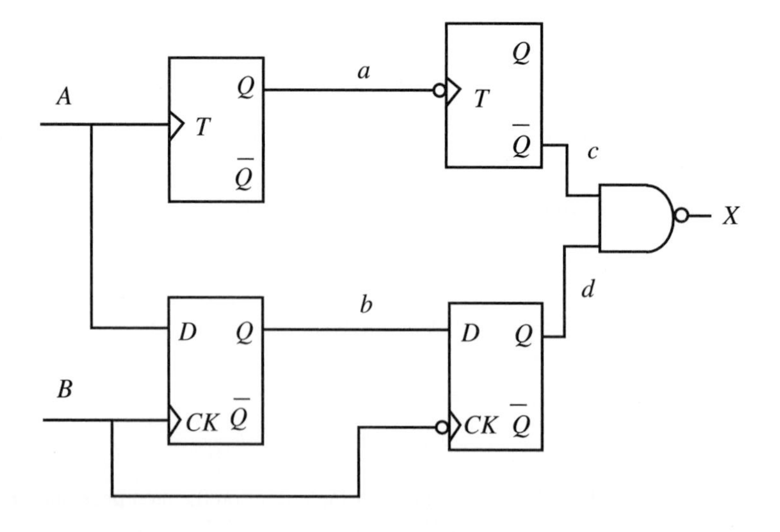

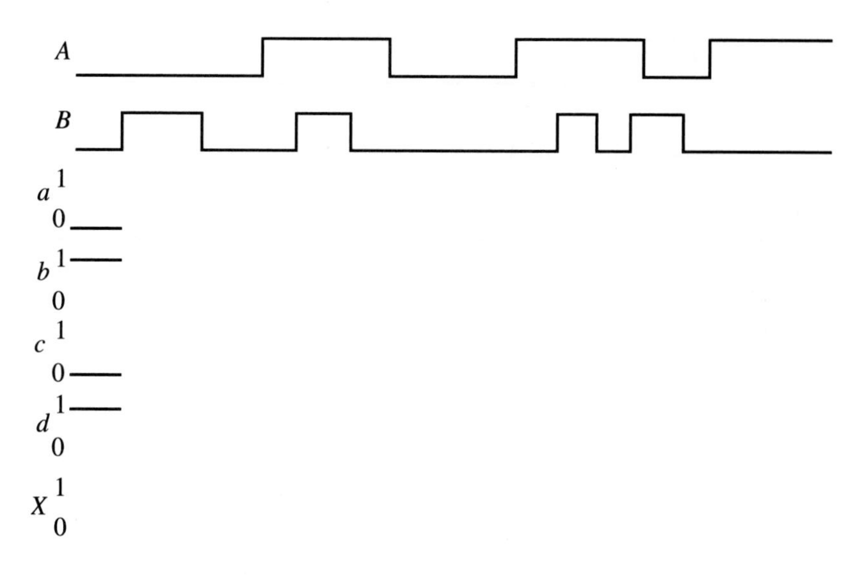


**6.16.** Complete the timing diagram for the circuit below assuming the JK flip-flops are positive edge triggered.

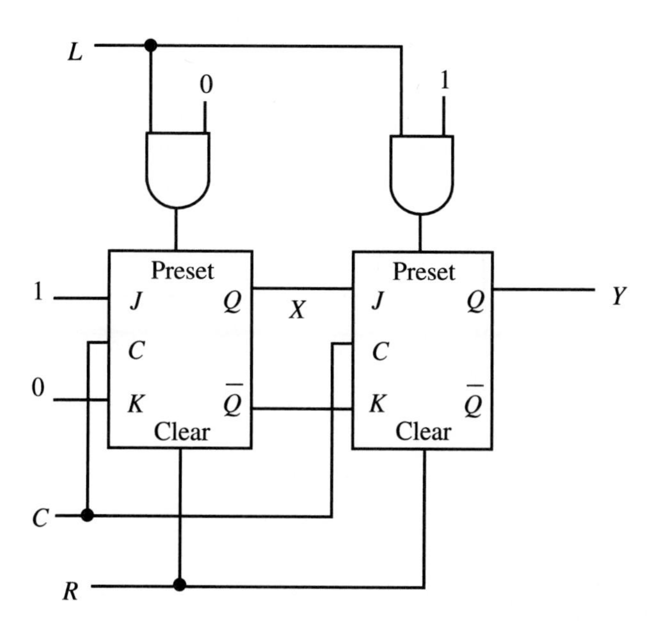

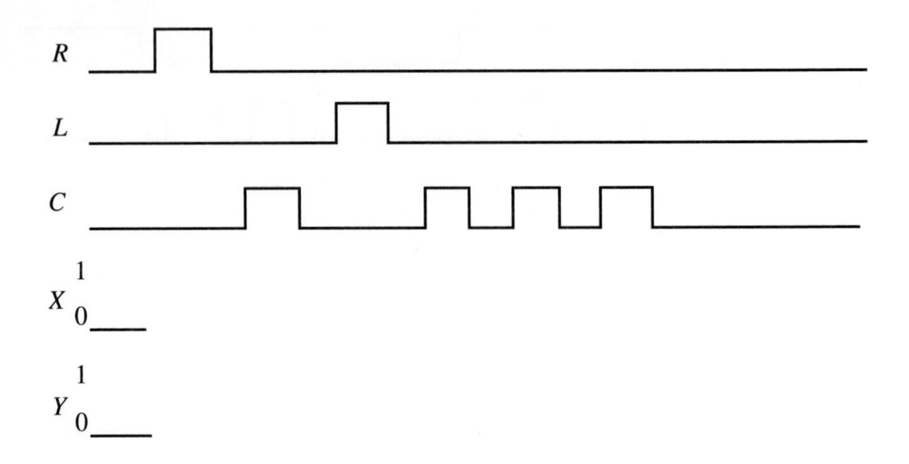


**6.17.** Complete the timing diagram for the circuit below.

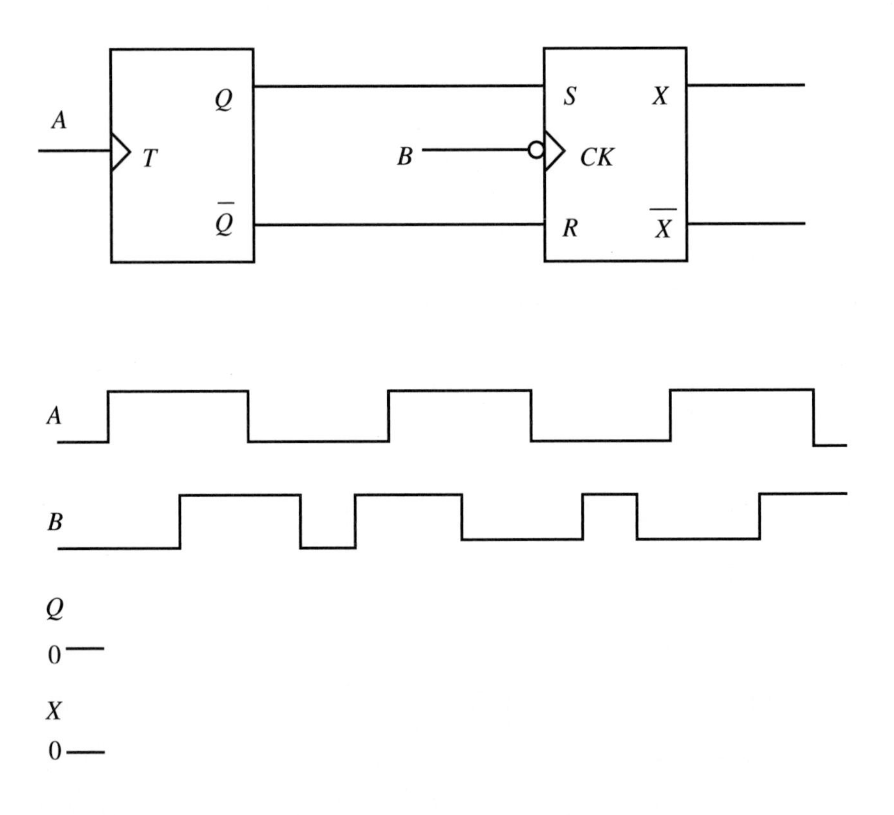


**6.18.** Using the information in a TTL data book, draw a complete schematic for a 7490 Decade Counter with BCD output. Include a reset feature using the four *R* lines in the timing diagram on the next page. Also, complete the timing diagram.

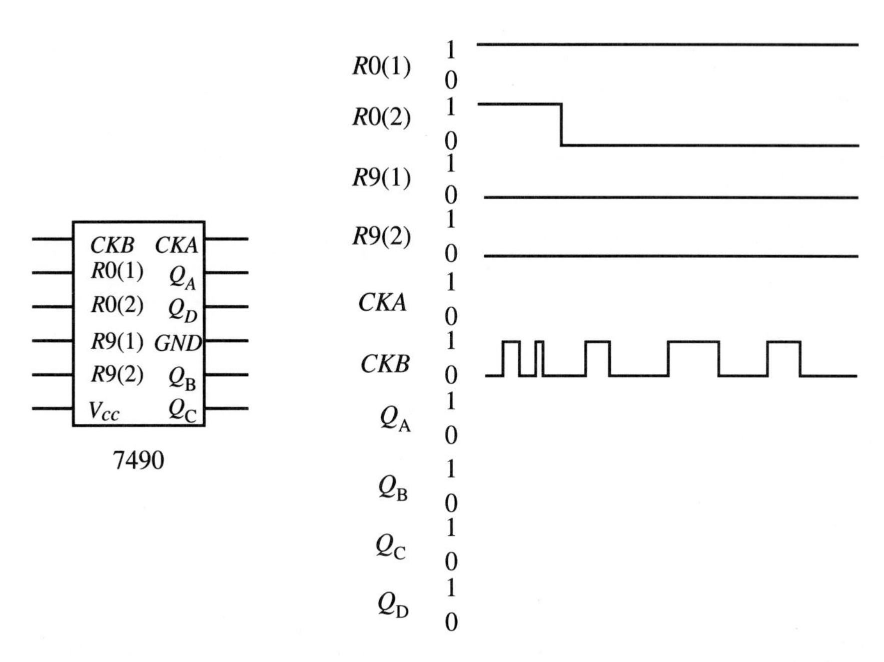


**6.19.** Determine the exact pulse width as a function of the resistance and capacitance in a one-shot circuit (see Equation 6.35).

## BIBLIOGRAPHY

Gibson, G. and Liu, Y., *Microcomputers for Engineers and Scientists,* Prentice-Hall, Englewood Cliffs, NJ, 1980.

Horowitz, P. and Winfield, H., *The Art of Electronics,* 2nd Edition, Cambridge University Press, New York, 1989.

Mano, M., *Digital Logic and Computer Design,* Prentice-Hall, Englewood Cliffs, NJ, 1979.

McWhorter, G. and Evans, A., *Basic Electronics,* Master Publishing, Inc., Richardson, TX, 1994.

Mims, F., *Getting Started in Electronics,* Radio Shack Archer Catalog No. 276-5003A, 1991.

Mims, F., *Engineers Mini-Notebook: 555 Circuits,* Radio Shack Archer Catalog No. 276-5010, 1984.

Mims, F., *Engineer's Mini-Notebook: Digital Logic Circuits,* Radio Shack Archer Catalog No. 276-5014, 1986.

Motorola Technical Summary, "MC68HC11EA9/MC68HC711EA9 8-bit Microcontrollers," Document number: MC68HC11EA9TS/D, Motorola Advanced Microcontroller Division, Austin, TX, 1994.

Peatman, J., *Design with Microcontrollers,* McGraw-Hill, New York, 1988.

Stiffler, A., *Design with Microprocessors for Mechanical Engineers,* McGraw-Hill, New York, 1992.

Texas Instruments, *TTL Linear Data Book,* Dallas, TX, 1992.

Texas Instruments, *TTL Logic Data Book,* Dallas, TX, 1988.

# 7

# DATA ACQUISITION

**OBJECTIVES:** After you read, discuss, study, and apply ideas in this chapter, you will be able to:

- Understand how to properly sample a signal for digital processing
- Understand how to code data that has been digitized
- Understand the important components of an A/D converter

## 7.1 INTRODUCTION

Microprocessors, microcontrollers, single board computers, and personal computers are in widespread use in mechatronic and measurement systems today, and it is increasingly important for engineers to understand how to directly access information and analog data from the surrounding environment with these devices. As an example, consider a signal from a sensor as illustrated by the analog signal in Figure 7.1. One could record the signal with an analog device such as a chart recorder, which physically plots the signal on paper, or display it with an oscilloscope. Another option is to store the data using a microprocessor or computer. This process is called computer ***data acquisition***, and it provides more compact storage of the data (magnetic media vs. long rolls of paper), can result in greater data accuracy, allows us to use the data in a real-time control system, and enables us to perform data processing long after the events have occurred.

To be able to get analog data to a digital circuit or microprocessor, the analog data must be transformed into digital values compatible with the digital processor. The first step is to numerically evaluate the signal at discrete instants in time. This process is called ***sampling***, and the result is a ***digitized signal*** that is composed of discrete values corresponding to each sample, as illustrated in Figure 7.1. Thus a digitized signal is a sequence of numbers that is an approximation to an analog signal. The time relation between the numbers is an inherent property of the sampling process because time itself is not also recorded. The collection of sampled data points forms a data array, and although this representation is no longer continuous, if properly executed it can retain the same signal information in the original analog signal.

An important question is how fast or often the signal should be sampled to obtain an accurate representation. The naive answer might be: as fast as you possibly can. The problems with this conclusion are that specialized high-speed hardware is required and large amounts of computer memory will be needed to store the data. A better answer is to select a minimal sampling rate for a given application that retains all important signal information.

The ***Sampling Theorem,*** also called ***Shannon's Sampling Theorem,*** states that we need to sample a signal at a rate at least two times the maximum frequency component occurring in a signal in order to retain all frequency components contained in the signal. In other words, to faithfully represent the analog signal, the digital samples must be taken at a frequency $f_s$ such that

$$f_s > 2f_{\max} \tag{7.1}$$

where $f_{\max}$ is the frequency of the highest frequency component in the input analog signal. $f_s$ is referred to as the ***sampling rate,*** and the minimum required rate ($2f_{\max}$) is called the ***Nyquist frequency***. If we approximate a signal by a truncated Fourier series, the maximum frequency component is the highest harmonic. The time interval between the digital samples is

$$\Delta t = 1/f_s \tag{7.2}$$

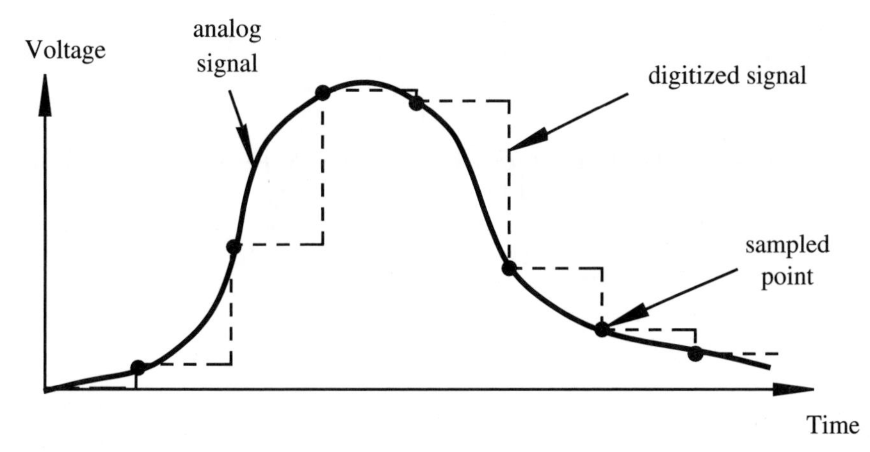


**FIGURE 7.1**
Analog signal and sampled equivalent.

As an example, if the sampling rate is 5000 Hz, the time interval between samples would be 0.2 ms.

If a signal is sampled at less than two times its maximum frequency component, **aliasing** can result. Figure 7.2 illustrates an analog sine wave sampled regularly at the points shown. Twelve samples are taken over ten cycles of the original signal. Therefore, the sampling frequency is $1.2f_0$ where $f_0$ is the frequency of the original sine wave. Since the sampling frequency is not more than $2f_0$, we do not correctly obtain the frequency in the original signal. Furthermore, the apparent frequency in the sampled signal is $0.2f_0$, which is incorrect. You can think of this as a "phantom" frequency, which is an alias of the true frequency. Therefore, undersampling not only results in errors but creates information that is not really there!

The method of Fourier decomposition presented in Chapter 4 provides a means to determine the frequency components of an arbitrary analog signal. Representing a signal in terms of its frequency components allows us to identify the bandwidth and correctly apply the Sampling Theorem.

▼▼▼ ***CLASS DISCUSSION ITEM 7.1. Wagon Wheels and the Sampling Theorem.*** Relate the sampling theorem and its implications about correct and incorrect data representation to the use of a movie camera to film a rotating wagon wheel in a western movie. The camera shutter can be considered a sampling device where the shutter speed is the sampling rate. Typically, the sampling rate is 30 Hz to provide a flicker-free image to the human eye. What effects would you expect in the movie, when viewed, if the wagon wheel is turning rapidly?

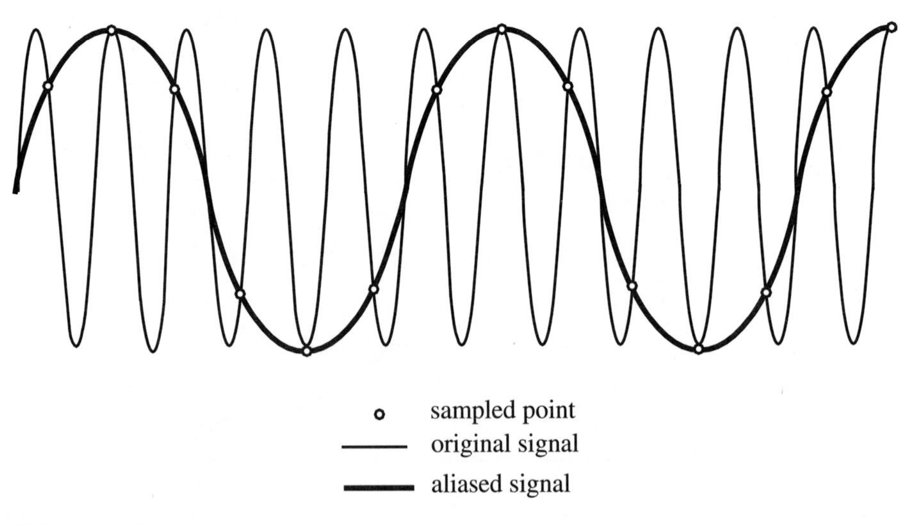


**FIGURE 7.2**
Aliasing.

▼ ***EXAMPLE 7.1. Sampling Theorem and Aliasing.*** Consider the function

$$F(t) = \sin(at) + \sin(bt)$$

Using a trigonometric identity for the sum of two sinusoidal functions, we can rewrite $F(t)$ as the following product:

$$F(t) = \left[2\cos\left(\frac{a-b}{2}t\right)\right] \cdot \sin\left(\frac{a+b}{2}t\right)$$

If frequencies $a$ and $b$ are close in value, the bracketed term has a very low frequency in comparison to the sinusoidal term on the right. Therefore, the bracketed term modulates the amplitude of the higher frequency sinusoidal term. The resulting waveform exhibits what is called a ***beat frequency*** that is common in optics, mechanics, and acoustics when two waves close in frequency add.

In order to illustrate aliasing associated with improper sampling, the waveform is plotted below using two different sampling frequencies. The function frequencies are chosen as

$$a = 2\pi\frac{\text{rad}}{\text{sec}} \qquad b = 0.9a$$

The frequency of the amplitude modulated sine wave is the average of the two input frequencies:

$$\frac{(a+b)}{2} = 0.95a = 0.95 \text{ Hz}$$

The first data set is plotted at 100 Hz ($\Delta t = 0.01$ sec), providing an excellent representation of the waveform. The second data set is plotted at 1.33 Hz ($\Delta t = 0.75$ sec), which is less than twice the maximum frequency of the waveform (1.9 Hz). Therefore, the signal is undersampled and aliasing results. The sampled waveform is an incorrect representation, and its observed maximum frequency appears to be approximately 0.4 Hz since there are approximately 4 cycles over 10 seconds.

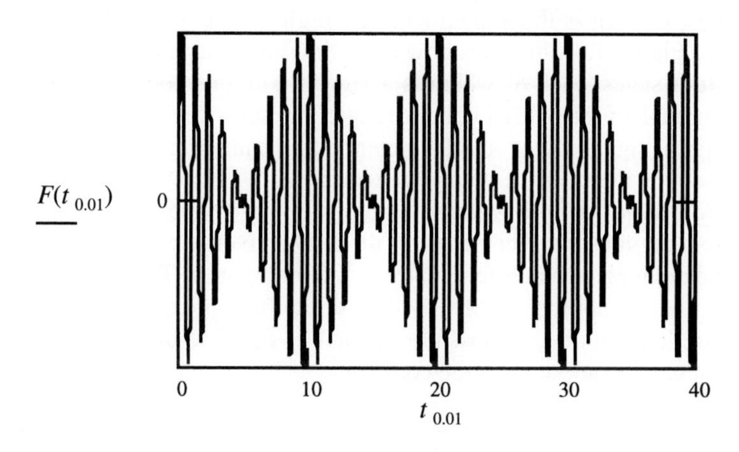

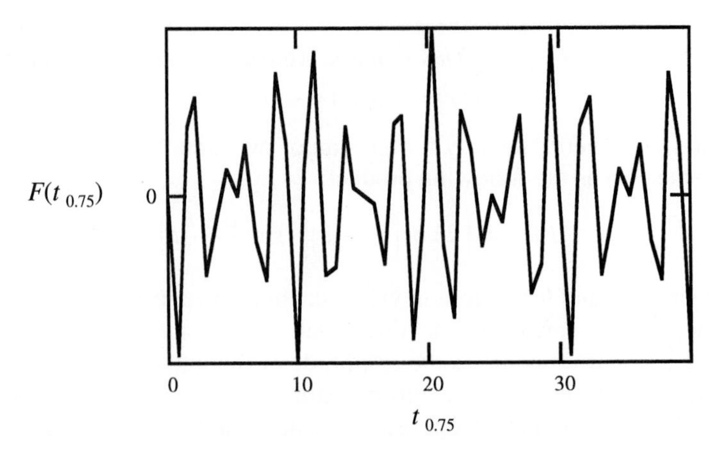


You might argue that neither plot is perfect. However, unlike the second plot, the first plot does retain all of the frequency information in the analog signal.

---

▼▼▼ ***CLASS DISCUSSION ITEM 7.2. Sampling a Beat Signal.*** What is the minimum sampling rate required to adequately represent the signal in Example 7.1?

## 7.2 QUANTIZING THEORY

Now we will look at the process required to change a sampled analog voltage into digital form. The process, called ***analog to digital (A/D) conversion***, conceptually is a two-step process requiring quantizing and coding. ***Quantizing*** is defined as the transformation of a continuous analog input into a set of data represented by discrete output states. ***Coding*** is the assignment of a digital code word or number to each output state. Figure 7.3 illustrates how a continuous voltage range is divided into discrete output states, each of which is assigned a unique code. Each output state covers a subrange of the overall voltage range. The stair-step signal represents the states of a digital signal that would be generated by sampling a linear ramp of an analog signal occurring over the voltage range shown.

An ***analog to digital (A/D) converter*** is an electronic device that converts an analog voltage to a digital code. The output of the A/D converter can be directly interfaced to digital devices such as microcontrollers or computers. The ***resolution*** of an A/D converter is the number of bits used to digitally approximate the analog value of the input. The number of possible states $N$ is equal to the number of bit combinations that can be output from the converter:

$$N = 2^n \tag{7.3}$$

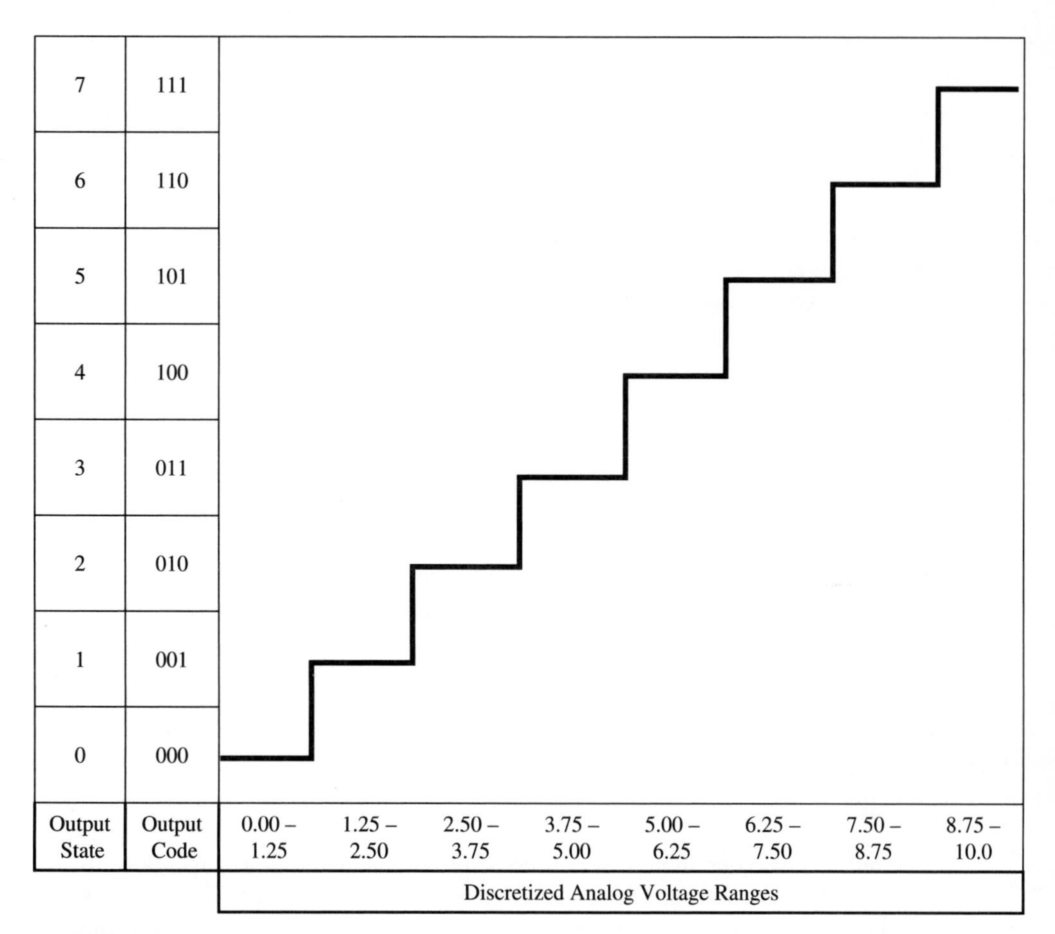


**FIGURE 7.3**
Analog to digital conversion.

where $n$ is the number of bits. For the example illustrated in Figure 7.3, the 3-bit device has $2^3$ or 8 output states as listed in the first column. The output states are usually numbered consecutively from 0 to $(N - 1)$. The corresponding code word for each output state is listed in the second column. Most commercial A/D converters are 8-, 10-, or 12-bit devices that resolve 256, 1024, and 4096 output states, respectively.

The number of analog ***decision points*** that occur in the process of quantizing is $(N - 1)$. In Figure 7.3, the decision points occur at 1.25 V, 2.50 V, . . . , and 8.75 V. The ***analog quantization size*** $Q$ is defined as the full scale range of the A/D converter divided by the number of output states:

$$Q = (V_{max} - V_{min}) / N \tag{7.4}$$

It is a measure of the analog change that can be resolved by the converter. Although the term *resolution* is defined as the number of output bits from an A/D converter, sometimes it is used to refer to the analog quantization size. For our example, the analog quantization size is 10/8 V = 1.25 V. This means that the amplitude of the digitized signal will have an error of at most 1.25 V.

▼▼▼ *CLASS DISCUSSION ITEM 7.3.* ***Laboratory A/D Conversion.*** Why is a 12-bit A/D converter satisfactory for most laboratory measurements?

## 7.3 HARDWARE FOR ANALOG TO DIGITAL CONVERSION

To properly convert an analog voltage signal for computer processing, the following hardware elements must be properly selected and used in this sequence:

1. Buffer amplifier
2. Low-pass filter
3. Sample and hold amplifier
4. Analog to digital (A/D) converter
5. Computer

The buffer amplifier should be chosen to provide a signal in a range close to but not exceeding the full input voltage range of the A/D converter. The low-pass filter is necessary to remove any undesirable high-frequency components in the signal that could produce aliasing. The cutoff frequency of the low-pass filter should be no greater than 1/2 the sampling rate. The sample and hold amplifier maintains a fixed input value (from an instantaneous sample) during the short conversion time of the A/D converter. The converter is chosen with a resolution and analog quantization size appropriate to the system and signal. The computer must be properly interfaced to the A/D converter system in order to store and process the data. The computer must also have sufficient memory and permanent storage media to hold all of the data. The arrangement of the necessary components along with an illustration of their respective outputs is shown in Figure 7.4.

The A/D system components described above can be found packaged on a variety of PC plug-in boards called data acquisition and control (DAC) cards. These cards usually support various language programming interfaces, including C, FORTRAN, and BASIC. Software libraries with various software function calls are provided that give easy high-level access to the board's functionality. Acquiring data from the world and storing it on the computer is a simple matter of calling a function from a program or from an object-oriented visual programming interface. Usually, other input and output features are also provided on the cards, in-

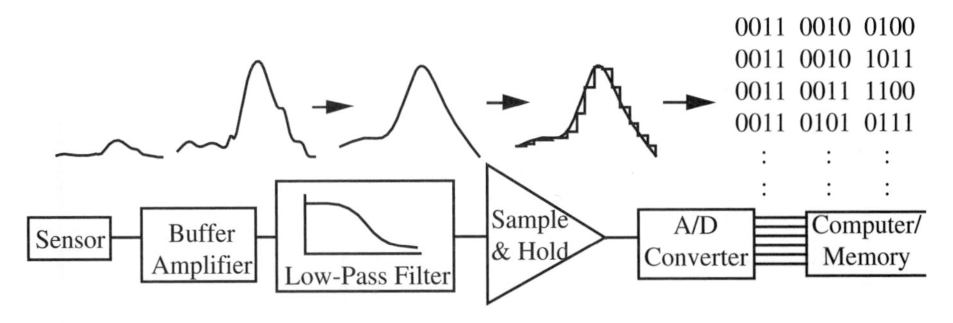


**FIGURE 7.4**
Components used in A/D conversion.

cluding binary (TTL) I/O, D/A conversion, counter/timer features, and limited signal conditioning circuitry. Important features to be aware of when purchasing a DAC card include the A/D and D/A resolution (number of bits) and the maximum sampling rate. These features are key to accuracy and reliability in computer control applications.

An example DAC card that can be inserted into a PC bus is shown in Figure 7.5. Its internal architecture is illustrated in Figure 7.6. The card's functions include two 12-bit D/A converters, one 12-bit A/D converter, 24 lines of digital I/O, and three 16-bit counter/timers. It also comes with software that is easily used for customizing applications.

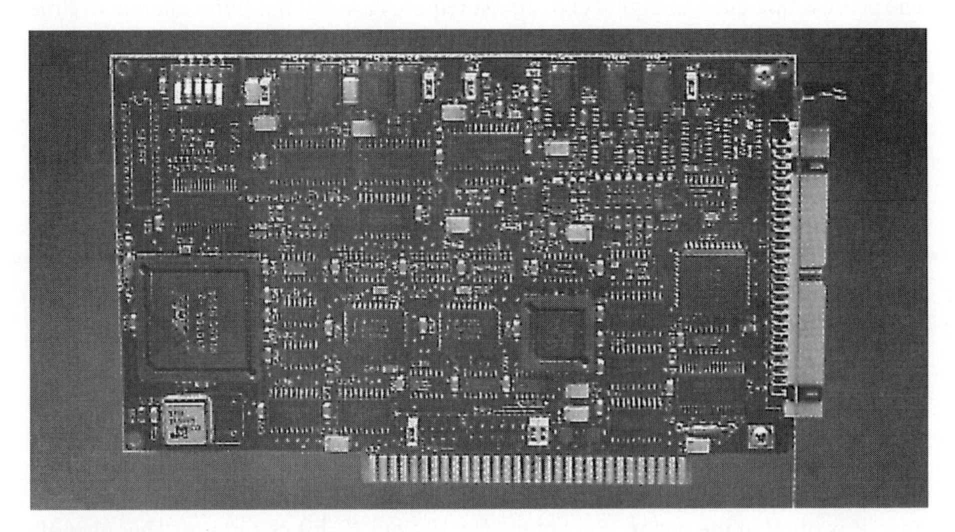

**FIGURE 7.5**
Example DAC card. *(Courtesy of National Instruments, Austin, TX)*

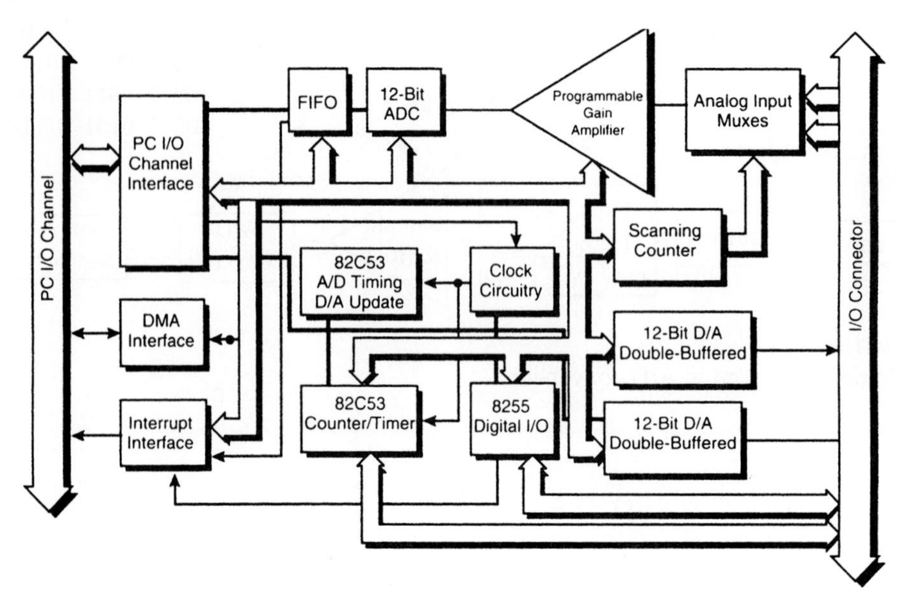


**FIGURE 7.6**
Example DAC card architecture. *(Courtesy of National Instruments, Austin, TX)*

## 7.4 ANALOG TO DIGITAL (A/D) CONVERSION

The process of analog to digital conversion requires a small but finite interval of time that must be taken into consideration when assessing the accuracy of the results. Conversion time is dependent on the design of the converter, the method used for conversion, and the speed of the components used in the electronic design. Since analog signals change continuously, the uncertainty about when in the time window the conversion occurs causes corresponding uncertainty in the digital value. The term ***aperture time*** defines the duration of the time window and is associated with any error in the digital output due to changes in the input during this time. The relationship between the aperture time and the uncertainty in the input amplitude is shown in Figure 7.7. During the aperture time $\Delta T_a$, the input signal changes by $\Delta V$ where

$$\Delta V \approx \frac{dV(t)}{dt}\Delta T_a \tag{7.5}$$

We conclude that sampling at or above the Nyquist frequency will yield the correct frequency components in a signal. However, to also obtain accurate amplitude resolution, we must have an A/D converter with a sufficiently small aperture time. It is often in the nanosecond range for 10- and 12-bit resolution. The other alternative is to use a sample and hold circuit (Chapter 5) with a high-quality capacitor to maintain a constant signal during A/D conversion.

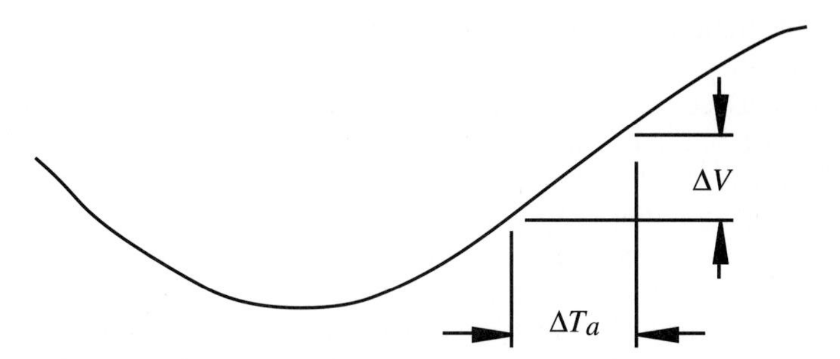


**FIGURE 7.7**
A/D conversion aperture time.

---

▼ ***EXAMPLE 7.2. Aperture Time.*** Consider a sinusoidal signal $A \sin(\omega t)$ as an input to an A/D converter. The time rate of change of the signal $A\omega \cos(\omega t)$ has a maximum value of $A\omega$. Using Equation 7.5, the maximum change in the input voltage during an aperture time $\Delta T_a$ is

$$\Delta V = A\omega\Delta T_a$$

We can compare the amplitude of the uncertainty caused by the aperture time to the peak-to-peak value of the input signal as follows:

$$\frac{\Delta V}{2A} = \frac{\omega\Delta T_a}{2}$$

If we assume that $\Delta V$ is no greater than the analog quantization size,

$$\frac{2A}{N} = \Delta V$$

where $N$ is the number of output states. Therefore,

$$\frac{\Delta V}{2A} = \frac{1}{N} = \frac{\omega\Delta T_a}{2}$$

Suppose we want to convert a signal to 10-bit resolution, which isn't extraordinarily high. This means the number of output states is $2^{10}$ or 1024. If the signal were speech (from a microphone) with a bandwidth of 10 kHz, $\Delta T_a$ would have to be less than

$$\Delta T_a = \frac{2}{2\pi(10,000)1024} = 3.2 \times 10^{-8} = 32\,ns$$

This is a very short time interval, much smaller than the required minimum sample period given by 1/2(10 kHz) = 50 μs.

---

There are a number of different principles on which A/D converters are designed: successive approximation, flash or parallel encoding, single-slope and dual-slope integration, switched capacitor, and delta sigma. We will consider the first two since they occur most often in commercial designs. The ***successive approximation*** A/D converter is very widely used because it is relatively fast and cheap. As shown in Figure 7.8, it uses a D/A converter (DAC) in a feedback loop. DACs will be described in the next section. When the start signal is applied, the sample and hold amplifier (S&H) latches the analog input. Then the control unit begins an iterative process where the digital value is approximated, converted to an analog value with the D/A converter, and compared to the analog input with the comparator. When the D/A output equals the analog input, the end signal is set by the control unit and the correct digital output is available at the output.

If $n$ is the resolution of the A/D converter, it takes $n$ steps to complete the conversion. More specifically, the input is compared to standard values in a decreasing binary sequence defined by 1, 1/2, 1/4, 1/8, . . . , $1/n$ of the full scale (FS) value of the A/D converter. The successive approximation register controls the D/A converter by proceeding through the binary sequence. The register first turns on the most significant bit (MSB) of the DAC, leaving all lesser bits at 0, and the comparator tests the DAC output against the analog input. If the analog input exceeds the DAC output, the MSB is left on (high); otherwise it is reset to zero. This procedure is then applied to the next lesser significant bit and the comparison is made again. After $n$ comparisons have occurred, the converter is down to the least significant bit (LSB). The output of the DAC then produces the best digital approximation to the analog input. The process terminates on the LSB, and the control unit sets the end signal signifying the end of the conversion.

As an example, a 4-bit successive approximation procedure is illustrated graphically in Figure 7.9. The MSB is 1/2 FS, which in this case is greater than the signal; therefore, the bit is turned off. The second bit is 1/4 FS and is less than the signal, so it is left on. The third bit gives 1/4 + 1/8 of FS, which is still less than the analog signal, so the third bit is left on. The fourth provides 1/4 + 1/8 + 1/16 of FS and is greater than the signal, so the fourth bit is turned off and the conversion is complete. The digital result is 0110. Higher resolution procedures would produce higher accuracy.

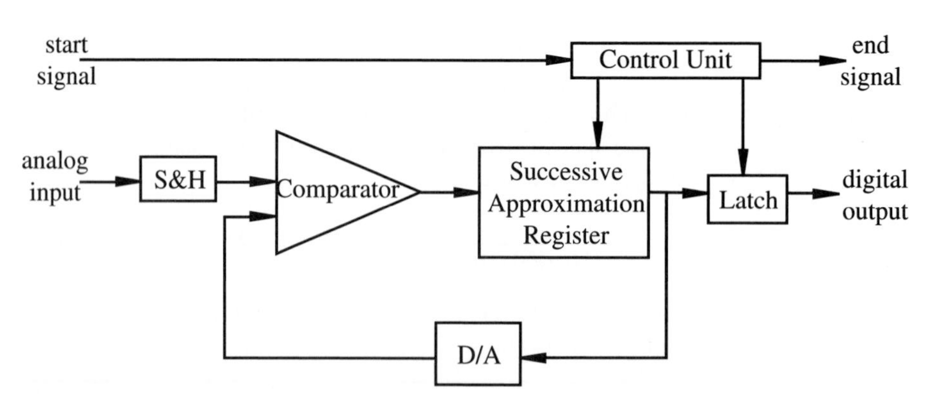


**FIGURE 7.8**
Successive approximation A/D converter.

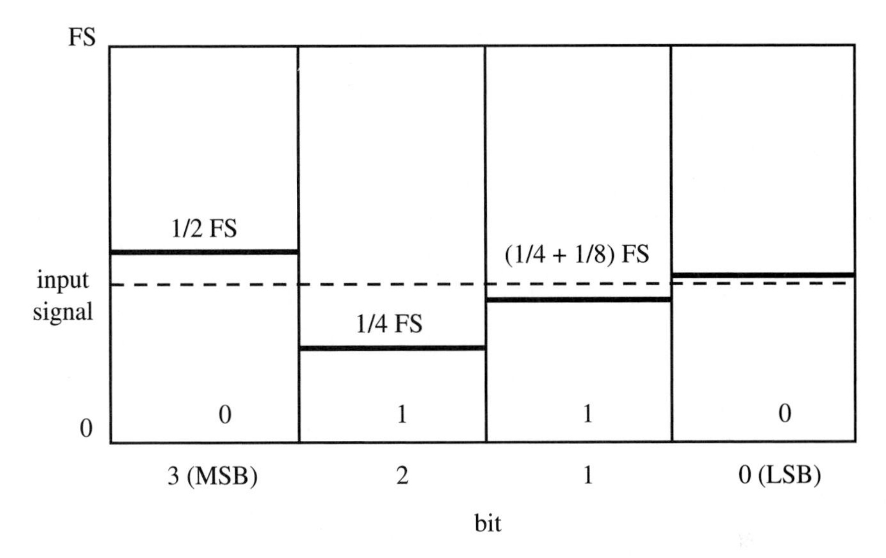


**FIGURE 7.9**
4-Bit successive approximation A/D conversion.

An $n$-bit A/D converter has a conversion time of $n\Delta T$, where $\Delta T$ is the cycle time for the D/A converter and control unit. Typical conversion times for 8-, 10-, and 12-bit A/D converters range from 1 to 100 μs. A 10-bit converter has 1024 output states, which results in a 0.1% precision over the output range.

▼▼▼ ***CLASS DISCUSSION ITEM 7.4. Selecting an A/D Converter.*** What are some reasons why a designer might select a 10-bit A/D converter instead of a 12-bit or higher resolution converter?

The fastest A/D converter is known as a ***flash converter***. As Figure 7.10 illustrates, it consists of a bank of input comparators acting in parallel to identify various levels of the signal. The output of the latches is in a coded form easily converted to the required binary output with combinational logic. The flash converter illustrated in Figure 7.10 is a 2-bit converter having a resolution of four output states. Table 7.1 lists the comparator output codes and corresponding binary outputs for each of the states, assuming an input voltage range of 0 to 4 V. The voltage range is set by the $V_{min}$ and $V_{max}$ supply voltages shown in Figure 7.10 (0 V and 4 V in this example). The code converter is a simple combinational logic circuit. For the 2-bit converter, the relationships between the code bits $G_i$ and the binary bits $B_i$ (see Question 7.5) are

$$B_0 = G_0 \cdot \overline{G_1} + G_2 \tag{7.6}$$

$$B_1 = G_1 \tag{7.7}$$

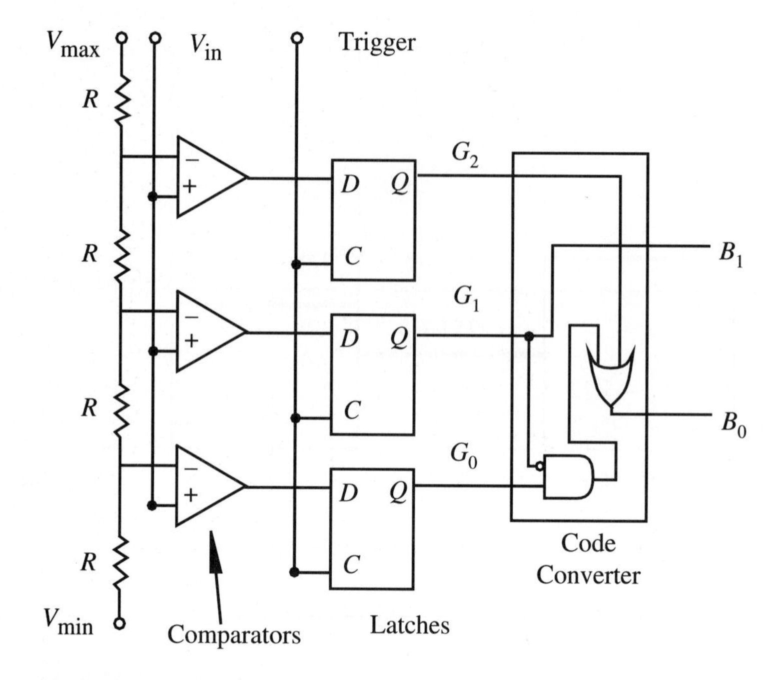


**FIGURE 7.10**
A/D flash converter.

**TABLE 7.1**
**2-Bit flash converter output**

| State | Code ($G_2G_1G_0$) | Binary ($B_1B_0$) | Voltage range |
|---|---|---|---|
| 0 | 000 | 00 | 0–1 |
| 1 | 001 | 01 | 1–2 |
| 2 | 011 | 10 | 2–3 |
| 3 | 111 | 11 | 3–4 |

Several analog signals can be digitized by a single A/D converter if the analog signals are multiplexed at the input to the A/D converter. An analog ***multiplexer*** simply switches between several analog inputs using transistors or relays and control signals. This can significantly reduce the cost of a system's design. Important parameters useful in selecting an A/D converter are the input voltage range, output resolution, and conversion time.

## 7.5 DIGITAL TO ANALOG (D/A) CONVERTERS

Often we need to reverse the process of A/D conversion by changing a digital value to an analog voltage. This is called ***digital to analog (D/A) conversion***. A D/A converter, also called a ***DAC***, allows a computer or other digital device to interface with external analog circuits and devices.

The simplest type of D/A converter is a resistor ladder network connected to an inverting summer op amp circuit as shown in Figure 7.11. This particular converter is a 4-bit R-2R resistor ladder network. It differs from other possible resistor ladder networks in that it only requires two precision resistance values ($R$ and $2R$). The digital input to the DAC is a 4-bit binary number represented by bits $b_0$, $b_1$, $b_2$, and $b_3$, where $b_0$ is the least significant bit (LSB) and $b_3$ is the most significant bit (MSB). Each bit in the circuit controls a switch between ground and the inverting input of the op amp. To understand how the analog output voltage $V_{out}$ is related to the input binary number, we can analyze the four different input combinations 0001, 0010, 0100, and 1000, and apply the principle of superposition for an arbitrary 4-bit binary number.

If the binary number is 0001, the $b_0$ switch is connected to the op amp, and the other bit switches are grounded. The resulting circuit is as shown in Figure 7.12. Since, the noninverting input of the op amp is grounded, the inverting input is at a virtual ground. The resistance between node $V_0$ and ground is $R$ since it is the parallel combination of two $2R$'s. Therefore, $V_0$ is the result of voltage division of $V_1$ across two series resistors of equal value $R$:

$$V_0 = \frac{1}{2}V_1 \tag{7.8}$$

Similarly, we can show that

$$V_1 = \frac{1}{2}V_2 \text{ and } V_2 = \frac{1}{2}V_3 \tag{7.9}$$

Therefore,

$$V_0 = \frac{1}{8}V_3 = \frac{1}{8}V_s \tag{7.10}$$

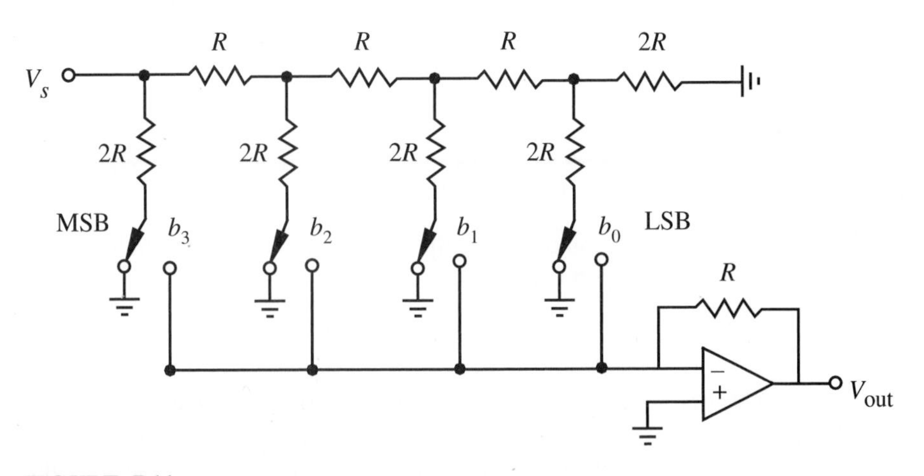


**FIGURE 7.11**
4-Bit resistor ladder D/A converter.

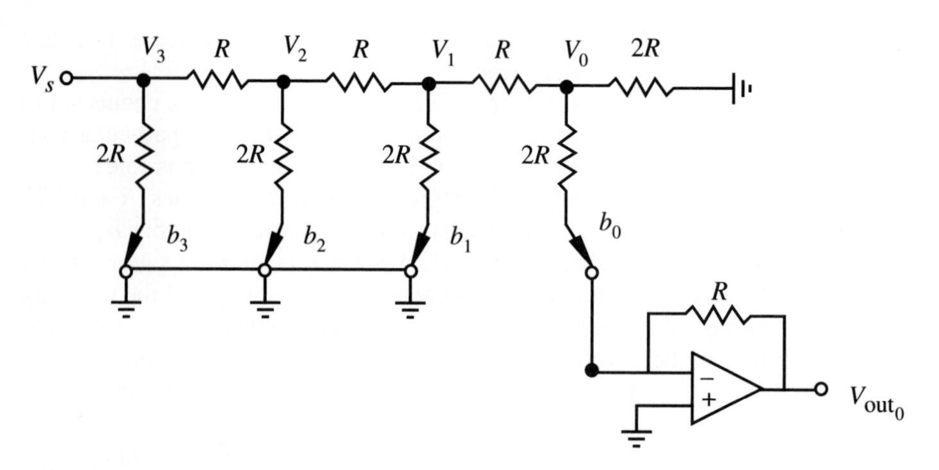


**FIGURE 7.12**
4-Bit resistor ladder D/A with digital input 0001.

$V_0$ is the input to the inverting amplifier circuit, which has a gain of

$$-\frac{R}{2R} = -\frac{1}{2} \tag{7.11}$$

Therefore, the analog output voltage corresponding to the binary input 0001 is

$$V_{out_0} = -\frac{1}{16}V_s \tag{7.12}$$

Similarly, we can show (see Question 7.7) that for the input 0010,

$$V_{out_1} = -\frac{1}{8}V_s \tag{7.13}$$

and for the input 0100,

$$V_{out_2} = -\frac{1}{4}V_s \tag{7.14}$$

and for the input 1000,

$$V_{out_3} = -\frac{1}{2}V_s \tag{7.15}$$

The output for any combination of bits comprising the input binary number can be found by the principle of superposition:

$$V_{out} = b_3V_{out_3} + b_2V_{out_2} + b_1V_{out_1} + b_0V_{out_0} \tag{7.16}$$

If $V_s$ is 10 V, the output ranges from 0 V to $-\frac{15}{16}$ 10 V for the 4-bit binary input, which has 16 values ranging from 0000 (0) to 1111 (15). A negative reference voltage $V_s$ can be used to produce a positive output voltage range. Either case yields a ***unipolar*** output, which is either positive or negative, but not both.

A ***bipolar*** output, which ranges over negative and positive values, can be produced by replacing all ground references in the circuit with a reference voltage of opposite sign as $V_s$.

▼▼▼ ***CLASS DISCUSSION ITEM 7.5. Bipolar 4-bit D/A Converter.*** If $V_s = 10$ V and the ground reference is replaced by a –10 V reference, what output voltage would correspond to each possible binary input applied to a 4-bit D/A *R*-2*R* ladder network?

Figure 7.13 illustrates the roles that A/D and D/A converters play in a mechatronic control system. An analog voltage signal from a sensor (e.g., a thermocouple) is converted to a digital value; the computer uses this value in a control algorithm; and the computer outputs an analog signal to an actuator (e.g., an electric motor) to cause some change in the system being controlled. Sensors and actuators are the topics of the next two chapters.

▼▼▼ ***CLASS DISCUSSION ITEM 7.6. Audio CD Technology.*** A compact disk (CD) stores music (an analog signal) in a digital format. How is this done? How is the digital data reconverted to the music you hear? Given that audible frequencies range from 20 Hz to 20 kHz, what is an appropriate sampling frequency? For this sampling rate, how many bits must stored on a CD to produce 45 minutes of listening pleasure?

▼▼▼ ***CLASS DISCUSSION ITEM 7.7. Digital Guitar.*** A digital guitar is a standard electric guitar equipped with additional components that send MIDI signals to a digital synthesizer. ***MIDI*** stands for *M*usical *I*nstrument *D*igital *I*nterface, and a MIDI signal consists of digital bytes containing amplitude and frequency codes for musical notes. What system components are required to do this?

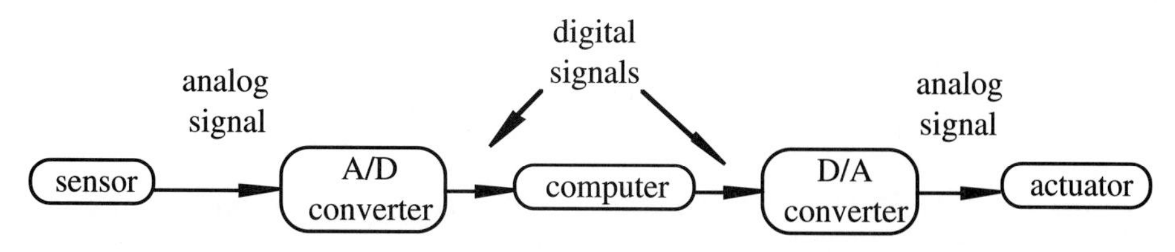


**FIGURE 7.13**
Computer control hardware.

## QUESTIONS AND EXERCISES

**7.1.** Given a 12-bit A/D converter operating over a voltage range from –5 V to 5 V, how much does the input voltage have to change, in general, in order to detect the change?

**7.2.** A signal's values range over ±5 V, and you wish to make measurements with an analog quantization size of 5 mV. What resolution A/D converter is required to execute this task?

**7.3.** Suppose you want to digitize the raw signals coming from the following sources:

(*a*) Thermocouple sensing room temperature

(*b*) The clock signal from a 120 MHz Pentium computer

(*c*) Stereo amplifier output

What is the minimum sampling frequency that you would choose in each of these cases?

**7.4.** Generate plots for the function in Example 7.1 with sample intervals of 0.5 sec, 1 sec, and 10 sec. Explain the results in light of the Sampling Theorem.

**7.5.** Derive Equations 7.6 and 7.7 from the truth table in Table 7.1.

**7.6.** Generate a plot similar to Figure 7.9 for a 5-bit converter if the input signal is 2.25 V and the input range is –5 V to 5 V. What is the correct digital output?

**7.7.** Prove Equations 7.13 through 7.15.

**7.8.** As a mechatronic designer moves progressively from an 8-bit to 10-bit to 12-bit A/D converter, what problems might be incurred in the design of the data acquisition system?

**7.9.** Using the Internet, look up the specifications of a simple A/D converter such as the National Semiconductor (www.national.com) ADC0800 8-bit A/D converter. Determine the maximum sampling rate and the method for performing the conversion. Also determine what the inputs and output are.

## BIBLIOGRAPHY

Datel Intersil, *Data Acquisition and Conversion Handbook*, Mansfield, MA, 1980.

Gibson, G. and Liu, Y., *Microcomputers for Engineers and Scientists,* Prentice-Hall, Englewood Cliffs, NJ, 1980.

Horowitz, P. and Winfield, H., *The Art of Electronics,* 2nd Edition, Cambridge University Press, New York, 1989.

O'Connor, P., *Digital and Microprocessor Technology,* 2nd Edition, Prentice-Hall, Englewood Cliffs, NJ, 1989.

# 8

# SENSORS

**OBJECTIVES:** After you read, discuss, study, and apply ideas in this chapter, you will be able to:

- Understand the fundamentals of simple electromechanical sensors, including proximity sensors and switches, potentiometers, linear variable differential transformers, optical encoders, strain gages, load cells, thermocouples, and accelerometers
- Describe how natural and binary codes are used to encode linear and rotational position in digital encoders
- Apply engineering mechanics principles to interpret data from a single strain gage or strain gage rosette
- Make accurate temperature measurements using thermocouples
- Measure acceleration and understand the frequency response limitations of accelerometers

## 8.1 INTRODUCTION

A ***sensor*** is an element in a mechatronic or measurement system that acquires a physical parameter and changes it into a signal that can be processed by the system. Often the active element of a sensor is referred to as a ***transducer***. Monitoring and control systems require sensors to measure physical quantities such as position, distance, force, strain, temperature, vibration, and acceleration. The following sections present the theory of some measurements and devices for measuring a variety of physical quantities.

Sensor and transducer design always involves the application of some law or principle of physics or chemistry that relates the quantity of interest to some measurable event. Appendix 2 summarizes many of the physical laws and principles that have potential application in sensor and transducer design. Some examples of applications are also provided. This list is useful to a transducer designer who is searching for a method to measure a physical quantity. Practically every transducer applies one or more of these principles in its operation.

## 8.2 POSITION AND SPEED MEASUREMENT

Other than electrical measurements (e.g., voltage, current, resistance), the most common measured quantity in mechatronic systems is position. We often need to know where various parts of our system are in order to control the system's behavior. Section 8.2.1 presents proximity sensors and limit switches that are a subset of position sensors and detect whether or not something is close or has reached a limit of travel. Section 8.2.2 presents the potentiometer, which is an inexpensive analog device for measuring rotary or linear position. Section 8.2.3 presents the linear variable differential transformer (LVDT), which is an analog device capable of measuring linear displacement. Finally, Section 8.2.4 presents the digital encoder, which is useful for measuring position in digital form suitable for directly interfacing to a computer or other digital system.

Because many applications involve measuring and controlling shaft rotation [e.g., in robot joints, numerically controlled (NC) lathe and mill axes, motors, and generators], and because linear motion can often be easily converted to rotary motion, rotary position sensors are more common than linear sensors.

Speed measurements can be obtained by taking consecutive position measurements at known time intervals and computing the time rate of change of the position values. A tachometer is an example of a speed sensor that does this for a rotating shaft.

### 8.2.1 Proximity Sensors and Switches

A ***proximity sensor*** consists of an element that either changes its state or changes an analog signal when it is close to, but often not actually touching, an object. Magnetic, electrical capacitance, inductance, and eddy current methods are particularly suited to the design of a proximity sensor. A photoemitter-detector pair represents another approach where interruption or reflection of a beam of light is used to detect an object in a noncontact manner. The emitter and detector are usually a phototransistor and a photodiode. Various configurations for photoemitter-detector pairs are illustrated in Figure 8.1. In the opposed and retroreflective configurations, the object interrupts the beam, and in the proximity configuration, the object reflects the beam. Figure 8.2 shows a commercial sensor that can be used in the retroreflective or proximity configuration. Common applications for proximity sensors and limit switches are in counting moving objects and in limiting the traverse of a mechanism.

There are many designs for limit switches, including push-button and levered microswitches. The switches can open or close connections as illustrated in Figure 8.3. Switches are characterized by the number of poles (P) and throws (T) and whether connections are normally open (NO) or normally closed (NC). A pole is a moving element in the switch that makes or breaks connections, and a throw is a contact point for a pole. The SPST switch is a single pole (SP), single throw (ST) device that opens or closes a single connection. The SPDT switch changes the pole between two different throw positions. There are many variations on the pole and throw configurations of switches, but their function is easily understood given this terminology.

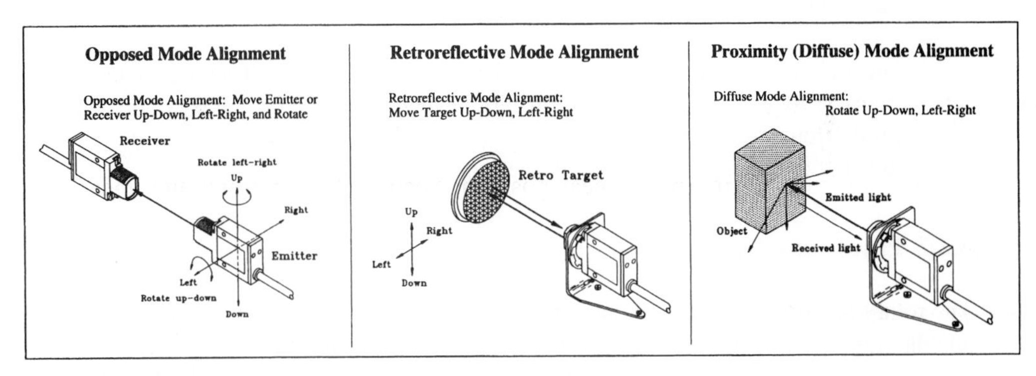


**FIGURE 8.1**
Various configurations for photoemitter-detector pairs. *(Courtesy of Banner Engineering, Minneapolis, MN)*

**FIGURE 8.2**
Example of a photoemitter-detector pair in a single housing. *(Courtesy of Banner Engineering, Minneapolis, MN)*

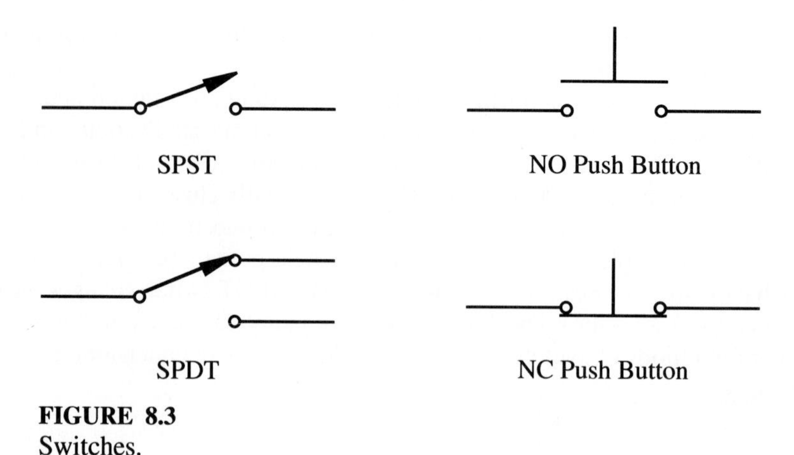


**FIGURE 8.3**
Switches.

When mechanical switches are opened or closed they exhibit switch bounce where many voltage transitions occur during the switching. If a switch is connected to a digital circuit that requires a single transition, the switch output must be debounced using a circuit as described in Section 6.14.1.

### 8.2.2 Potentiometer

The rotary ***potentiometer*** is a variable resistance device that can be used to measure angular position. It consists of a wiper that makes contact with a resistive element, and as this point of contact moves, the resistance between the wiper and end leads of the device changes in proportion to the angular displacement. Figure 8.4 illustrates the form and internal schematic for a typical rotary potentiometer. Through voltage division, the change in resistance can be used to create an output voltage that is directly proportional to the input displacement. This relationship was derived in Section 4.8.

### 8.2.3 Linear Variable Differential Transformer

The ***linear variable differential transformer*** (***LVDT***) is a transducer for measuring linear displacement. As illustrated in Figure 8.5, it consists of primary and secondary windings and a movable iron core. It functions much like a transformer where voltages are induced in the secondary coil in response to excitation in the primary

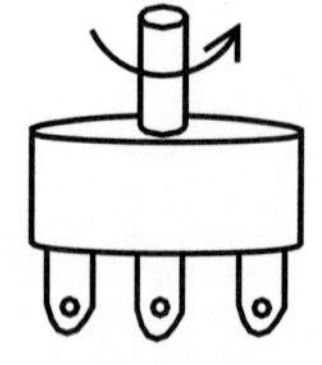

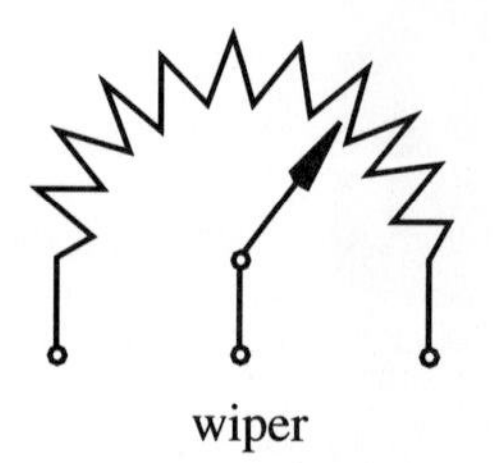


**FIGURE 8.4**
Potentiometer.

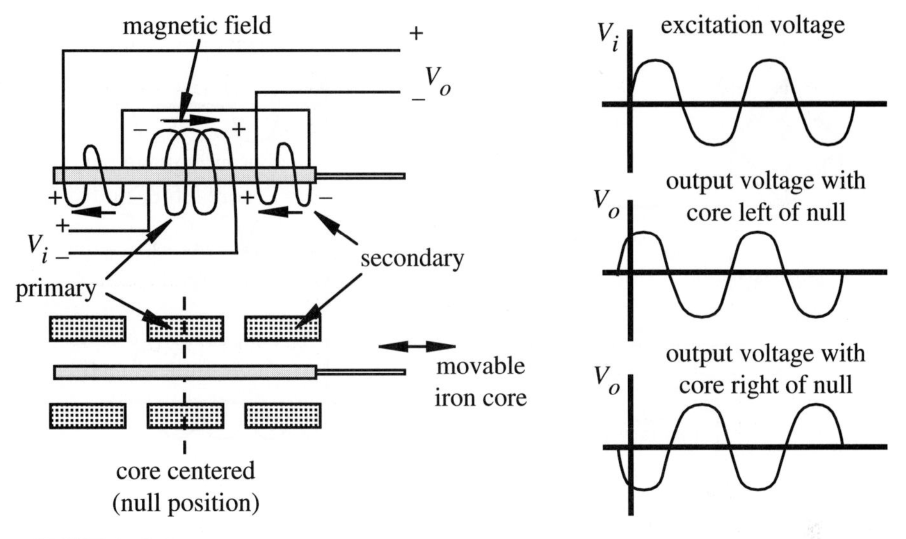


**FIGURE 8.5**
Linear variable differential transformer.

coil. The LVDT must be excited by an ac signal to induce an ac response in the secondary. The core position can be determined by measuring the secondary response.

With two secondary coils connected in the series-opposing configuration as shown, the output signal includes both the magnitude and direction of the core motion. The primary ac excitation $V_i$ and the output signal $V_o$ for two different core positions are shown in Figure 8.5. There is a midpoint in the core's position where the voltage induced in each coil will be of the same amplitude and 180° out of phase, producing a "null" output. As the core moves from the null position, the output amplitude will increase a proportional amount over a linear range around the null as shown in Figure 8.6. Therefore, by measuring the output voltage amplitude, we can easily and accurately determine the magnitude of the core displacement.

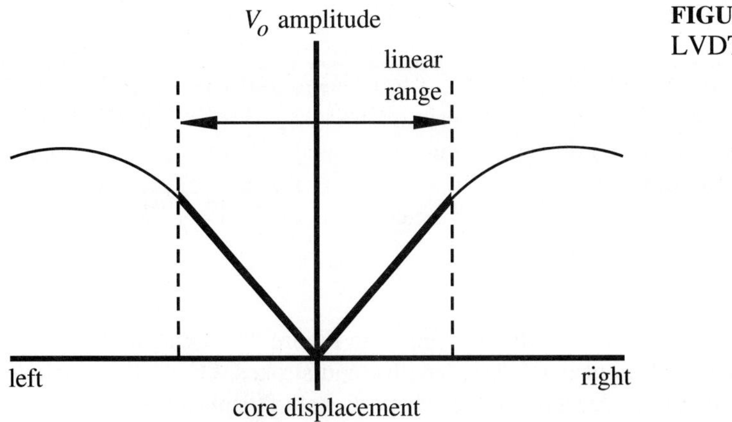


**FIGURE 8.6**
LVDT linear range.

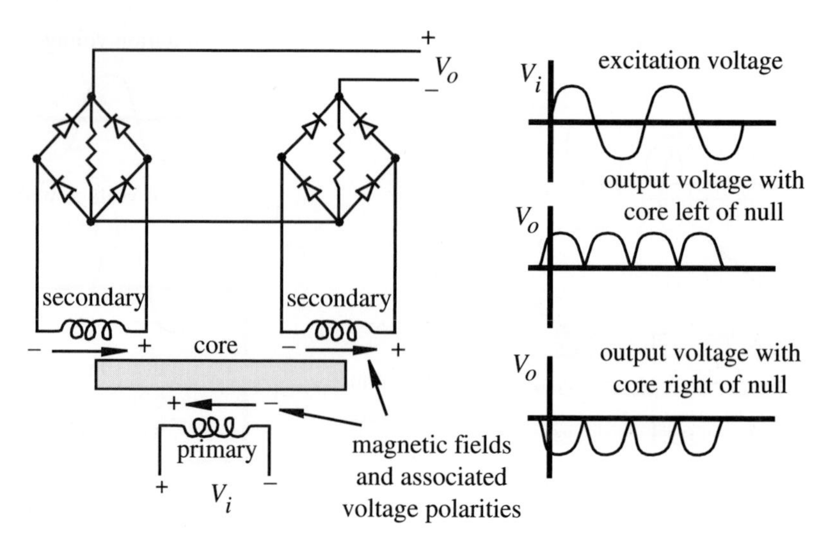


**FIGURE 8.7**
LVDT demodulation.

In order to determine the direction of the core displacement, the secondary coils can be included in a demodulation circuit as shown in Figure 8.7. The diode bridges in this circuit produce a positive or negative rectified sine wave depending on what side of the null position the core is on (see Class Discussion Item 8.1).

As illustrated in Figure 8.8, a low-pass filter may also be used to convert the rectified output into a smoothed signal that tracks the core position. The cutoff frequency of this low-pass filter must be chosen carefully to filter out the high frequencies in the rectified wave but not filter out frequency components associated with the core motion. The excitation frequency is usually chosen to be at least 10 times the maximum expected frequency of the core motion to yield a good representation of the time-varying displacement.

▼▼▼ ***CLASS DISCUSSION ITEM 8.1. LVDT Demodulation.*** Trace the currents through the diodes in the demodulation circuit shown in Figure 8.7 for different core positions (null, left of null, and right of null) and explain why the output voltage behaves as shown. Assume ideal diodes. Also, explain why the output is zero when the core is in the null or center position.

Commercial LVDTs, such as the one shown in Figure 8.9, are available in cylindrical forms with different diameters, lengths, and strokes. Often, they include internal circuitry that provides a dc voltage proportional to displacement.

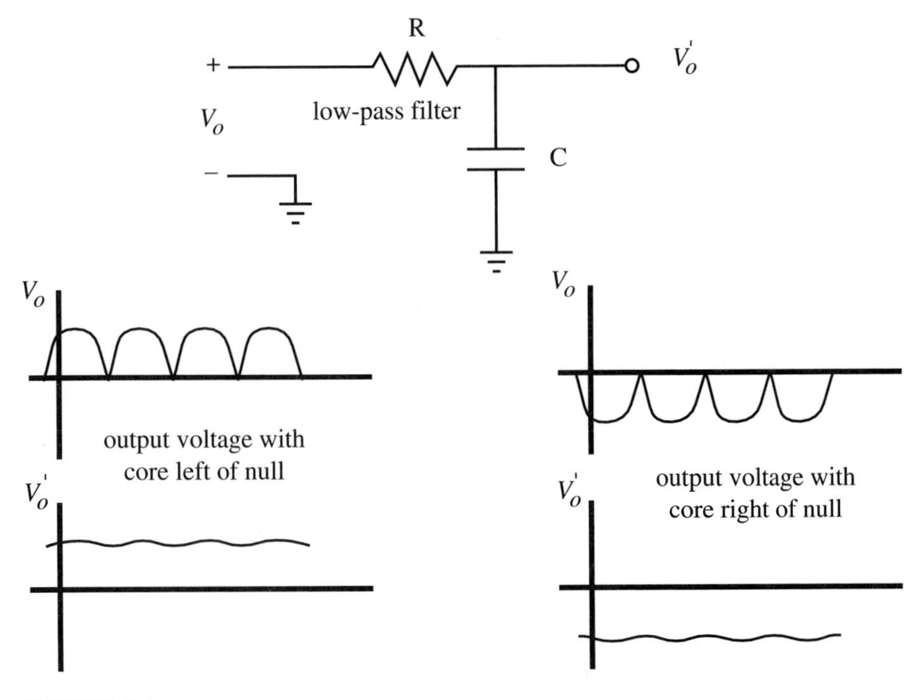


**FIGURE 8.8**
LVDT output filter.

The advantages of the LVDT are accuracy over the linear range and an analog output that may not require amplification. Also, it is less sensitive to wide ranges in temperature than other position transducers (e.g., potentiometers, encoders, and semiconductor devices). The LVDT's disadvantages include limited range of motion and limited frequency response. The overall frequency response is limited by inertial effects associated with core motion and by the choice of the primary excitation frequency and the filter cutoff frequency.

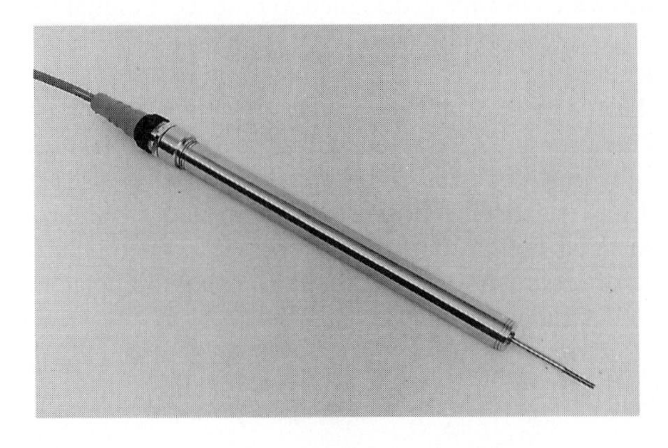

**FIGURE 8.9**
Commercial LVDT.
*(Courtesy of Sensotec, Columbus, OH)*

▼▼▼ ***CLASS DISCUSSION ITEM 8.2. LVDT Signal Filtering.*** Given the spectrum of a time-varying core displacement, what effect does the choice of the primary excitation frequency have, and how should the low-pass filter be designed to produce an output most representative of the displacement?

A ***resolver*** is an analog rotary position sensor that operates very much like the LVDT. It consists of a rotating shaft (rotor) with a primary winding and a stationary housing (stator) with two secondary windings offset by 90°. When the primary is excited with an ac signal, ac voltages are induced in the secondary coils, which are proportional to the sine and cosine of the shaft angle. Because of this, the resolver is useful in applications where trigonometric functions of position are required without use of a microprocessor.

### 8.2.4 Digital Optical Encoder

A digital optical encoder is a device that converts motion into a sequence of digital pulses. By counting a single bit or by decoding a set of bits, the pulses can be converted to relative or absolute position measurements. Encoders have both linear and rotary configurations, but the most common type is rotary. Rotary encoders are manufactured in two basic forms: the absolute encoder where a unique digital word corresponds to each rotational position of the shaft, and the incremental encoder, which produces digital pulses as the shaft rotates, allowing measurement of relative position of the shaft. As illustrated in Figure 8.10, most rotary encoders are composed of a glass or plastic code disk with a photographically deposited radial pattern organized in tracks. As radial lines in each track interrupt the beam between a photoemitter-detector pair, digital pulses are produced.

The optical disk of the ***absolute encoder*** is designed to produce a digital word that distinguishes $N$ distinct positions of the shaft. For example, if there are 8 tracks, the encoder is capable of producing $2^8 = 256$ distinct positions or an angular resolution of 1.406° (360°/256). The most common types of numerical encoding used in the absolute encoder are gray and natural binary codes. To illustrate the action of an absolute encoder, the gray code and natural binary code disk track patterns for a simple 4-track (4-bit) encoder are illustrated in Figures 8.11 and 8.12. The linear patterns and associated timing diagrams are what the photodetectors sense as the code disk circular tracks rotate with the shaft. The output bit codes for both coding schemes are listed in Table 8.1.

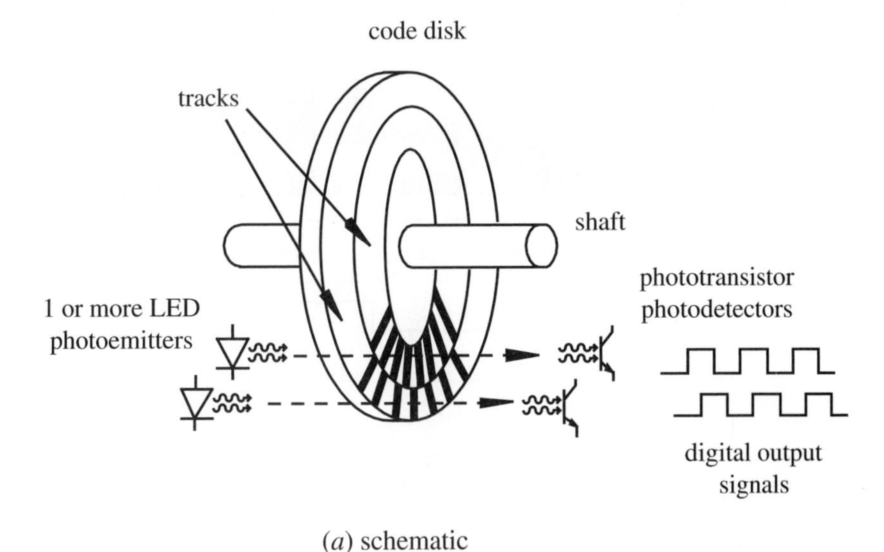


(*a*) schematic

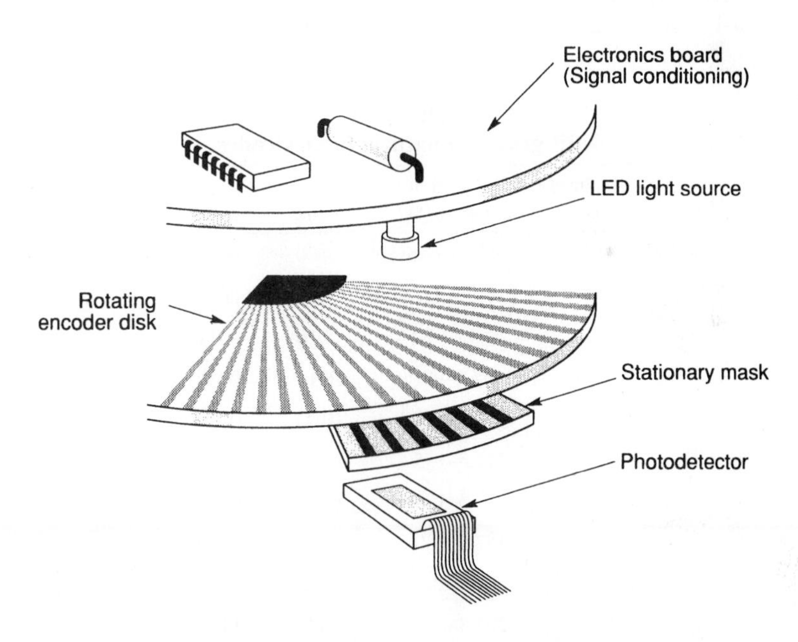


(*b*) typical construction *(Courtesy of Lucas Ledex Inc., Vandalia, OH)*

**FIGURE 8.10**
Components of an optical encoder.

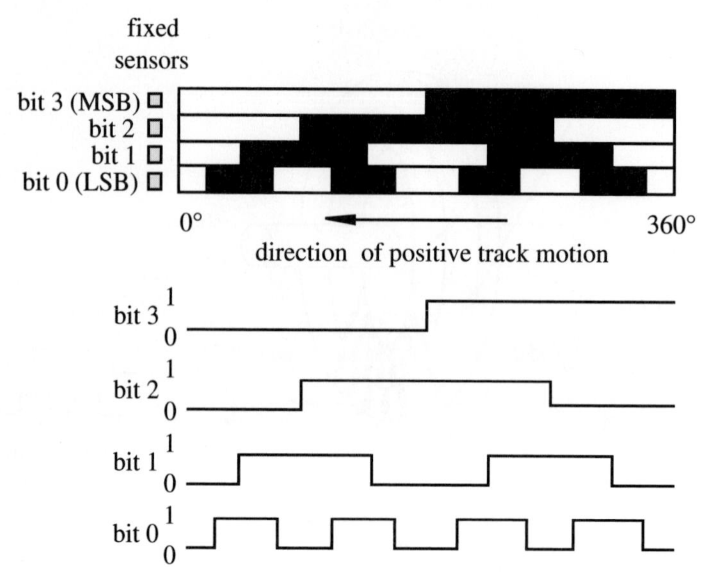


**FIGURE 8.11**
4-Bit gray code absolute encoder disk track patterns.

**TABLE 8.1**
**4-Bit gray and natural binary codes**

| Decimal code | Rotation range (°) | Binary code | Gray code |
|---|---|---|---|
| 0 | 0–22.5 | 0000 | 0000 |
| 1 | 22.5–45 | 0001 | 0001 |
| 2 | 45–67.5 | 0010 | 0011 |
| 3 | 67.5–90 | 0011 | 0010 |
| 4 | 90–112.5 | 0100 | 0110 |
| 5 | 112.5–135 | 0101 | 0111 |
| 6 | 135–157.5 | 0110 | 0101 |
| 7 | 157.5–180 | 0111 | 0100 |
| 8 | 180–202.5 | 1000 | 1100 |
| 9 | 202.5–225 | 1001 | 1101 |
| 10 | 225–247.5 | 1010 | 1111 |
| 11 | 247.5–270 | 1011 | 1110 |
| 12 | 270–292.5 | 1100 | 1010 |
| 13 | 292.5–315 | 1101 | 1011 |
| 14 | 315–337.5 | 1110 | 1001 |
| 15 | 337.5–360 | 1111 | 1000 |

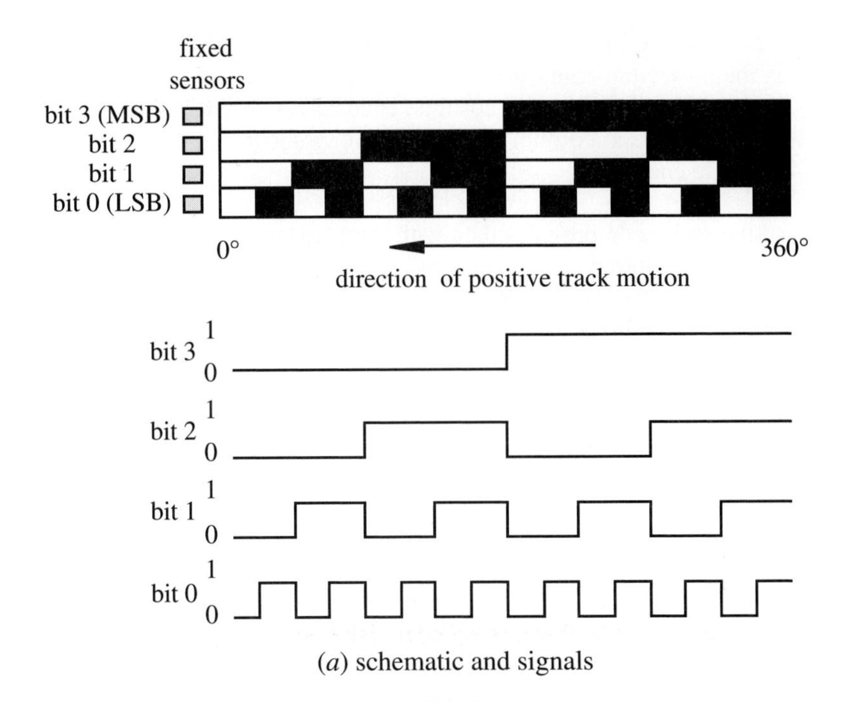


(*a*) schematic and signals

(*b*) actual disk *(Courtesy of Parker Compumotor Division, Rohnert Park, CA)*

**FIGURE 8.12**
4-Bit natural binary absolute encoder disk track patterns.

The gray code is designed so that only one track (one bit) will change state for each count transition, unlike the binary code where multiple tracks (bits) change at certain count transitions. This effect can be seen clearly in Figures 8.11 and 8.12 and in the last two columns of Table 8.1. For the gray code, the uncertainty during a transition is only one count, unlike with the binary code, where the uncertainty could be multiple counts.

▼▼▼ ***CLASS DISCUSSION ITEM 8.3. Encoder Binary Code Problems.*** What is the maximum count uncertainty for a 4-bit gray code absolute encoder and a 4-bit natural binary absolute encoder? At what decimal code transitions does the maximum count uncertainty occur in a 4-bit natural binary absolute encoder?

Since the gray code provides data with the least uncertainty but the natural binary code is the preferred choice for direct interface to computers and other digital devices, a circuit to convert from gray to binary code is desirable. Figure 8.13 shows a simple circuit that utilizes exclusive OR gates (XOR) to perform this function.

The Boolean expressions that relate the binary bits ($B_i$) to the gray code bits ($G_i$) are

$$\begin{aligned} B_3 &= G_3 \\ B_2 &= B_3 \oplus G_2 \\ B_1 &= B_2 \oplus G_1 \\ B_0 &= B_1 \oplus G_0 \end{aligned} \tag{8.1}$$

For a gray code to binary code conversion of any number of bits $N$ (e.g., $N = 4$ above), the most significant bits (MSB) of the binary and gray code are always identical ($B_{N-1} = G_{N-1}$), and for each other bit, the binary bit is the exclusive OR (XOR) combination of adjacent gray code bits ($B_i = B_{i+1} \oplus G_i$ for $i = 0$ to $N - 2$). This pattern can be easily seen in the 4-bit example above (Equations 8.1).

▼▼▼ ***CLASS DISCUSSION ITEM 8.4. Gray to Binary Code Conversion.*** Examine the validity of Equations 8.1 by applying them to the last two columns in Table 8.1.

The ***incremental encoder***, sometimes called a ***relative encoder***, is simpler in design than the absolute encoder. It consists of two tracks and two sensors whose outputs are called channels A and B. As the shaft rotates, pulse trains occur on these

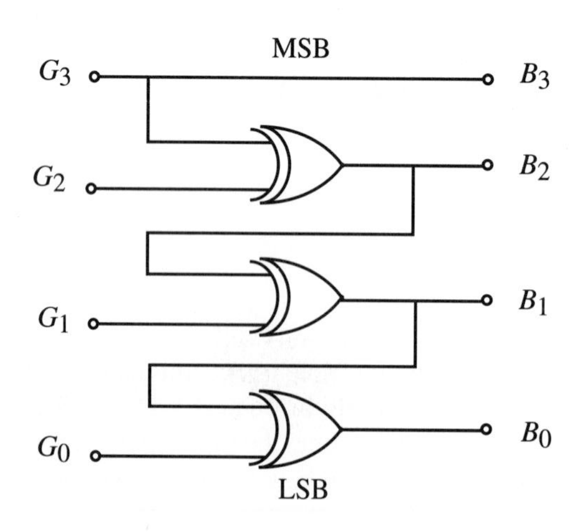


**FIGURE 8.13**
Gray code to binary code conversion.

channels at a frequency proportional to the shaft speed, and the phase relationship between the signals yields the direction of rotation. The code disk pattern and output signals A and B are illustrated in Figure 8.14. By counting the number of pulses and knowing the resolution of the disk, the angular motion can be measured. The A and B channels are used to determine the direction of rotation by assessing which channel "leads" the other. The signals from the two channels are a 1/4 cycle out of phase with each other and are known as ***quadrature signals***. Often a third output channel, called INDEX, yields one pulse per revolution, which is useful in counting full revolutions. It also useful as a reference to define a home base or zero position.

Figure 8.14a illustrates two separate tracks for the A and B channels, but a more common configuration uses a single track (see Figure 8.14b) with the A and B sensors offset a 1/4 cycle on the track to yield the same signal pattern. A single-track code disk is simpler and cheaper to manufacture.

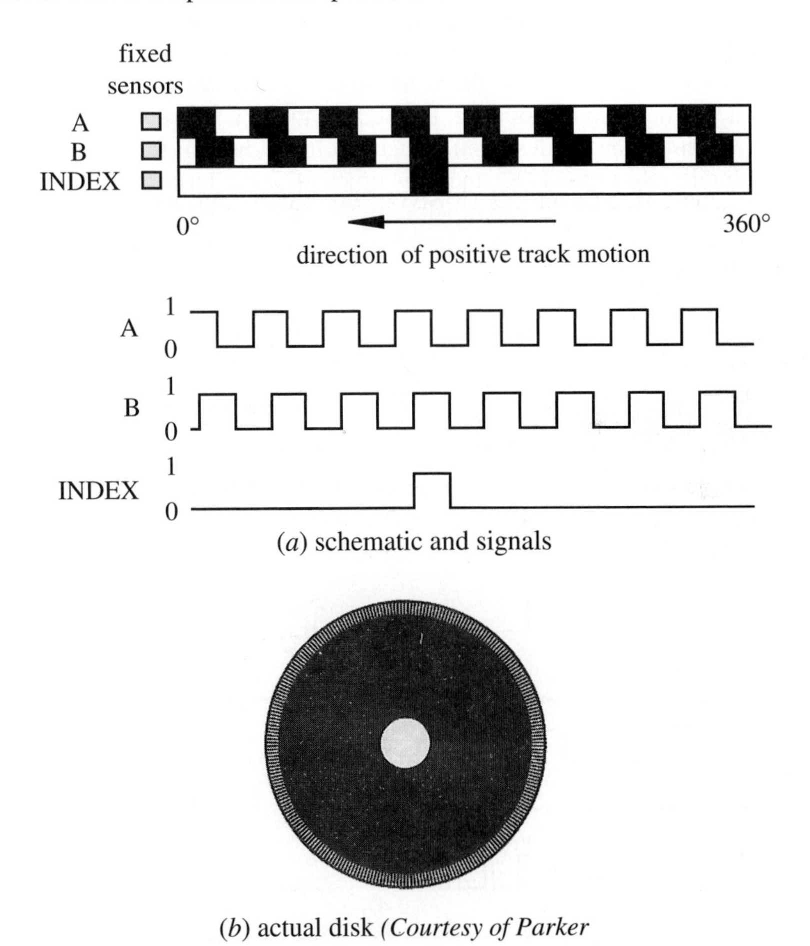


(*a*) schematic and signals

(*b*) actual disk *(Courtesy of Parker Compumotor Division, Rohnert Park, CA)*

**FIGURE 8.14**
Incremental encoder disk track patterns.

The quadrature signals A and B can be decoded to yield the direction of rotation as shown in Figure 8.15. Decoding transitions of A and B by using sequential logic circuits in different ways can provide three different resolutions of the output pulses: 1X, 2X, and 4X. 1X resolution only provides a single pulse for each cycle in one of the signals A or B. 4X resolution provides a pulse at every edge transition in the two signals A and B providing four times the 1X resolution. The direction of rotation (CW or CCW) is determined by the level of one signal during an edge transition of the second signal. For example, in the 1X mode, $A = \downarrow$ with $B = 1$ implies a CW pulse, and $B = \downarrow$ with $A = 1$ implies a CCW pulse. If we only had a single output channel A or B, it would be impossible to determine the direction of rotation. Furthermore, shaft jitter around an edge transition in the single signal would result in erroneous pulses.

Figure 8.16 shows a circuit that will yield the base 1X resolution by creating and counting pulses at specific negative edges of the quadrature signals. The D flip-flops decode whether the shaft is rotating clockwise or counterclockwise, and this information is used to drive an up-down counter to keep the current pulse count for the encoder rotation. In addition to the edges detected for the 1X resolution, circuits can be designed to detect other edges in the quadrature signals resulting in two times (2X) and four times (4X) the base resolution (1X) as illustrated in Figure 8.15. These quadrature decoder circuits can be constructed with discrete components, but they are also available on ICs (e.g., Hewlett Packard's HCTL-2016).

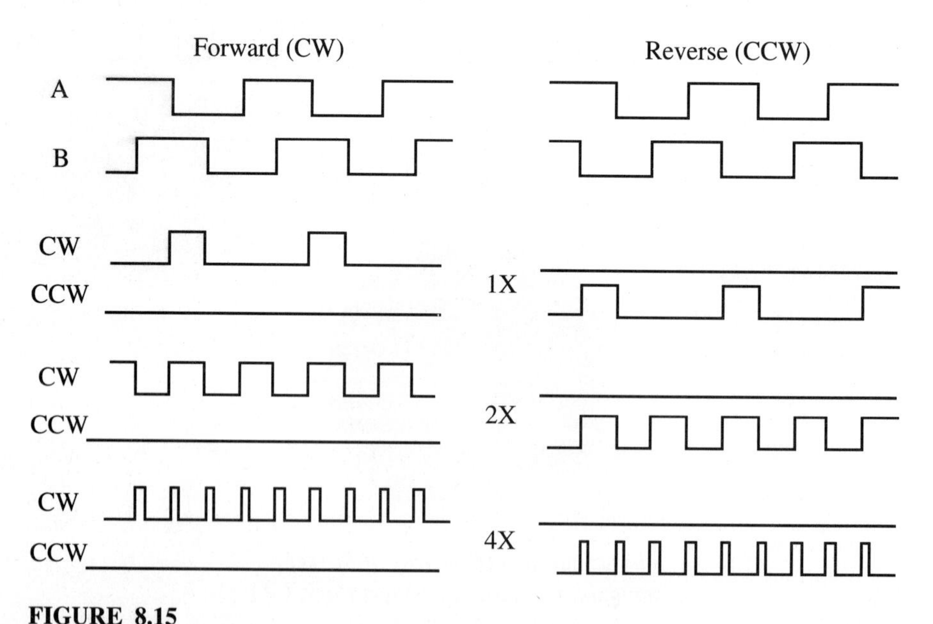


**FIGURE 8.15**
Quadrature direction sensing and resolution enhancement.

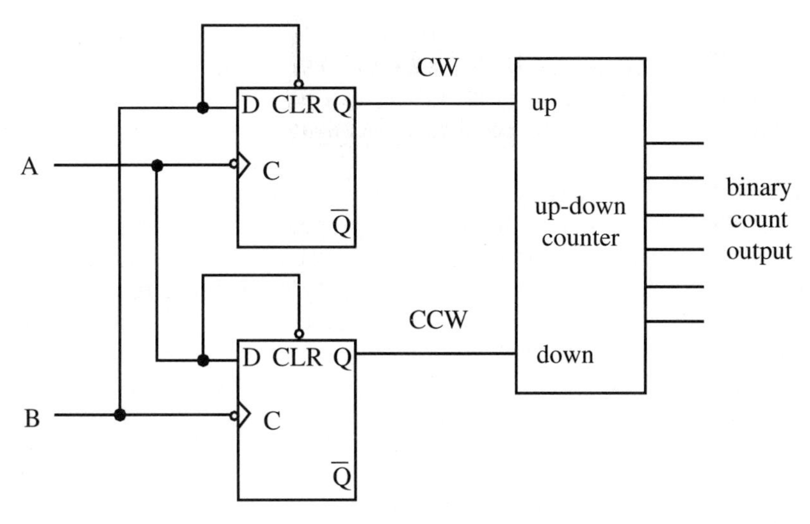


**FIGURE 8.16**
1X quadrature decoder circuit.

▼▼▼ ***CLASS DISCUSSION ITEM 8.5. Encoder 1X Circuit with Jitter.*** An incremental encoder connected to a 1X quadrature decoder circuit is experiencing a small rotational vibration with an amplitude roughly equivalent to one quadrature pulse width. During this vibration, you observe many pulses on both the CW and CCW lines but no net change in the output of the up-down counter. Explain how this can happen.

▼▼▼ ***CLASS DISCUSSION ITEM 8.6. Encoder Resolution.*** What is the angular resolution of a 2-channel incremental encoder with a 4X quadrature decoder circuit if the code disk track has 2500 radial lines?

Incremental encoders provide more resolution at lower cost than absolute encoders, but they only measure relative motion and do not provide absolute position directly. An incremental encoder can be used in conjunction with a limit switch to define absolute position relative to some home position defined by the switch. Absolute encoders are chosen in applications where establishing a reference position is impractical.

▼▼▼ ***CLASS DISCUSSION ITEM 8.7. Robotic Arm with Encoders.*** When a robotic arm with absolute encoders on its joints is powered up, the robot knows exactly where its arm is relative to its base. If the absolute encoders were replaced with incremental encoders, would this remain true? If not, how would the robot establish a home or reference position for the arm?

## 8.3 STRESS AND STRAIN MEASUREMENT

Measurement of stress in a mechanical component is important when assessing whether or not it is or will be subjected to safe load levels. The most common transducer used to measure stress is the electrical resistance strain gage. As we will see, stress values can be determined from strain measurements using principles of solid mechanics. Stress and strain measurements can also be used to indirectly measure other physical quantities such as force (by measuring strain of a flexural element), pressure (by measuring strain in a flexible diaphragm), and temperature (by measuring thermal expansion of a material).

Basic stress and strain relations and planar stress analysis techniques are presented in Appendix 2 for your review if necessary.

### 8.3.1 Electrical Resistance Strain Gage

The most common transducer for experimentally measuring strain in a mechanical component is the bonded metal foil strain gage illustrated in Figure 8.17. It consists of a thin foil of metal, usually constantan, etched in a grid pattern onto a thin plastic backing material, usually polyimide. The foil pattern is terminated at both ends with large metallic pads, which are usually copper-coated to allow leadwires to be attached easily with solder. The entire gage is usually very small, typically 5 mm to 15 mm long.

To measure strain on the surface of a mechanical component, the gage is adhesively bonded directly to the component, usually with epoxy or cyanoacrylate. The backing makes the foil gage easy to handle and provides a good bonding surface that also electrically insulates the metal foil from the component. Leadwires are then soldered to the solder tabs on the gage. When the component is loaded, the resistance of the metal foil changes in a predictable way, and if this resistance change is measured accurately, the strain on the surface of the component can be determined. These strain measurements allow us to determine the state of stress on the surface of the component, where stresses typically have their highest values. Knowing stresses at critical locations on a component under load can help a designer verify analytical results (e.g., from a finite element analysis) and verify that stress levels remain below safe limits for the material (e.g., the yield strength). It is important to note that because strain gages are finite in size, a measurement actually reflects an average of the strain over a small area. Hence, making measurements where stress gradients are large (e.g., at stress concentrations) can yield poor results.

It is important to note that experimental stress analysis (e.g., with strain gages) and analytical or numerical stress analysis (e.g., with finite element analysis) are both important to reliable design of mechanical parts. The two approaches should be considered complements to each other and not replacements. Finite element analysis involves many assumptions about material properties and boundary conditions that may not accurately model the actual component when it is manufactured and loaded. Strain gage measurements may also have some inaccuracies due to poor bonding and alignment on the component surface and due to uncompensated

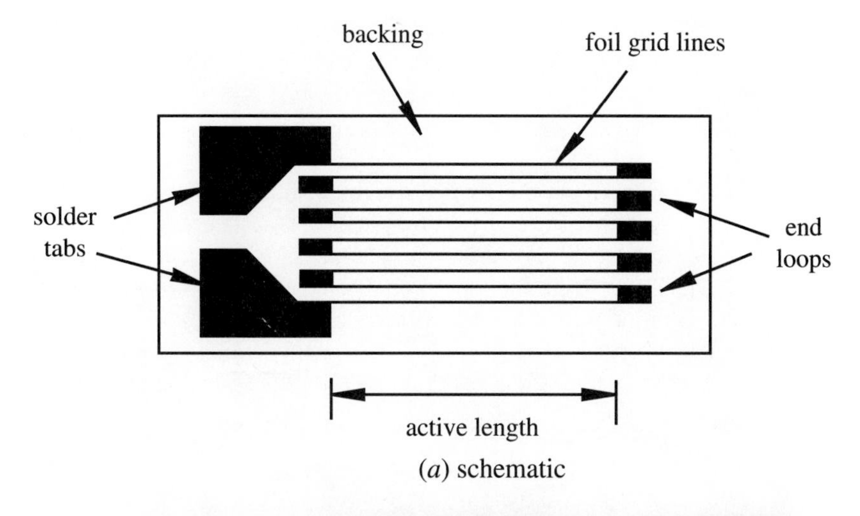


(*a*) schematic

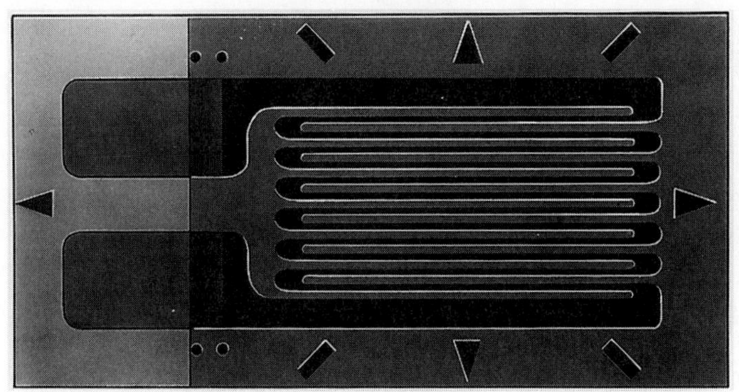

(*b*) actual *(Courtesy of Measurements Group Inc., Raleigh, NC)*

**FIGURE 8.17**
Metal foil strain gage construction.

temperature effects. Also, only specific locations can be checked with strain gages since space and access on the component can be limiting factors.

Effects that are easily measured with strain gages but difficult to model with finite element analysis include stresses resulting from mechanical assembly of components and from complex loading and boundary conditions. These and other effects are often difficult to predict and model with analytical and numerical methods.

Experimental stress applications usually involve mounting a large number of strain gages on a mechanical component, as illustrated in Figure 8.18, before it is loaded. Experimental strain values are usually acquired through an automated data acquisition system. The strain data can be converted to stresses in the object under different loading conditions, and the stresses can be compared to analytical and numerical finite element analysis results.

**FIGURE 8.18**
Strain gage application. *(Courtesy of Measurements Group Inc., Raleigh, NC)*

To understand how a strain gage is used to measure strain, we first look at how the resistance of the foil changes when deformed. The connected metal foil grid lines in the active portion of the gage can be approximated by a single rectangular conductor as illustrated in Figure 8.19, whose total resistance is given by

$$R = \frac{\rho L}{A} \tag{8.2}$$

where $\rho$ is the foil metal resistivity, $L$ is the total length of the grid lines, and $A$ is the grid line cross-sectional area. The gage end loops and solder tabs have negligible effects on the gage resistance since they typically have a much larger cross section than the foil lines.

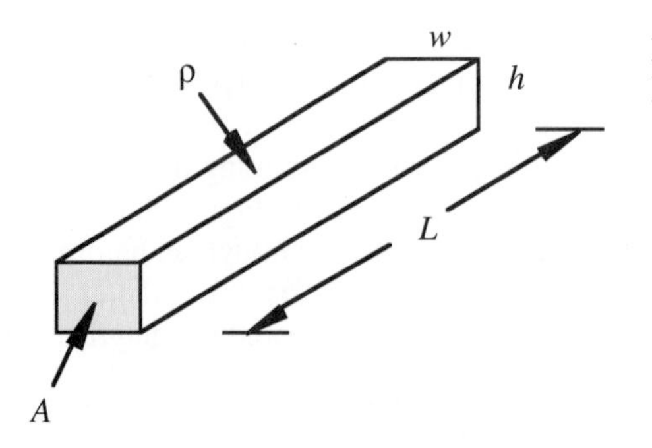


**FIGURE 8.19**
Rectangular conductor.

To see how the resistance changes under deformation, we need to take the differential of Equation 8.2. We can simplify this operation by first taking the natural logarithm, giving

$$\ln R = \ln \rho + \ln L - \ln A \tag{8.3}$$

Taking the differential yields the following expression for the change in resistance given material property and geometry changes in the conductor:

$$dR/R = d\rho/\rho + dL/L - dA/A \tag{8.4}$$

As we would expect, the signs in this equation imply that the resistance of the conductor increases ($dR > 0$) with increased resistivity and increased length and decreases with increased cross-sectional area. Since the cross-sectional area of the conductor is

$$A = wh \tag{8.5}$$

the area differential term is

$$\frac{dA}{A} = \frac{w \cdot dh + h \cdot dw}{w \cdot h} = \frac{dh}{h} + \frac{dw}{w} \tag{8.6}$$

From the definition of Poisson's Ratio (see Appendix 2),

$$\frac{dh}{h} = -\nu\frac{dL}{L} \tag{8.7}$$

and

$$\frac{dw}{w} = -\nu\frac{dL}{L} \tag{8.8}$$

so

$$\frac{dA}{A} = -2\nu\frac{dL}{L} = -2\nu\varepsilon_{\text{axial}} \tag{8.9}$$

where $\varepsilon_{\text{axial}}$ is the axial strain in the conductor (see Appendix 2). When the conductor is elongated ($\varepsilon_{\text{axial}} > 0$) the cross-sectional area decreases ($dA/A < 0$), causing the resistance to increase.

Using Equation 8.9, Equation 8.4 can be expressed as

$$dR/R = \varepsilon_{\text{axial}}(1 + 2\nu) + d\rho/\rho \tag{8.10}$$

Dividing through by $\varepsilon_{\text{axial}}$ gives

$$\frac{dR/R}{\varepsilon_{\text{axial}}} = 1 + 2\nu + \frac{d\rho/\rho}{\varepsilon_{\text{axial}}} \tag{8.11}$$

The first two terms on the right-hand side, 1 and $2\nu$, represent the change in resistance due to increased length and decreased area. The last term $(d\rho/\rho)/(\varepsilon_{\text{axial}})$ represents the ***piezoresistive effect*** in the material, which explains how the resistivity of the material changes with strain. This term is constant over the operating range of typical strain gage metal foils.

Commercially available strain gage specifications usually report a constant ***gage factor F*** to represent the right-hand side of Equation 8.11. This factor represents the material characteristics of the gage that relate the gage's change in resistance to strain:

$$F = \frac{\Delta R/R}{\varepsilon_{axial}} \tag{8.12}$$

Thus when a gage of known resistance $R$ and gage factor $F$ is bonded to the surface of a component and the component is then loaded, we can determine the strain in the gage $\varepsilon_{axial}$ simply by measuring the change in resistance of the gage $\Delta R$:

$$\varepsilon_{axial} = \frac{\Delta R/R}{F} \tag{8.13}$$

This gage strain is the strain experienced on the surface of the loaded component in the direction of the gage long (grid) axis.

For the bonded metal foil strain gage, the gage factor $F$ is commonly 2 and the gage resistance $R$ is 120 Ω. Strain gage suppliers also report a ***transverse sensitivity*** for the gage, which is a measure of the resistance changes in the end loops and grid lines due to strain in the transverse direction. The transverse sensitivity for a bonded metal foil gage is usually approximately 1%. This number predicts the gage's sensitivity to transverse strains: those perpendicular to the measuring axis of the gage. A gage experiencing 50 με ($50 \times 10^{-6}$) in the axial direction and 100 με in the transverse direction with a transverse sensitivity of 1% will sense 51 με (50 + 1% of 100), not 50 με.

---

▼ ***EXAMPLE 8.1. Strain Gage Resistance Changes.*** If a 120 Ω strain gage with gage factor 2.0 is used to measure a strain of 100 με ($100 \times 10^{-6}$), how much does the resistance of the gage change from the unloaded state to the loaded state?

Equation 8.12 tells us that

$$\Delta R = R \cdot F \cdot \varepsilon$$

so the change in resistance would be

$$\Delta R = (120\ \Omega)(2.0)(0.000100) = 0.024\ \Omega$$

---

▼▼▼ *CLASS DISCUSSION ITEM 8.8. **Piezoresistive Effect in Strain Gages.*** For a typical metal foil strain gage with a gage factor of 2.0, how large is the piezoresistive effect in comparison to the effects of change in area and change in length?

### 8.3.2 Measuring Resistance Changes with a Wheatstone Bridge

In order to use strain gages to accurately measure strains experimentally, we need to be able to accurately measure small changes in resistance. The most common circuit used to measure small changes in resistance is the ***Wheatstone bridge,*** which consists of a four-resistor network excited by a dc voltage. Two different modes of operation of a Wheatstone bridge circuit are the static balanced mode and the dynamic unbalanced mode. The static mode is illustrated in Figure 8.20.

For the ***static balanced mode***, $R_2$ and $R_3$ are precision resistors, $R_4$ is a precision potentiometer (variable resistor) with an accurate scale for displaying the resistance value, and $R_1$ is the strain gage resistance for which the change is to be measured. To balance the bridge, the variable resistor is adjusted until the voltage between nodes A and B is zero. In the balanced state, the voltages at A and B must be equal so

$$i_1 R_1 = i_2 R_2 \tag{8.14}$$

Also, since the high-input impedance voltmeter between A and B is assumed to draw no current,

$$i_1 = i_4 = \frac{V_{ex}}{R_1 + R_4} \tag{8.15}$$

and

$$i_2 = i_3 = \frac{V_{ex}}{R_2 + R_3} \tag{8.16}$$

where $V_{ex}$ is the dc voltage applied to the bridge called the ***excitation voltage***. Substituting these expressions into Equation 8.14 and rearranging gives

$$\frac{R_1}{R_4} = \frac{R_2}{R_3} \tag{8.17}$$

If we know $R_2$ and $R_3$ accurately and we note the value for $R_4$, we can accurately calculate the unknown resistance $R_1$ as

$$R_1 = \frac{R_4 R_2}{R_3} \tag{8.18}$$

Note that this result is independent of the excitation voltage, $V_{ex}$ (see Class Discussion Item 8.9).

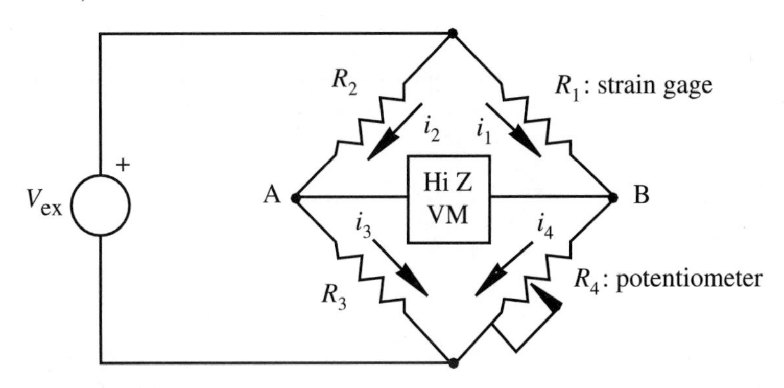


**FIGURE 8.20**
Static balanced bridge circuit.

As shown above, the static or balanced mode of operation can be used to measure an unknown resistance, but usually balancing is used only as a preliminary step to measuring changes in resistance. In ***dynamic deflection operation*** (see Figure 8.21), again with $R_1$ representing a strain gage and $R_4$ representing a potentiometer, the bridge is first balanced by adjusting $R_4$ until there is no output voltage. Then changes in the strain gage resistance $R_1$ that occur when the mechanical component is loaded can be determined from changes in the output voltage.

The output voltage can be expressed in terms of the resistor currents as

$$V_o = i_1 R_1 - i_2 R_2 = -i_1 R_4 + i_2 R_3 \tag{8.19}$$

and the excitation voltage can be related to the same currents as

$$V_{\text{ex}} = i_1(R_1 + R_4) = i_2(R_2 + R_3) \tag{8.20}$$

Eliminating the currents from these equations results in

$$V_o = V_{\text{ex}}\left(\frac{R_1}{R_1 + R_4} - \frac{R_2}{R_2 + R_3}\right) \tag{8.21}$$

When the bridge is balanced, $V_o$ is zero and $R_1$ has a known value. When $R_1$ changes value, as the strain gage deforms, Equation 8.21 can be used to relate this voltage change $\Delta V_o$ to the change in resistance $\Delta R_1$. To find this relation, we can replace $R_1$ by its new resistance $R_1 + \Delta R_1$ and $V_o$ by the output deflection voltage $\Delta V_o$. Then Equation 8.21 gives

$$\frac{\Delta V_o}{V_{\text{ex}}} = \frac{R_1 + \Delta R_1}{R_1 + \Delta R_1 + R_4} - \frac{R_2}{R_2 + R_3} \tag{8.22}$$

Rearranging this equation gives us the desired relation between the change in resistance and the measured output voltage:

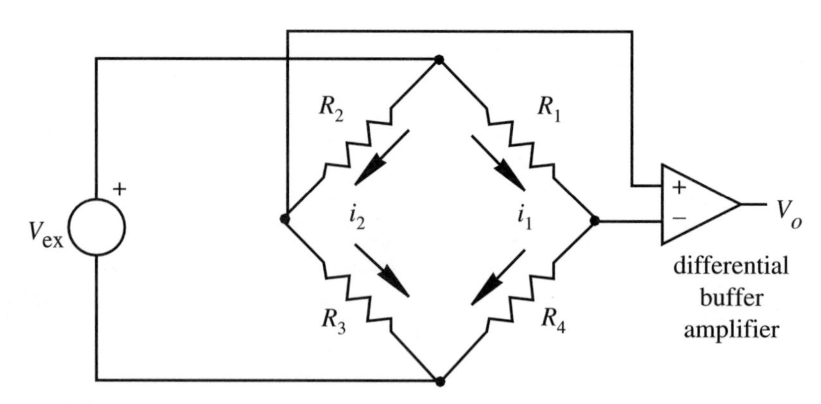


**FIGURE 8.21**
Dynamic unbalanced bridge circuit.

$$\frac{\Delta R_1}{R_1} = \frac{\frac{R_4}{R_1}\left(\frac{\Delta V_o}{V_{\text{ex}}} + \frac{R_2}{R_2 + R_3}\right)}{\left(1 - \frac{\Delta V_o}{V_{\text{ex}}} - \frac{R_2}{R_2 + R_3}\right)} - 1 \tag{8.23}$$

By measuring the change in the output voltage $\Delta V_o$, we can determine the gage resistance change $\Delta R_1$ from Equation 8.23 and therefore the gage strain from Equation 8.13.

▼▼▼ ***CLASS DISCUSSION ITEM 8.9. Wheatstone Excitation Voltage.*** What undesirable effects can the magnitude of the excitation voltage have on the resistance change measurements we are making with a Wheatstone bridge?

Figure 8.22 illustrates the effects of leadwires when using a strain gage located far from the bridge circuit. Figure 8.22a illustrates a 2-wire connection from a strain gage to a bridge circuit. With this configuration, each of the leadwire resistances $R'$ add to the resistance of the strain gage branch of the bridge. The problem with this is that if the leadwire temperature changes, it will cause changes in the resistance of the bridge branch. This effect can be substantial if the leadwires are long and extend through environments where the temperature changes. Figure 8.22b illustrates a 3-wire connection that solves this problem. With this configuration, equal leadwire resistances are added to adjacent branches in the bridge so the effects of changes in the leadwire resistances offset each other. The third leadwire is connected to the high-input impedance voltage measuring circuit, and its resistance has a negligible effect since it carries negligible current.

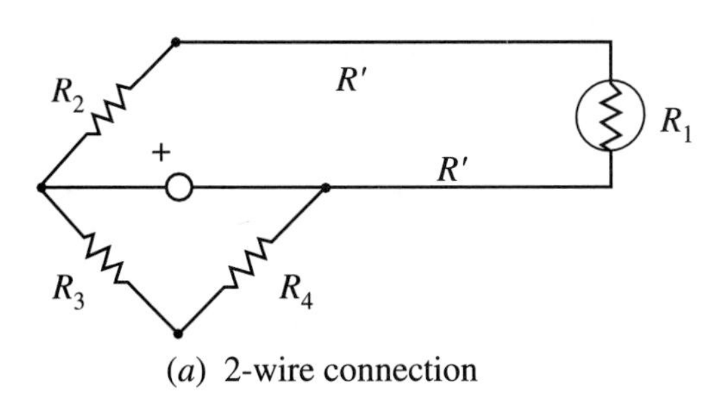


(*a*) 2-wire connection

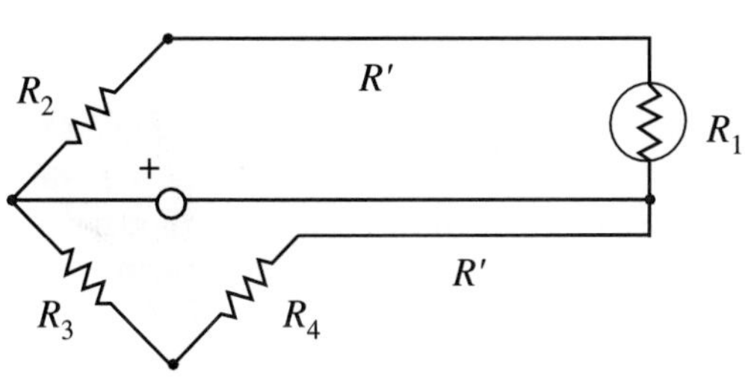


(*b*) 3-wire connection

**FIGURE 8.22**
Leadwire effects in 1/4 bridge circuits.

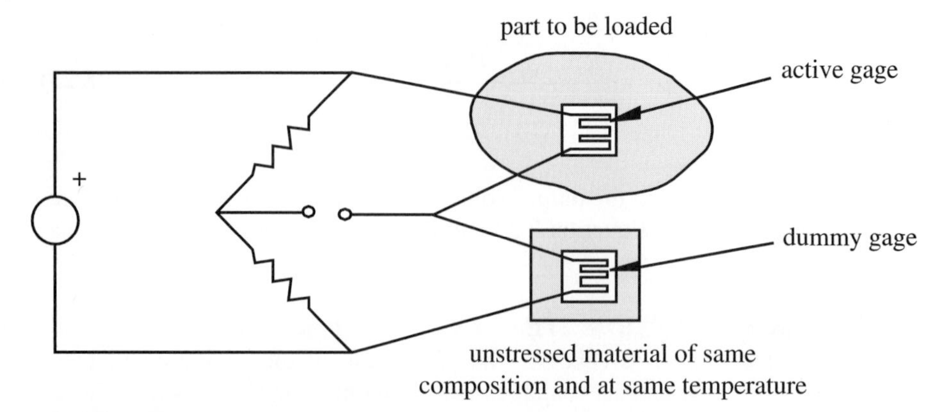


**FIGURE 8.23**
Temperature compensation with dummy gage in 1/2 bridge.

▼▼▼ ***CLASS DISCUSSION ITEM 8.10. Bridge Resistances in 3-Wire Bridges.*** What must be true about the bridge resistance $R_4$ for the 3-wire configuration shown in Figure 8.22b to be effective? (Hint: Use Equation 8.21.) Does your conclusion hold true when the gage experiences strain and its resistance changes? Is this a problem?

In addition to temperature effects in leadwires, temperature changes in the actual strain gage can cause significant changes in resistance, which would lead to erroneous readings. A convenient method for eliminating this effect is to use a 1/2 bridge circuit where two of the four bridge legs contain strain gages (see Figure 8.23). The gage in the top branch is the active gage used to measure surface strains on a component to be loaded. The second "dummy" gage is mounted to an unloaded sample of material identical in composition to the component. If this sample is kept at the same temperature as the component by keeping it in close proximity, the resistance changes in the two gages due to temperature will cancel since they are in adjacent branches of the bridge circuit. Therefore, the bridge will generate an unbalanced voltage only in response to strain in the active gage.

### 8.3.3 Measuring Different States of Stress with Strain Gages

Mechanical components may have complex shapes and are often subjected to complex loading conditions. In these cases, it is difficult to predict the orientation of principal stresses at arbitrary points on the component. However, with some geometries and loading conditions the principal axes are known and measuring the state of stress is easier.

If a component is loaded uniaxially (i.e., loaded only in one direction in tension or compression), the state of stress in the component can be determined with a single gage mounted in the direction of the load. Figure 8.24 illustrates a bar in tension and the associated state of stress. By measuring the strain $\varepsilon_x$, the stress is known from Hooke's Law (see Appendix 2) to be

$$\sigma_x = E\varepsilon_x \tag{8.24}$$

The axial stress in the bar $\sigma_x$ is given by

$$\sigma_x = \frac{P}{A} \tag{8.25}$$

where $A$ is the bar's cross-sectional area. Therefore, the force $P$ applied to the bar can be determined from the strain gage measurement:

$$P = AE\varepsilon_x \tag{8.26}$$

If a component is known to be loaded biaxially (i.e., loaded in two orthogonal directions in tension or compression), the state of stress in the component can be determined with two gages aligned with the stress directions. Figure 8.25 illustrates a pressurized tank and the associated state of stress. By measuring the strains $\varepsilon_x$ and $\varepsilon_y$, the stresses in the tank shell can be determined from Hooke's Law generalized to two dimensions:

$$\varepsilon_x = \frac{\sigma_x}{E} - \nu\frac{\sigma_y}{E} \tag{8.27}$$

$$\varepsilon_y = \frac{\sigma_y}{E} - \nu\frac{\sigma_x}{E} \tag{8.28}$$

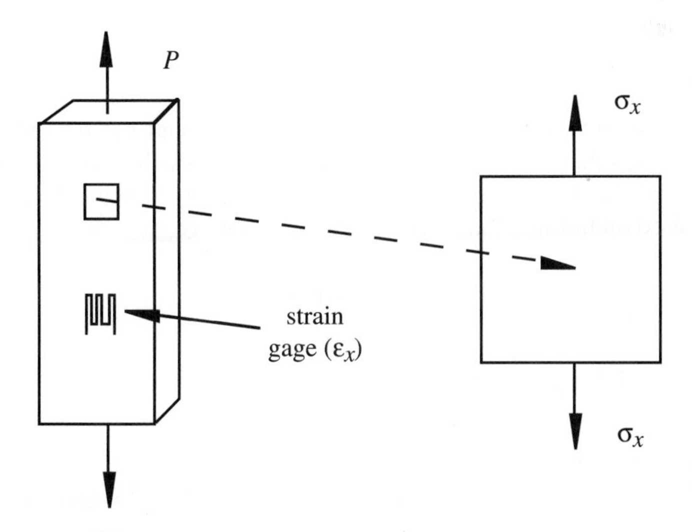


**FIGURE 8.24**
Bar under uniaxial stress.

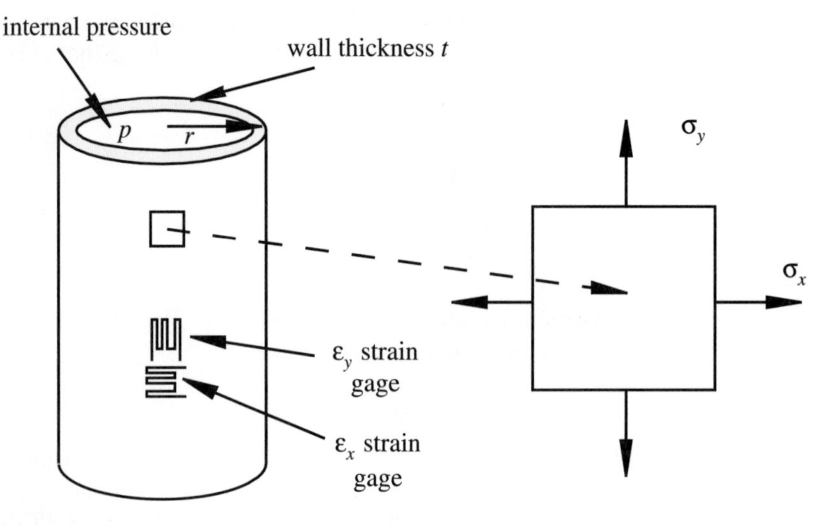


**FIGURE 8.25**
Biaxial stress in a long thin-walled pressure vessel.

Solving for the stress components gives

$$\sigma_x = \frac{E}{1-\nu^2}(\varepsilon_x + \nu\varepsilon_y) \tag{8.29}$$

$$\sigma_y = \frac{E}{1-\nu^2}(\varepsilon_y + \nu\varepsilon_x) \tag{8.30}$$

For a thin-walled pressure vessel (i.e., $t/r < 1/10$), the theoretical stresses are given by

$$\sigma_x = \frac{pr}{t} \qquad \sigma_y = \frac{pr}{2t} \tag{8.31}$$

where $p$ is the internal pressure, $t$ is the wall thickness, and $r$ is the radius of the vessel. $\sigma_x$ is called the transverse or hoop stress, and $\sigma_y$ is called the axial or longitudinal stress. Either Equation 8.29 or 8.30 can be used to compute the pressure in the vessel based on the strain gage measurements yielding

$$p = \frac{t\sigma_x}{r} = \frac{tE}{r(1-\nu^2)}(\varepsilon_x + \nu\varepsilon_y) \tag{8.32}$$

or

$$p = \frac{2t\sigma_y}{r} = \frac{2tE}{r(1-\nu^2)}(\varepsilon_y + \nu\varepsilon_x) \tag{8.33}$$

Either expression would yield the correct pressure value for an ideal thin-walled vessel and error-free measurements. In this example, the strain gage is serving as a pressure transducer.

For uniaxial and biaxial loading, we already know the directions of principal stresses in the component; hence we only need one or two gages respectively to de-

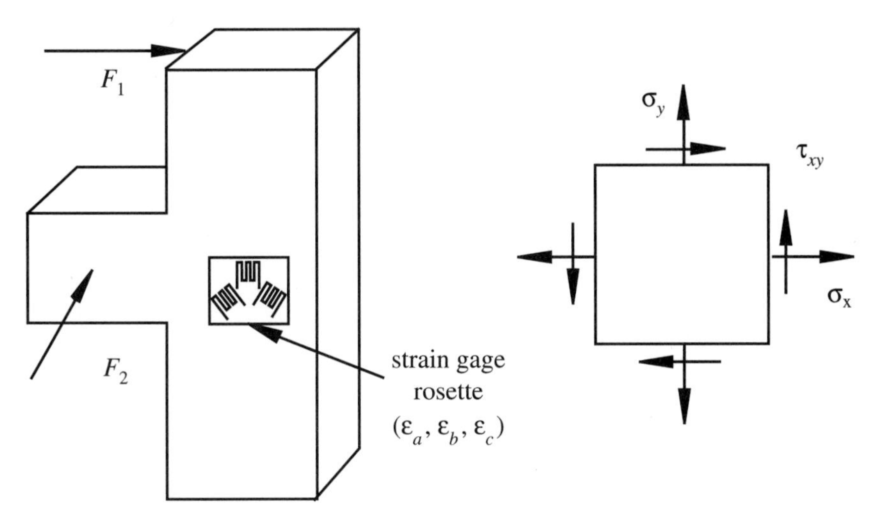


**FIGURE 8.26**
General state of planar stress on the surface of a component.

termine the stress magnitudes. However, when the loading is more complex or when the geometry is more complex, which is usually the case in mechanical design, we have to use three gages mounted in three different directions as illustrated in Figure 8.26.

An assembly of strain gages is referred to as a ***strain gage rosette***. There is a wide variety of commercially available rosettes with two or more grid patterns accurately oriented on a single backing in close proximity. An assortment of rosettes and single element gages illustrating the variety of shapes and sizes is shown in Figure 8.27.

The most common rosette patterns for measuring a general state of planar stress are illustrated in Figure 8.28, where the grids are shown as single lines labeled by letters. Of these, the rectangular strain gage is the most common configuration, where the strain gages are positioned 45° apart (see Figure 8.29). Figure 8.30 shows several commercially available 3-gage rosettes.

**FIGURE 8.27**
Assortment of different strain gage and rosette configurations. *(Courtesy of Measurements Group Inc., Raleigh, NC)*

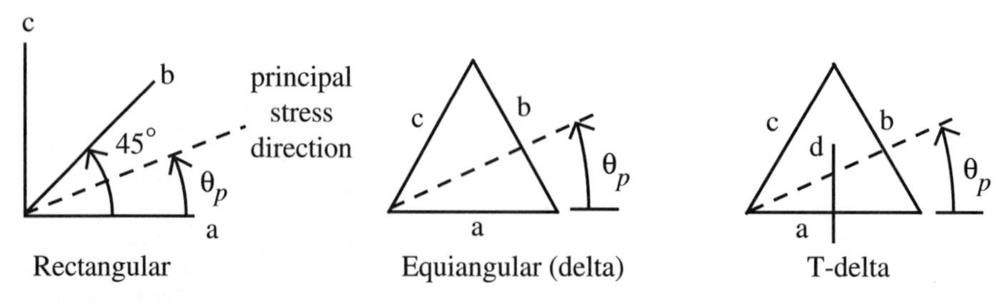


**FIGURE 8.28**
Most common strain gage rosette configurations.

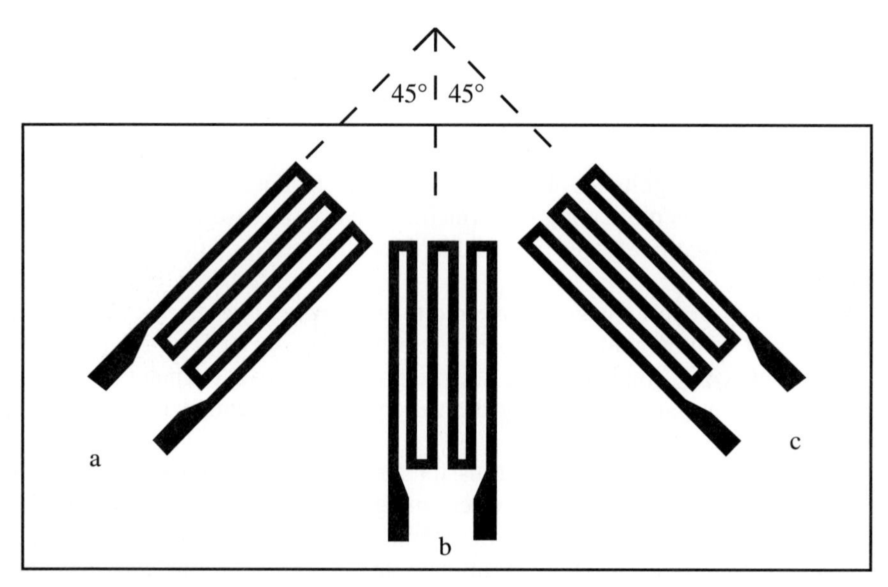


**FIGURE 8.29**
Rectangular strain gage rosette.

Using principles of solid mechanics, the magnitude and direction of the principal stresses can be determined directly from three simultaneous strain measurements using any of the rosette patterns shown in Figure 8.28. The magnitude and direction of the principal stresses for the rectangular rosette are

$$\sigma_{\text{max, min}} = \frac{E}{2}\left[\frac{\varepsilon_a + \varepsilon_c}{1-\nu} \pm \frac{1}{1+\nu}\sqrt{2(\varepsilon_a - \varepsilon_b)^2 + 2(\varepsilon_b - \varepsilon_c)^2}\right] \tag{8.34}$$

Three-Element 60°(Delta) Rosettes

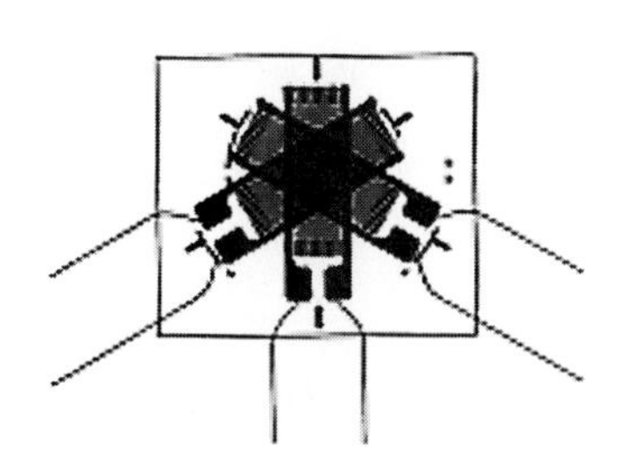

Stacked Type 250WY

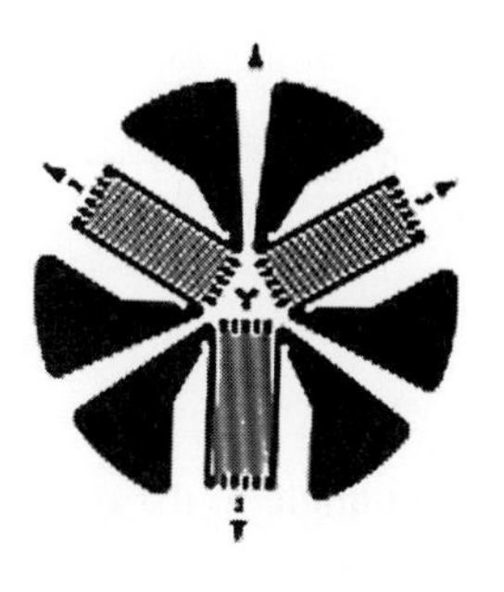

Single-Plane Type 250YA

Three-Element 45°(Rectangular) Rosettes

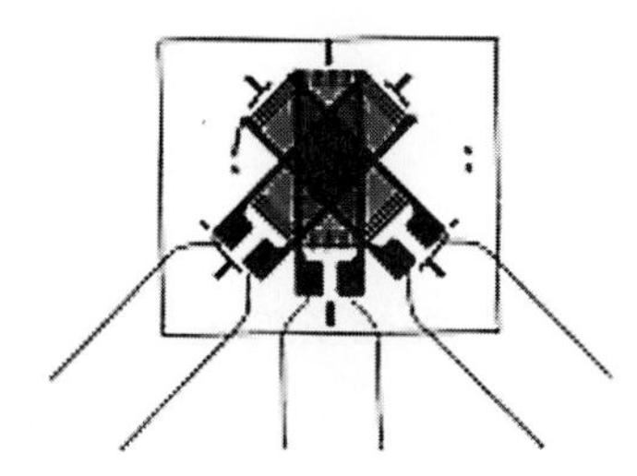

Stacked Type 250WR

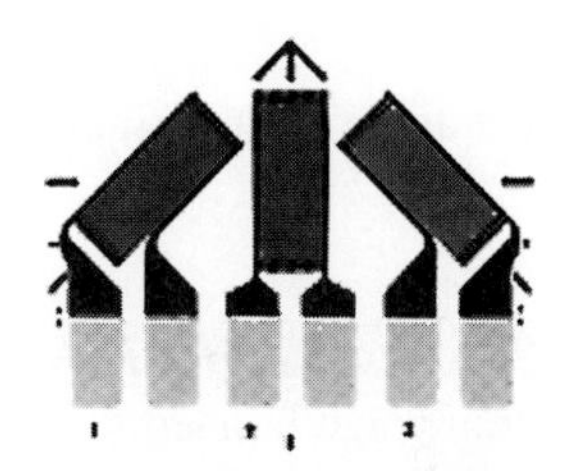

Single-Plane Type 250UR

**FIGURE 8.30**
Various 3-gage commercial rosettes. *(Courtesy of Measurements Group Inc., Raleigh, NC)*

$$\tau_{\max} = \frac{E}{2(1+\nu)}\sqrt{2(\varepsilon_a - \varepsilon_b)^2 + 2(\varepsilon_b - \varepsilon_c)^2} \tag{8.35}$$

$$\tan 2\theta_p = \frac{2\varepsilon_b - \varepsilon_a - \varepsilon_c}{\varepsilon_a - \varepsilon_c} \tag{8.36}$$

where $\varepsilon_a$, $\varepsilon_b$, and $\varepsilon_c$ are the strains in each of the rosette gages. The correct angle to the principal axis is $0 < \theta_p < 90°$ for $\varepsilon_b > (\varepsilon_a + \varepsilon_c)/2$. The relations for the equiangular (delta) rosette are

$$\sigma_{\max,\min} = \frac{E}{3}\left[\frac{\varepsilon_a+\varepsilon_c+\varepsilon_c}{1-\nu} \pm \frac{1}{1+\nu}\sqrt{2(\varepsilon_a-\varepsilon_b)^2+2(\varepsilon_b-\varepsilon_c)^2+2(\varepsilon_c-\varepsilon_a)^2}\right] \tag{8.37}$$

$$\tau_{\max} = \frac{E}{3(1+\nu)}\sqrt{2(\varepsilon_a-\varepsilon_b)^2+2(\varepsilon_b-\varepsilon_c)^2+2(\varepsilon_c-\varepsilon_a)^2} \tag{8.38}$$

$$\tan 2\theta_p = \frac{\sqrt{3}(\varepsilon_c-\varepsilon_b)}{2\varepsilon_a-\varepsilon_b-\varepsilon_c} \tag{8.39}$$

The correct angle to the principal axis is $0 < \theta_p < 90°$ for $\varepsilon_c > \varepsilon_b$. The relations for the T-delta rosette, which has four gages, are

$$\sigma_{\max,\min} = \frac{E}{2}\left[\frac{\varepsilon_a+\varepsilon_d}{1-\nu} \pm \frac{1}{1+\nu}\sqrt{(\varepsilon_a-\varepsilon_d)^2+\frac{4}{3}(\varepsilon_b-\varepsilon_c)^2}\right] \tag{8.40}$$

$$\tau_{\max} = \frac{E}{2(1+\nu)}\sqrt{(\varepsilon_a-\varepsilon_d)^2+\frac{4}{3}(\varepsilon_b-\varepsilon_c)^2} \tag{8.41}$$

$$\tan 2\theta_p = \frac{2(\varepsilon_c-\varepsilon_b)}{\sqrt{3}(\varepsilon_a-\varepsilon_d)} \tag{8.42}$$

The correct angle to the principal axis is $0 < \theta_p < 90°$ for $\varepsilon_c > \varepsilon_b$.

▼▼▼ ***CLASS DISCUSSION ITEM 8.11. Strain Gage Bond Effects.*** Does a strain gage bonded to a component have any influence on the stresses that are being measured? If so, how? In what cases are these effects significant?

### 8.3.4 Force Measurement with Load Cells

A ***load cell*** is a sensor used to measure a force. It contains an internal flexural element, usually with several strain gages mounted to its surface. The flexural element's shape is designed so that the strain gage outputs can be related to the applied force. The load cell is usually connected to a bridge circuit to yield a voltage proportional to the load. Two commercial load cells, which are used to measure uniaxial force, are shown in Figure 8.31. An example of the application of load cells is in commercially available laboratory materials testing machines for measuring forces applied to a test specimen. Load cells are used in weight scales, and they are sometimes included as integral parts of mechanical structures to monitor forces in the structures.

(*a*) *Courtesy of MTS Systems Corp., Minneapolis, MN*

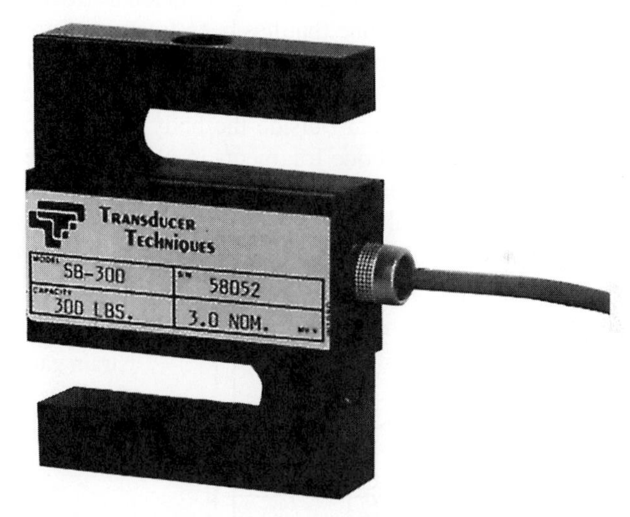


(*b*) *Courtesy of Transducer Techniques, Temecula, CA*

**FIGURE 8.31**
Typical axial load cells.

---

▼▼ ***DESIGN EXAMPLE 8.1. A Strain Gage Load Cell for an Exteriorized Skeletal Fixator.*** Orthopedic biomechanics is a field that includes the analysis of the loading of the skeletal system, engineering approaches to understanding the mechanical properties of biological tissues, and the choice of appropriate systems to replace the tissues when they fail. It is a major growth industry in the health care field and offers many opportunities to engineers interested in problems that have a biological or medical flavor. There are many interesting engineering design problems associated with the choice of implantable materials, the long-term strength of the materials in the hostile environment of the body, and the mechanical design of engineered materials that must attach securely to biological materials. Replacement joints are among the most successful designs to date, and you probably know someone who has had a joint replaced with a stainless steel prosthesis.

The authors have been involved in a number of exciting bioengineering research projects. One of the most interesting involved the analysis of loading in the limbs in patients who had suffered severely broken bones. When one suffers a multiple fracture in a leg bone, the application of a simple plaster or fiberglass cast is not sufficient to allow the bone to remodel and heal. Bone is a living tissue and is constantly being remodeled or replaced. In fact, stress is important in initiating and maintaining the healing process. A case in point: astronauts in a weightless space environment for extended periods exhibit bone loss due to the lack of a gravitational stress.

In a severely broken bone, one that has been fractured into a number of pieces, an interesting biomechanical invention is often used. It is known as an exteriorized skeletal fixator. As illustrated in the figure below, it consists of a structural bar outside the body fastened to stainless steel pins that pierce and hold bone fragments in place until the bone tissue heals. The engineered structure therefore carries the load of the body when the animal or person walks.

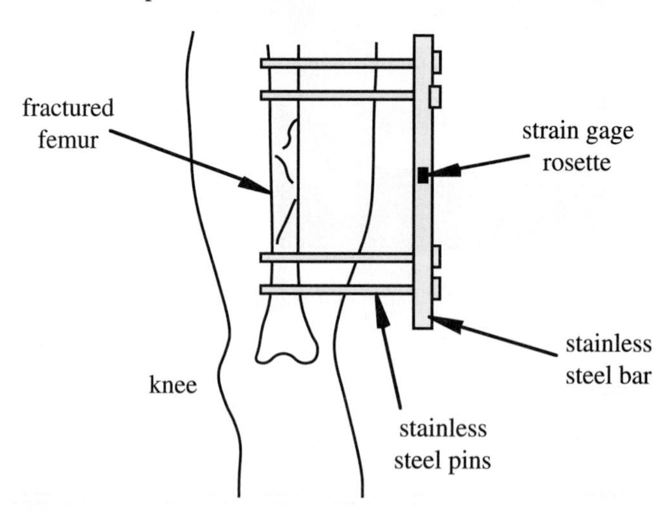


An important question involves the loading of the fixator while the patient walks. This information is necessary in order to size the fixator so that it does not deform too much and prevent bone healing. One of the subtle features of the healing process is that a very small amount of relative motion is necessary for healing; if all of the stress is removed from the bone and no relative motion occurs between the fragments, the bone will not heal. But too much relative motion hinders healing too.

To measure the magnitude of loading for a verification study, we need to design a load cell as part of a skeletal fixator to monitor the complex loading profiles that occur during walking. We are interested in measuring the load profiles in the cylindrical stainless steel bar that supports the broken leg. Loads subjected to the bone during walking are fairly complex and will consists of axial stress, bending stress, and torsional stress.

Our objective is to design a load cell that will allow us to easily and reliably determine the axial, bending, and torsional loads while the subject is walking. We can do this by mounting three rectangular strain gage rosettes on the bar 90° apart. The center (*b*) gage of each rosette (see Figure 8.29) will be aligned with the axis of the bar.

If the principal stresses are calculated at each rosette, the axial, bending, and torsion loads can be determined. The principal stresses in the axial direction can be used to determine the axial load and bending loads in two directions. The maximum shear stresses at each rosette can be averaged to determine the torsional load. Thus the nine strain gages (from three rosettes) will yield simultaneous measurements of axial load, bending load, and torsional load.

The nine strain gages now must be connected to a strain gage bridge in a quarter-gage configuration. A bridge such as the Measurements Group 2100 will provide up to 10 bridges for simultaneous measurement of unbalanced strain. The analog output of the bridge then must be digitized at an appropriate sampling frequency to yield the real-time stress profiles in the biomechanical load cell.

## 8.4 TEMPERATURE MEASUREMENT

Since temperature is such an important variable in many engineering systems, an engineer should be familiar with the basic methods of measuring it. Temperature sensors appear in buildings, chemical process plants, engines, transportation vehicles, appliances, computers, and many other devices that require the monitoring and control of temperature.

Since many physical phenomena depend on temperature, we can often use this dependence to indirectly measure temperature by measuring quantities such as pressure, volume, electrical resistance, and strain and converting the values using physical relations.

The temperature scales used to express temperature are:

- Celsius (°C): Common SI unit of relative temperature.
- Kelvin (K): Standard SI unit of absolute thermodynamic temperature. Note the absence of the degree symbol.

- Fahrenheit (°F): English System unit of relative temperature.
- Rankine (°R): English System unit of absolute thermodynamic temperature.

The relationships between these scales are summarized below:

$$T_C = T_K - 273.15 \tag{8.43}$$

$$T_F = (9/5)T_C + 32 \tag{8.44}$$

$$T_R = T_F + 459.67 \tag{8.45}$$

where $T_C$ is temperature in degrees Celsius, $T_K$ is temperature in Kelvin, $T_F$ is temperature in degrees Fahrenheit, and $T_R$ is temperature in degrees Rankine.

### 8.4.1 Liquid-in-Glass Thermometer

A simple nonelectrical temperature-measuring device is the liquid-in-glass thermometer. It typically uses alcohol or mercury as the working fluid, which expands and contracts relative to the glass container. The upper range is usually on the order of 600°F. When taking readings in a liquid, the depth of immersion is important as it can result in different readings. Since readings are taken visually and since there can be a meniscus at the top of the working fluid, readings must be taken carefully and consistently.

### 8.4.2 Bimetallic Strip

Another nonelectrical temperature-measuring device used in simple control systems is the bimetallic strip. As illustrated in Figure 8.32, it is composed of two or more metal layers having different coefficients of thermal expansion. Since these layers are permanently bonded together, the structure will deform when the temperature changes. This is due to the difference in the thermal expansions of the two metal layers. The deflection $\delta$ can be related to the temperature of the strip. Bimetallic strips are used in household and industrial thermostats where the mechanical motion of the strip makes or breaks an electrical contact to turn a heating or cooling system on or off.

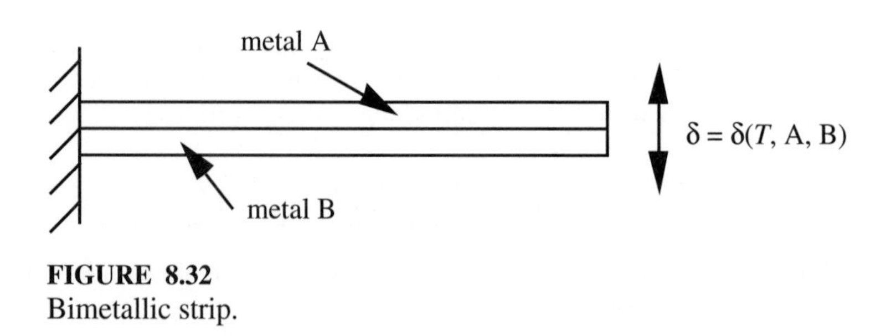


**FIGURE 8.32**
Bimetallic strip.

### 8.4.3 Electrical Resistance Thermometer

A ***resistance temperature device***, called an ***RTD***, is constructed of metallic wire wound around a ceramic or glass core and hermetically sealed. The resistance of the metallic wire increases with temperature. The resistance-temperature relationship is usually approximated with the following linear expression:

$$R = R_0[1 + \alpha(T - T_0)] \tag{8.46}$$

where $T_0$ is a reference temperature, $R_0$ is the resistance at the reference temperature, and $\alpha$ is a calibration constant. The sensitivity ($dR/dT$) is $R_0\alpha$. The reference temperature is usually the ice point of water (0°C). The most common metal used in RTDs is platinum because of its high melting point, resistance to oxidation, predictable temperature characteristics, and stable calibration values. The operating range for a typical platinum RTD is –220°C to 750°C. Lower cost nickel and copper types are also available, but they have narrower operating ranges.

A ***thermistor*** is a semiconductor device whose resistance changes exponentially with temperature. Its resistance-temperature relationship is usually expressed in the form

$$R = R_0 e^{\left[\beta\left(\frac{1}{T} - \frac{1}{T_0}\right)\right]} \tag{8.47}$$

where $T_0$ is a reference temperature, $R_0$ is the resistance at the reference temperature, and $\beta$ is a calibration constant called the characteristic temperature of the material. A well-calibrated thermistor can be accurate to within 0.01°C or better, which is better than typical RTD accuracies. However, thermistors have much narrower operating ranges than RTDs.

### 8.4.4 Thermocouple

Two dissimilar metals in contact (see Figure 8.33) form a thermoelectric junction that produces a voltage proportional to the temperature of the junction. This is known as the ***Seebeck effect***.

Since an electrical circuit must form a closed loop, thermoelectric junctions occur in pairs, resulting in a device called a ***thermocouple***. We can represent a thermoelectric circuit containing two junctions as illustrated in Figure 8.34. Here we have wires of metals A and B forming junctions at different temperatures $T_1$ and $T_2$, resulting in a potential $V$ that can be measured. The thermocouple voltage $V$ depends on the metal properties of A and B and the difference between the junction

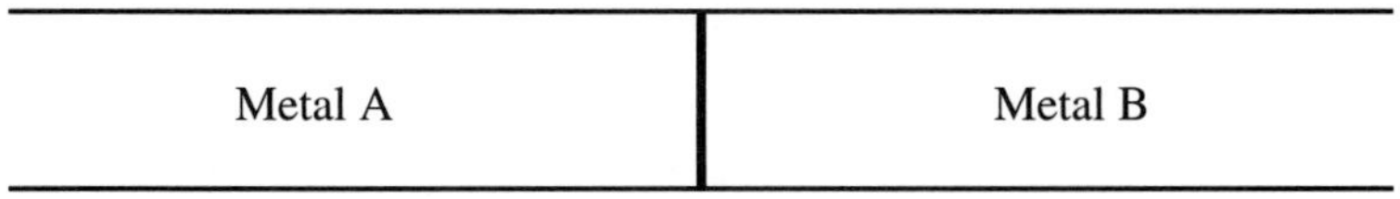

| Metal A | Metal B |
|---|---|

**FIGURE 8.33**
Thermoelectric junction.

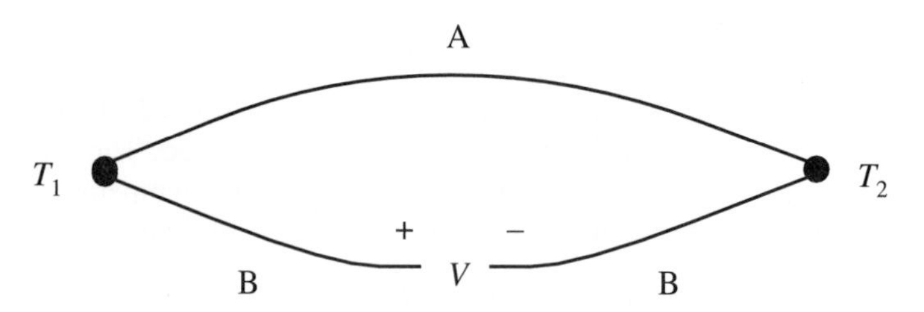


**FIGURE 8.34**
Thermocouple circuit.

temperatures $T_1$ and $T_2$. The thermocouple voltage is directly proportional to the junction temperature difference:

$$V = \alpha(T_1 - T_2) \tag{8.48}$$

where $\alpha$ is called the ***Seebeck coefficient***. As we will see later in this section, the relationship between voltage and temperature difference is not exactly linear. However, over a small temperature range, $\alpha$ is nearly constant.

There are secondary thermoelectric effects known as the Peltier and Thompson effects, which are associated with current flow in the thermocouple circuit, but these are usually negligible in measurement systems when compared to the Seebeck effect. When the current is large in a thermocouple circuit, these other effects become significant. The Peltier effect relates the current flow to heat flow into one junction and out of the other. This effect forms the basis of a thermoelectric refrigerator.

In order to properly design systems using thermocouples for temperature measurement, it is necessary to understand the basic laws that govern their application. The five basic laws of thermocouple behavior along with figures illustrating their meaning follow.

1. *Law of Leadwire Temperatures.* The thermoelectric voltage due to two junctions in a circuit consisting of two different conducting metals depends only on the junction temperatures $T_1$ and $T_2$. As illustrated in Figure 8.35, the temperature environment of the leads away from the junctions ($T_3$, $T_4$, $T_5$) does not influence the measured voltage. Therefore, we do not need to be concerned about shielding the leadwires from environmental conditions.

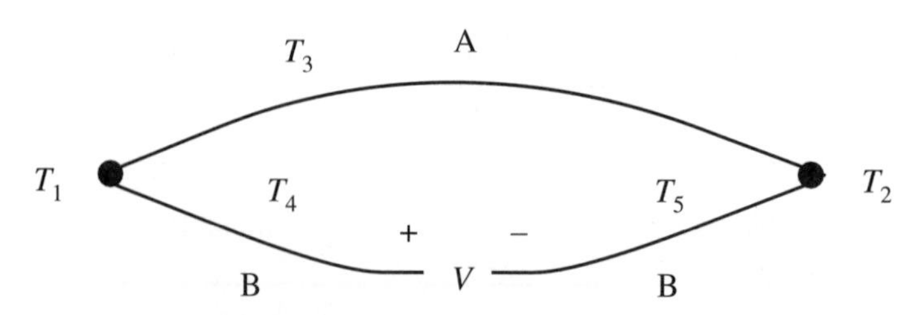


**FIGURE 8.35**
Law of leadwire temperatures.

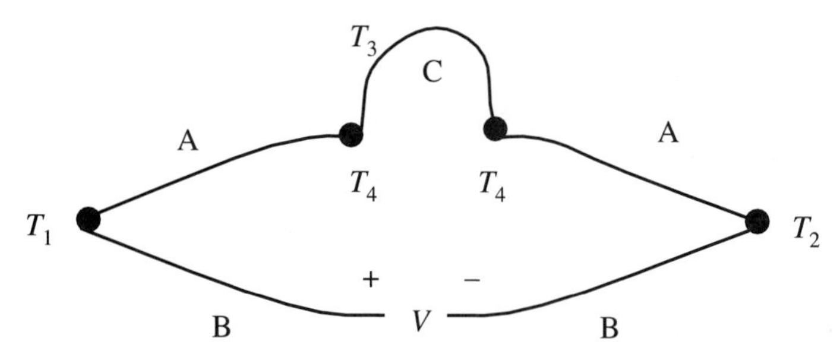


**FIGURE 8.36**
Law of intermediate leadwire metals.

2. *Law of Intermediate Leadwire Metals.* As illustrated in Figure 8.36, a third metal C introduced in the circuit comprising the thermocouple has no influence on the resulting voltage as long as the temperatures of the two new junctions (A-C and C-A) are the same ($T_4$). As a result of this law, a voltage measurement device that creates two new junctions can be inserted into the thermocouple circuit without altering the resulting voltage.
3. *Law of Intermediate Junction Metals.* As illustrated in Figure 8.37, if a third metal is used within a junction creating two new junctions (A-C and C-B), the measured voltage will not be affected as long as the two new junctions are at the same temperature ($T_1 = T_3$). Therefore, although soldered or brazed joints introduce thermojunctions, they have no resulting effect on the measured voltage. If $T_1 \neq T_3$, the sensed temperature at C will be the average of the two temperatures ($(T_1 + T_3)/2$).
4. *Law of Intermediate Temperatures.* Junction pairs at $T_1$ and $T_3$ produce the same voltage as two sets of junction pairs spanning the same temperature range ($T_1$ to $T_2$ and $T_2$ to $T_3$); therefore, as illustrated in Figure 8.38,

$$V_{13} = V_{12} + V_{23} \tag{8.49}$$

This result supports the use of a reference junction to allow accurate measurement of an unknown temperature based on a fixed reference temperature.

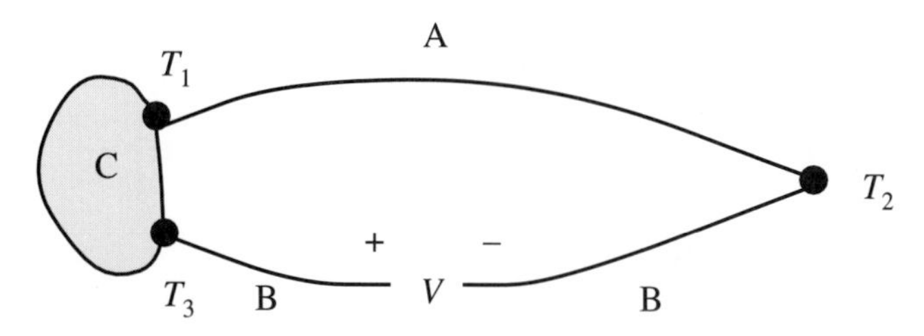


**FIGURE 8.37**
Law of intermediate junction metals.

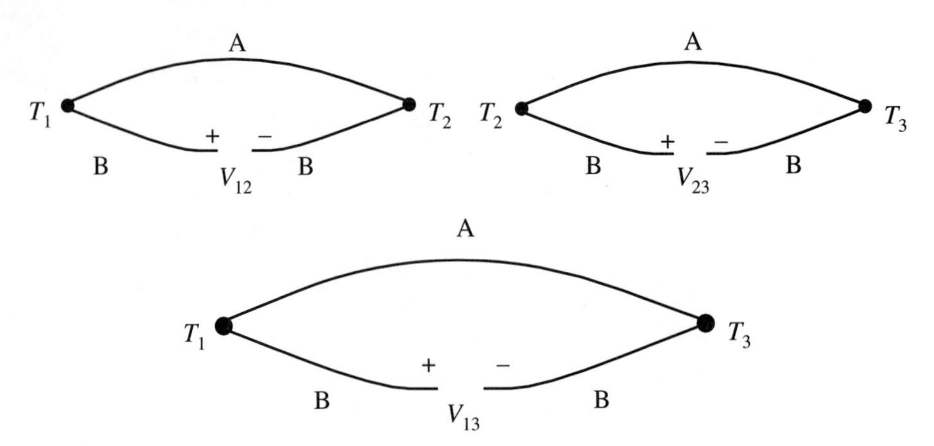


**FIGURE 8.38**
Law of intermediate temperatures.

5. *Law of Intermediate Metals.* As illustrated in Figure 8.39, the voltage produced by two metals A and B is the same as the sum of the voltages produced by each metal (A and B) relative to a third metal C. This result supports the use of a reference metal to be used as a basis to calibrate all other metals.

A standard configuration for thermocouple measurements is shown in Figure 8.40. It consists of wires of two metals, A and B, attached to a voltage-measuring device with terminals made of metal C. The reference junction is used to establish a temperature reference for one of the junctions to ensure accurate temperature measurements at the other junction relative to the reference. A convenient refer-

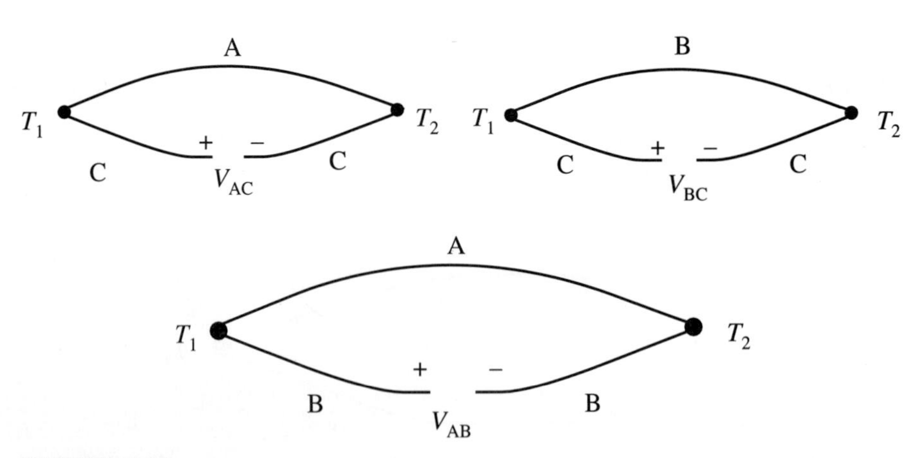


**FIGURE 8.39**
Law of intermediate metals.

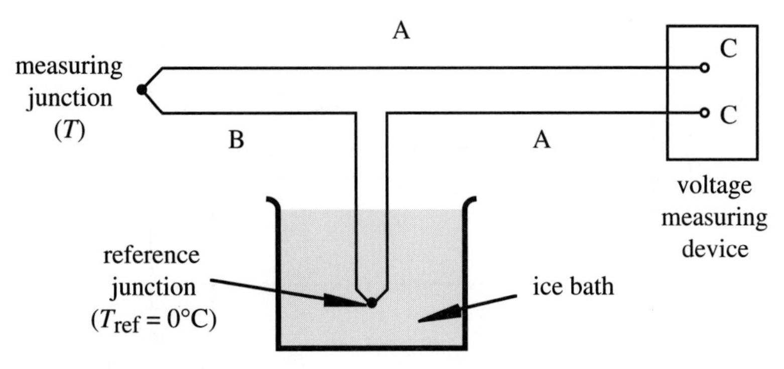


**FIGURE 8.40**
Standard thermocouple configuration.

ence temperature is 0°C since this temperature can be accurately established and maintained with a distilled water ice bath (i.e., ice-water mixture). If the terminals of the voltage measuring device are at the same temperature, the law of intermediate leadwire metals ensures that the measuring device terminal metal C has no effect on the measurement. For a given pair of thermocouple metals and a reference temperature, a standard reference table can be compiled for converting voltage measurements to temperatures.

An important alternative to using an ice bath is a semiconductor reference, which electrically establishes the reference temperature based on solid state physics principles. These reference devices are usually included in thermocouple instrumentation to eliminate the need for an external reference temperature.

Figure 8.41 illustrates a two-reference junction configuration, which allows independent choice of the leadwire metal. Copper is a good choice for this metal

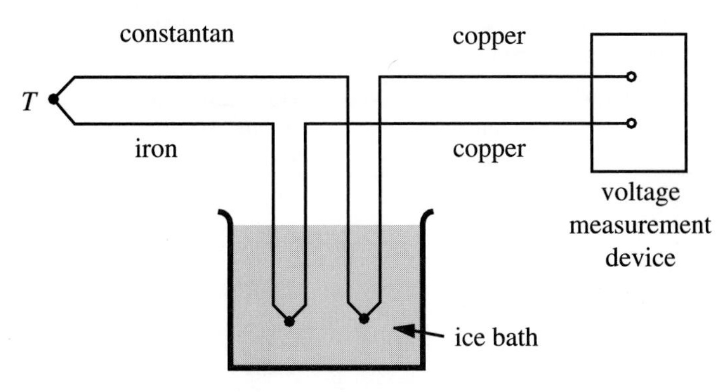


**FIGURE 8.41**
Attaching leadwires of selected metal.

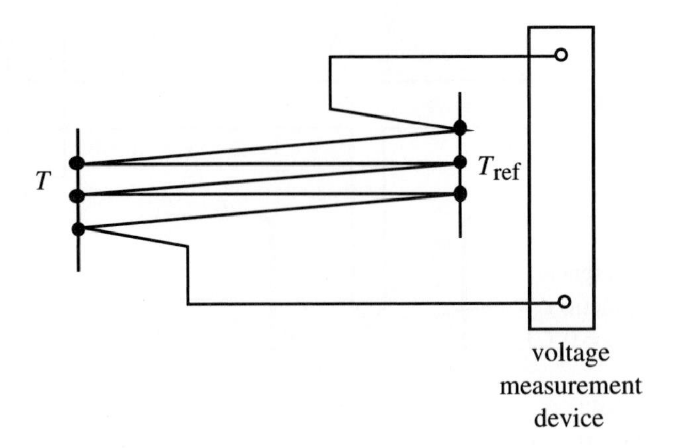


**FIGURE 8.42**
Thermopile.

since copper leadwires are readily available and since no junctions are introduced at the voltmeter connections, which are usually copper.

Figure 8.42 illustrates the configuration for a ***thermopile,*** which combines $N$ pairs of junctions resulting in a voltage signal multiplied by a factor $N$. In the example shown, the resulting voltage would be three times that of a single thermocouple.

---

▼ ***EXAMPLE 8.2. Thermocouple Configuration with Nonstandard Reference.*** A standard 2-junction thermocouple configuration is being used to measure the temperature in a wind tunnel. The reference junction is being held at a constant temperature of 10°C. We only have a thermocouple table referenced to 0°C. A portion of the table is listed below. We want to determine what the output voltage would be when the measuring junction is exposed to an air temperature of 100°C.

| Junction temperature (°C) | Output voltage (mV) |
|---|---|
| 0 | 0 |
| 10 | 0.507 |
| 20 | 1.019 |
| 30 | 1.536 |
| 40 | 2.058 |

*(continued)*

| Junction temperature (°C) | Output voltage (mV) |
|---|---|
| 50 | 2.585 |
| 60 | 3.115 |
| 70 | 3.649 |
| 80 | 4.186 |
| 90 | 4.725 |
| 100 | 5.268 |

By applying the law of intermediate temperature for this example, we can write

$$V_{100/0} = V_{100/10} + V_{10/0}$$

We wish to find $V_{100/10}$, the voltage measured for a temperature of 100°C relative to a reference junction at 10°C. We can get the other voltages, $V_{100/0}$ and $V_{10/0}$, in the equation from the table since they are both referenced to 0°C. Therefore,

$$V_{100/10} = V_{100/0} - V_{10/0} = (5.268 - 0.507)\ \text{mV} = 4.761\ \text{mV}$$

---

The six most commonly used thermocouple pairs are denoted by the letters E, J, K, R, S, and T. The 0°C reference junction calibration for each of the types is nonlinear and can be approximated with a polynomial. The metals in the junction pair, the thermoelectric polarity, the commonly used color code, the operating range, the accuracy, and the polynomial order and coefficients are shown for each type in Table 8.2. The general form for the polynomial using the coefficients in the table is

$$\begin{aligned} T &= \sum_{i=0}^{9} c_i V^i \\ &= c_0 + c_1 V + c_2 V^2 + c_3 V^3 + c_4 V^4 + c_5 V^5 + c_6 V^6 + c_7 V^7 + c_8 V^8 + c_9 V^9 \end{aligned} \tag{8.50}$$

where $V$ is the thermoelectric voltage measured in volts and $T$ is the measuring junction temperature in °C, assuming a 0°C reference junction. Figure 8.43 shows the sensitivity curves for some commercially available thermocouples. Even though we use a ninth order polynomial to represent the temperature voltage relation, providing an extremely close fit, the relationship is close to linear as predicted by the Seebeck effect.

**TABLE 8.2**
**Thermocouple data**

| | Type E | Type J | Type K | Type R | Type S | Type T |
|---|---|---|---|---|---|---|
| **Metal pair** | chromel (+) and constantan (–) | iron (+) and constantan (–) | chromel (+) and alumel (–) | 87% platinum, 13% rhodium (+) and platinum (–) | 90% platinum, 10% rhodium (+) and platinum (–) | copper (+) and constantan (–) |
| **Color code** | purple | black | yellow | green | green | blue |
| **Operating range** | –100°C to 1000°C | 0°C to 760°C | 0°C to 1370°C | 0°C to 1000°C | 0°C to 1750°C | –160°C to 400°C |
| **Accuracy** | ±0.5°C | ±0.1°C | ±0.7°C | ±0.5°C | ±0.1°C | ±0.5°C |
| **Approximate sensitivity (mV/°C)** | 0.079 | 0.054 | 0.042 | 0.012 | 0.011 | 0.049 |
| **Polynomial order** | 9 | 5 | 8 | 8 | 9 | 7 |
| $c_0$ | 0.104967 | –0.0488683 | 0.226585 | 0.263633 | 0.927763 | 0.100861 |
| $c_1$ | 17,189.5 | 19,873.1 | 24,152.1 | 179,075. | 169,527. | 257,27.9 |
| $c_2$ | –282,639. | –218,615. | 67,233.4 | $-4.88403 \times 10^{7}$ | $-3.15684 \times 10^{7}$ | –767,346. |
| $c_3$ | $1.26953 \times 10^{7}$ | $1.15692 \times 10^{7}$ | $2.21034 \times 10^{6}$ | $1.90002 \times 10^{10}$ | $8.99073 \times 10^{9}$ | $7.80256 \times 10^{7}$ |
| $c_4$ | $-4.48703 \times 10^{8}$ | $-2.64918 \times 10^{8}$ | $-8.60964 \times 10^{8}$ | $-4.82704 \times 10^{12}$ | $-1.63565 \times 10^{12}$ | $-9.24749 \times 10^{9}$ |
| $c_5$ | $1.10866 \times 10^{10}$ | $2.01844 \times 10^{9}$ | $4.83506 \times 10^{10}$ | $7.62091 \times 10^{14}$ | $1.88027 \times 10^{14}$ | $6.97688 \times 10^{11}$ |
| $c_6$ | $-1.76807 \times 10^{11}$ | – | $-1.18452 \times 10^{12}$ | $-7.20026 \times 10^{16}$ | $-1.37241 \times 10^{16}$ | $-2.66192 \times 10^{13}$ |
| $c_7$ | $1.71842 \times 10^{12}$ | – | $1.38690 \times 10^{13}$ | $3.71496 \times 10^{18}$ | $6.17501 \times 10^{17}$ | $3.94078 \times 10^{14}$ |
| $c_8$ | $-9.19278 \times 10^{12}$ | – | $-6.33708 \times 10^{13}$ | $-8.03104 \times 10^{19}$ | $-1.56105 \times 10^{19}$ | – |
| $c_9$ | $2.06132 \times 10^{13}$ | – | – | – | $1.69535 \times 10^{20}$ | – |

*Source*: G. Burns, M. Scroger, and G. Strouse, "Temperature-Electromotive Force Reference Functions and Tables for the Letter-Designated Thermocouple Types Based on the ITS-90," NIST Monograph 175, April 1993.

| ANSI DESIGNATION | ALLOY (Generic or Trade Names) |
|---|---|
| JP | Iron |
| JN, EN, or TN | Constantan, Cupron, Advance |
| KP or EP | Chromega, Tophel, $T_1$, Thermokanthal KP |
| KN | Alomega, Nial, $T_2$, Thermokanthal KN |
| TP | Copper |
| RN or SN | Pure Platinum |
| RP | Platinum 13% Rhodium |
| SP | Platinum 10% Rhodium |

Trade Names: Advance T—Driver Harris Co., Chromega and Alomega—OMEGA Engineering, Inc., Cupron, Nial and Tophel—Wilbur B. Driver Co., Thermokanthal KP and Thermokanthal KN—The Kanthal Corporation.

ANSI LETTER DESIGNATIONS—Currently thermocouple and extension wire is ordered and specified by an ANSI letter designation. Popular generic and trade name examples are Chromega/Alomega—ANSI Type K: Iron/Constantan—ANSI Type J: Copper/Constantan—ANSI Type T: Chromega/Constantan—ANSI Type E: Platinum/Platinum 10% Rhodium—ANSI Type S: Platinum/Platinum 13% Rhodium—ANSI Type R. The positive and negative legs are identified by letter suffixes P and N, respectively, as listed in the tables.

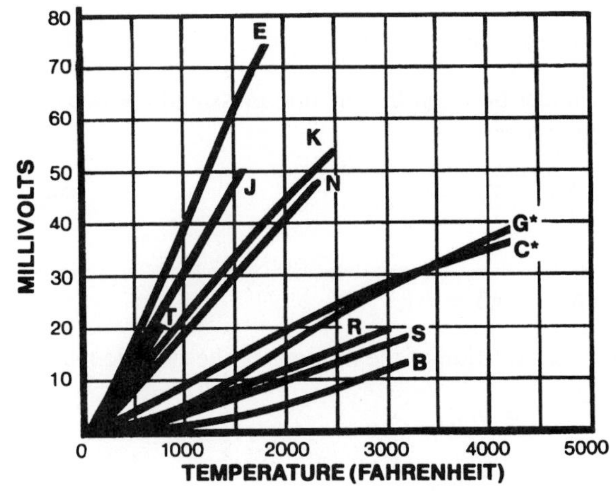


**ANSI Symbol**

- **T** Copper vs. Constantan
- **E** Chromega vs. Constantan
- **J** Iron vs. Constantan
- **K** Chromega vs. Alomega
- **N*** Omegalloy® (Nicrosil-Nisil)
- **G*** Tungsten vs. Tungsten 26% Rhenium
- **C*** Tungsten 5% Rhenium vs Tungsten 26% Rhenium
- **D*** Tungsten 3% Rhenium vs. Tungsten 25% Rhenium
- **R** Platinum 13% Rhodium vs. Platinum
- **S** Platinum 10% Rhodium vs. Platinum
- **B** Platinum 30% Rhodium vs. Platinum 6% Rhodium

*Not ANSI Symbol

© COPYRIGHT 1995 OMEGA ENGINEERING, INC. ALL RIGHTS RESERVED.
REPRODUCED WITH THE PERMISSION OF OMEGA ENGINEERING INC., STAMFORD, CT 06907

**FIGURE 8.43**
Thermocouple types and characteristics. *(Courtesy of OMEGA Engineering Inc., Stamford, CT)*

291

## 8.5 VIBRATION AND ACCELERATION MEASUREMENT

An accelerometer is a sensor designed to measure continuous mechanical vibration such as bearing vibration or aerodynamic flutter and transitory vibration such as shock waves, blasts, or impacts. Accelerometers are normally mechanically bonded to an object or structure for which acceleration is to be measured. The accelerometer detects acceleration along one axis and is insensitive in orthogonal directions. Strain gages or piezoelectric elements comprise the sensing element of an accelerometer, converting vibration into a voltage signal.

The design of an accelerometer is based on the inertial effects associated with a mass connected to a moving object through a spring, damper, and displacement sensor. Figure 8.44 illustrates the components of an accelerometer along with the displacement references, terminology, and free-body diagram. When the object accelerates, there will be relative motion between the housing and the mass. A displacement transducer senses the relative motion. Through a frequency response analysis of the second order system modeling the accelerometer, we can relate the displacement transducer output to either absolute position or acceleration of the object.

To determine the frequency response of the accelerometer, we first express the forces shown in the free-body diagram. To do this we define the relative displacement $x_r$ between the seismic mass and the object as

$$x_r = x_o - x_i \tag{8.51}$$

It is measured by a position transducer between the seismic mass and the housing. Therefore, the spring force can be expressed as

$$F_k = k(x_o - x_i) = kx_r \tag{8.52}$$

and the damper force as

$$F_b = b(\dot{x}_o - \dot{x}_i) = b\dot{x}_r \tag{8.53}$$

Applying Newton's Second Law, the equation of the motion for the seismic mass is

$$\sum F_{\text{ext}} = m\ddot{x}_o \tag{8.54}$$

or

$$-F_k - F_b = m\ddot{x}_o \tag{8.55}$$

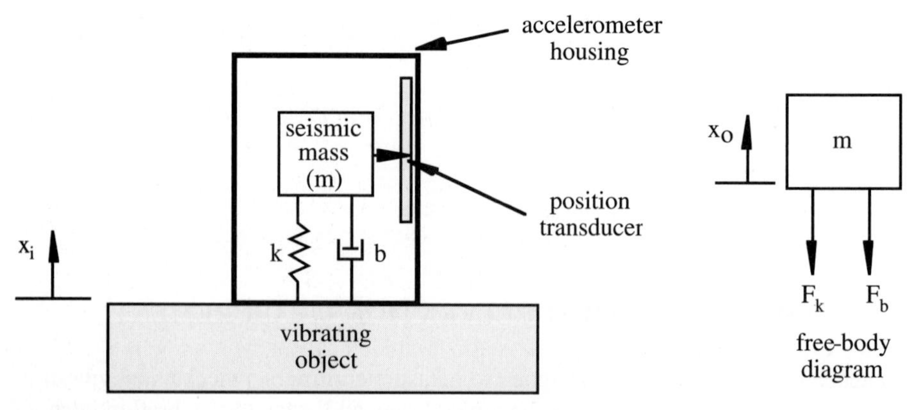


**FIGURE 8.44**
Accelerometer displacement references and free-body diagram.

The forces have negative signs in this equation since they are in an opposite direction to the reference direction $x_o$ in the free-body diagram. Substituting Equations 8.52 and 8.53, the result is

$$-kx_r - b\dot{x}_r = m\ddot{x}_o \tag{8.56}$$

Since the relative displacement $x_r$ is

$$x_r = x_o - x_i \tag{8.57}$$

we can replace $\ddot{x}_o$ with

$$\ddot{x}_o = \ddot{x}_r + \ddot{x}_i \tag{8.58}$$

Therefore, we can write Equation 8.56 as

$$-kx_r - b\dot{x}_r = m(\ddot{x}_r + \ddot{x}_i) \tag{8.59}$$

which can be rearranged as

$$m\ddot{x}_r + b\dot{x}_r + kx_r = -m\ddot{x}_i \tag{8.60}$$

This second order differential equation relates the measured relative displacement $x_r$ to the input displacement $x_i$. As in the analysis of a second order system in Chapter 4, we can rewrite this equation as

$$\frac{1}{\omega_n^2}\ddot{x}_r + \frac{2\zeta}{\omega_n}\dot{x}_r + x_r = -\frac{1}{\omega_n^2}\ddot{x}_i \tag{8.61}$$

where the natural frequency is

$$\omega_n = \sqrt{\frac{k}{m}} \tag{8.62}$$

and the damping ratio is

$$\zeta = \frac{b}{2\sqrt{km}} \tag{8.63}$$

▼▼▼ ***CLASS DISCUSSION ITEM 8.12. Effects of Gravity on an Accelerometer.*** The free-body diagram and resulting equation of motion for the accelerometer do not show a gravitational force explicitly. Explain why.

To determine the frequency response, the input displacement is assumed to be a sinusoid:

$$x_i(t) = X_i\sin(\omega t) \tag{8.64}$$

Because the system is linear, the resulting output displacement is also a sinusoid of the same frequency but different phase:

$$x_r(t) = X_r \sin(\omega t + \phi) \tag{8.65}$$

The analysis results in the amplitude ratio

$$\frac{X_r}{X_i} = \frac{(\omega/\omega_n)^2}{\left(\left[1-\left(\frac{\omega}{\omega_n}\right)^2\right]^2 + 4\zeta^2\left(\frac{\omega}{\omega_n}\right)^2\right)^{1/2}} \tag{8.66}$$

and the phase angle

$$\phi = -\tan^{-1}\left(\frac{2\zeta(\omega/\omega_n)}{1-\left(\frac{\omega}{\omega_n}\right)^2}\right) \tag{8.67}$$

To relate the output displacement signal $x_r$ to the input acceleration $\ddot{x}_i$, we differentiate Equation 8.64, resulting in

$$\ddot{x}_i(t) = -X_i\omega^2 \sin(\omega t) \tag{8.68}$$

Likewise, the relative acceleration is

$$\ddot{x}_r(t) = -X_r\omega^2 \sin(\omega t + \phi) \tag{8.69}$$

Note that the amplitude of the input acceleration is

$$X_i\omega^2 \tag{8.70}$$

and the amplitude of the output acceleration is

$$X_r\omega^2 \tag{8.71}$$

Rearranging Equation 8.66, we have

$$H_a(\omega) = \frac{X_r\omega_n^2}{X_i\omega^2} = \frac{1}{\left(\left[1-\left(\frac{\omega}{\omega_n}\right)^2\right]^2 + 4\zeta^2\left(\frac{\omega}{\omega_n}\right)^2\right)^{1/2}} \tag{8.72}$$

where $H_a(\omega)$ is used to represent the ratio $(X_r\omega_n^2)/(X_i\omega^2)$ as a function of frequency $\omega$. Figures 8.45 and 8.46 illustrate the amplitude ratio and phase angle relationships graphically for different values for the damping ratio.

The denominator of $H_a(\omega)$ is the input acceleration amplitude $X_i\omega^2$, and the numerator is the product of the output displacement amplitude $X_r$ and the square of the natural frequency $\omega_n^2$. Therefore, we can relate the measured output displacement amplitude to the input acceleration amplitude as

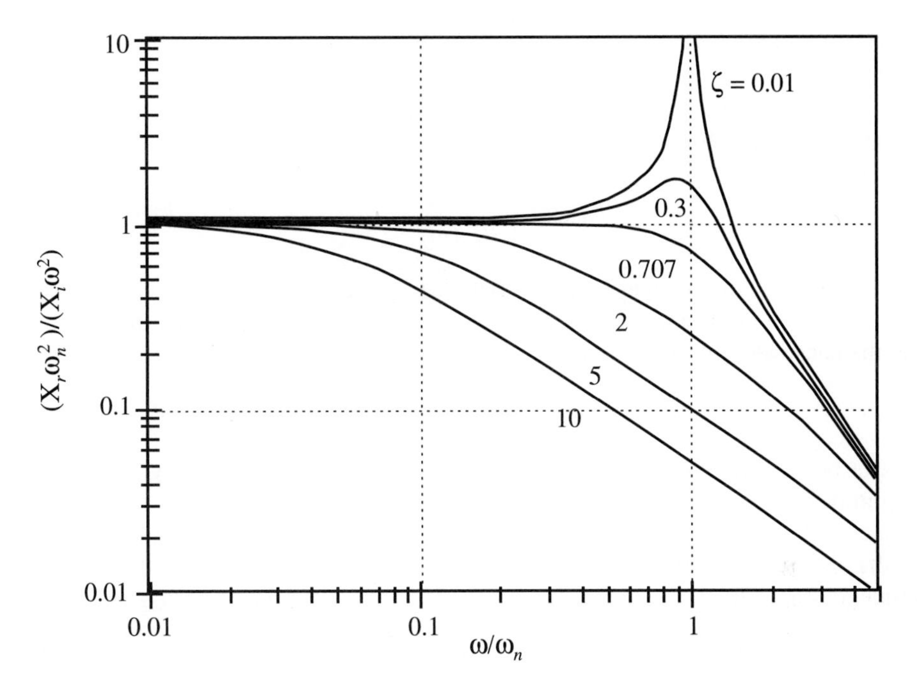


**FIGURE 8.45**
Ideal accelerometer amplitude response.

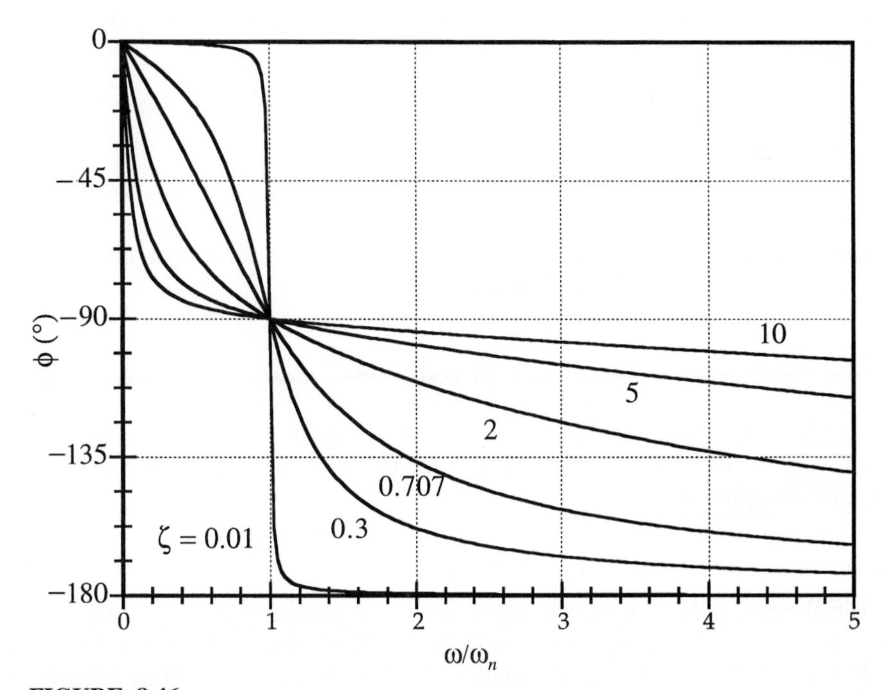


**FIGURE 8.46**
Ideal accelerometer phase response.

$$X_r = \left(\frac{1}{\omega_n^2}\right) H_a(\omega)(X_i\omega^2) \tag{8.73}$$

so the input acceleration amplitude can be expressed as

$$(X_i\omega^2) = \frac{X_r\omega_n^2}{H_a(\omega)} \tag{8.74}$$

If we design the accelerometer so that $H_a(\omega)$ is nearly 1 over a large frequency range, then the input acceleration amplitude is given directly in terms of the relative displacement amplitude scaled by a constant factor $\omega_n^2$:

$$(X_i\omega^2) = (\omega_n^2)X_r \tag{8.75}$$

As can be seen in Figure 8.45, the largest frequency range resulting in a unity amplitude ratio occurs when the damping ratio $\zeta$ is 0.707 and when the natural frequency $\omega_n$ is as large as possible. Also, as is clear in Figure 8.46, a $\zeta$ of 0.707 results in the best phase linearity for the system. We can make the natural frequency large by choosing a small seismic mass and a large spring constant. This is easy to accomplish in a very small package common to commercial accelerometers.

The same spring-mass-damper configuration used to measure acceleration can also be designed to measure displacement instead. This type of device is called a ***vibrometer***. The amplitude ratio given in Equation 8.66 provides us with the necessary relationship between the input and output displacements. As we did in the accelerometer analysis, we can now define the displacement ratio as

$$H_d(\omega) = \frac{X_r}{X_i} \tag{8.76}$$

Figure 8.47 illustrates the amplitude ratio to frequency relationship for different values of the damping ratio. The phase angle relationship is the same as for the accelerometer (see Figure 8.46). The input displacement amplitude $X_i$ is related to the measured relative displacement amplitude $X_r$ as

$$X_i = \frac{X_r}{H_d(\omega)} \tag{8.77}$$

If we design the vibrometer so that $H_d(\omega)$ is nearly 1 over a large frequency range, then the input displacement amplitude is given directly by the relative displacement amplitude:

$$X_i = X_r \tag{8.78}$$

As can be seen in Figure 8.47, the largest frequency range resulting in a unity amplitude ratio occurs when the damping ratio $\zeta$ is 0.707 and when the natural frequency $\omega_n$ is as small as possible. We can make the natural frequency small by choosing a

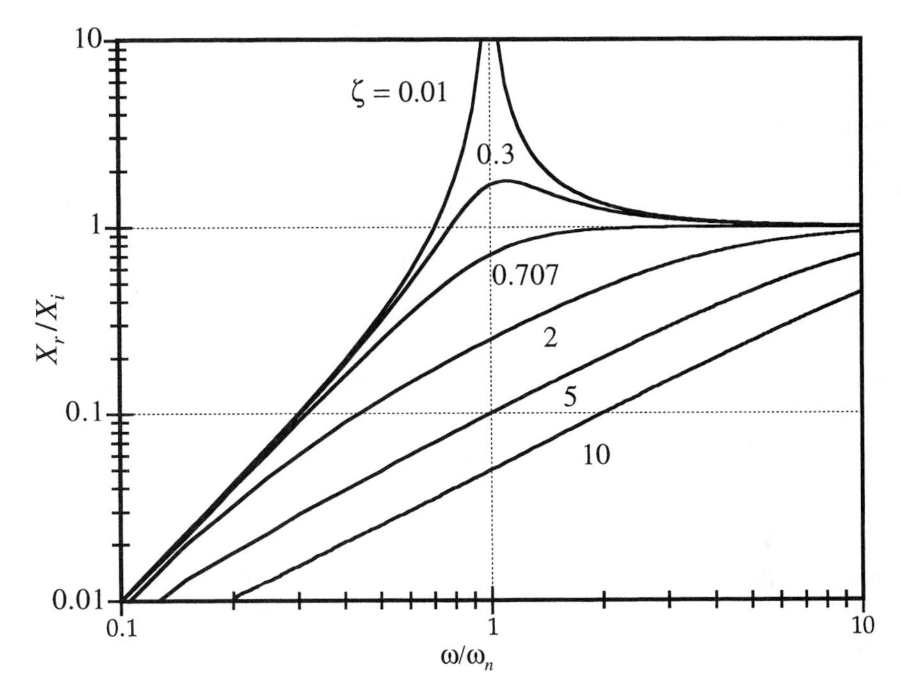


**FIGURE 8.47**
Vibrometer amplitude response.

large seismic mass and a small spring constant. This explains the large size of seismographs used to measure the earth's displacement during an earthquake.

## 8.5.1 Piezoelectric Accelerometers

The highest quality accelerometers are constructed using a ***piezoelectric crystal***, a material whose deformation results in charge polarization across the crystal. In a reciprocal manner, application of an electric field to a piezoelectric material results in deformation. As illustrated in Figure 8.48a, a piezoelectric accelerometer consists of a crystal in contact with a mass, supported in a housing by a spring. Figure 8.48b shows a commercially available piezoelectric accelerometer. In addition to natural damping properties inherent in the crystal and spring, additional damping is sometimes incorporated (e.g., by filling the housing with oil). When the supporting object experiences acceleration, relative displacement occurs between the case and the mass due to the inertia of the mass. The resulting strain in the piezoelectric crystal causes a displacement charge between the crystal conductive coatings as a result of the piezoelectric effect. An accelerometer using a piezoelectric crystal requires no external power supply. It is important to recognize that the accelerometer measures acceleration only in the direction in which it is mounted (i.e., along the axis of the spring, mass, and crystal).

A piezoelectric material produces a large output for its size. Naturally occurring piezoelectric materials are Rochelle salt, tourmaline, and quartz. Some crystalline materials can be artificially polarized to take on piezoelectric characteristics by heating and then slowly cooling them in a strong electric field. Such materials are barium titanate, lead zirconate (PZT), lead titanate, and lead metaniobate. These ferroelectric ceramics are more often used in accelerometers because the sensitivity can be controlled during manufacturing.

A simple equivalent circuit for a piezoelectric crystal is shown in Figure 8.49, implying that the crystal is effectively a capacitor and a charge source $Q$ that generates charge across the capacitor plates proportional to the deformation of the crystal. Representing the accelerometer by a Thevenin equivalent circuit (see Figure 8.50), the open circuit voltage $V$ is equal to the charge, typically in the picocoulomb range, divided by the capacitance, typically in the picofarad range:

$$V = \frac{Q}{C_p} \tag{8.79}$$

The sensitivity of the accelerometer is the ratio of the charge output to the acceleration of the housing expressed in pC/$g$, (rms pC)/(rms $g$), or (peak pC)/(peak $g$), where $g$ is the acceleration measured in units of the acceleration due to gravity. The output of the accelerometer is attached to a ***charge amplifier,*** which converts the charge on the crystal to a voltage that can be measured. Most accelerometers are calibrated in millivolts per $g$ for a specified charge amplifier.

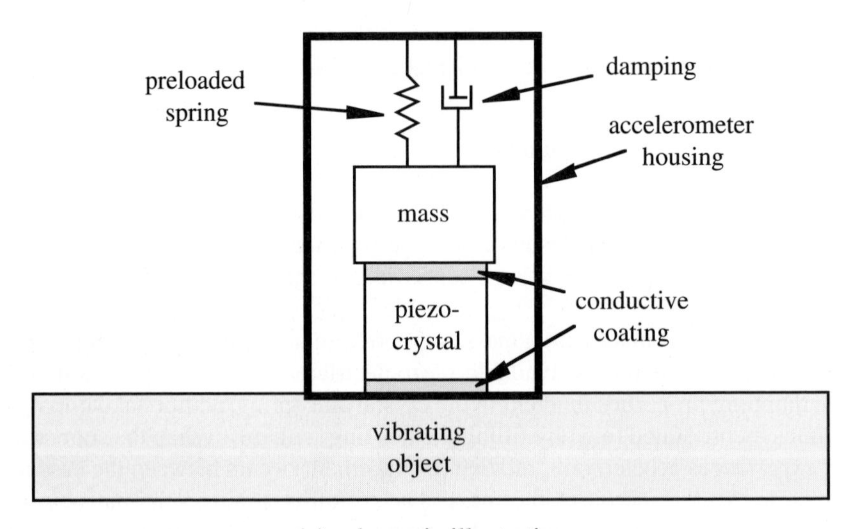


(*a*) schematic illustration

**FIGURE 8.48**
Piezoelectric accelerometer construction.

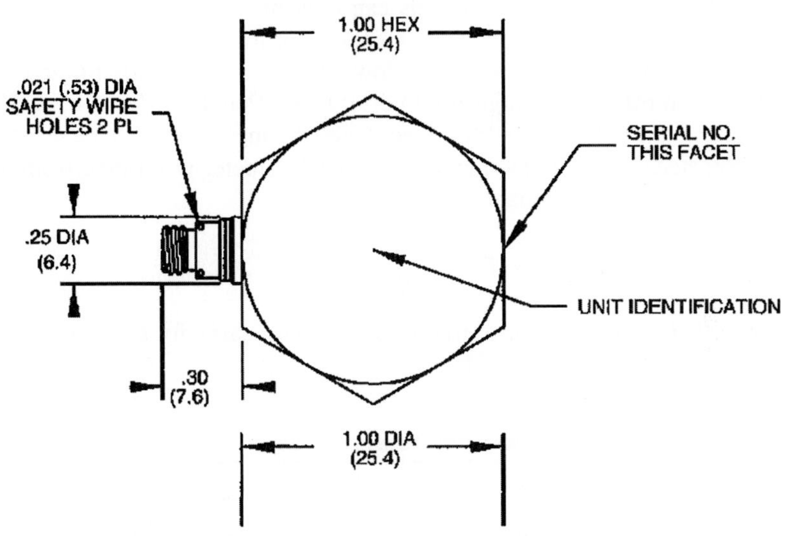


(*b*) actual device
*(Courtesy of Endevco, San Juan Capistrano, CA)*

**FIGURE 8.48**
*(continued)*

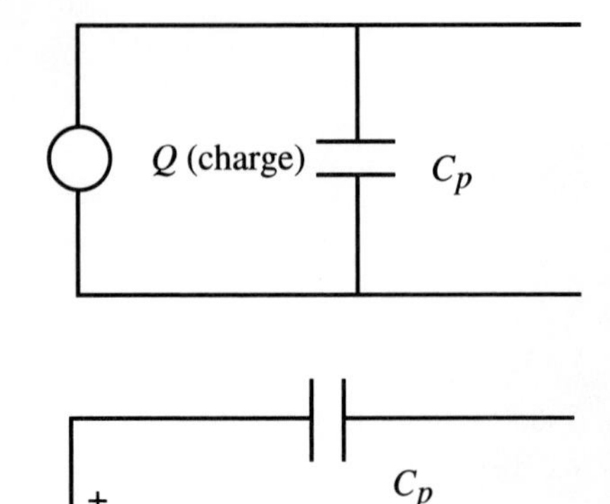


**FIGURE 8.49**
Equivalent circuit for piezoelectric crystal.

$C_p$

+

$V = Q/C_p$

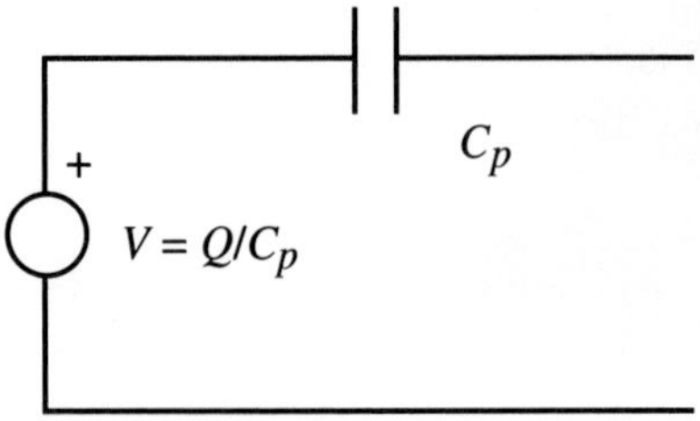

**FIGURE 8.50**
Thevenin equivalent of piezoelectric crystal.

In general, piezoelectric accelerometers cannot measure constant or slowly changing accelerations since the crystals can only measure a change in force by sensing a change in strain. But they are excellent for dynamic measurements such as vibration and impacts. The response at low frequencies is usually limited by the low-frequency cutoff of the charge amplifier. It may be on the order of a few Hertz. The response at high frequencies is a function of the mechanical characteristics of the accelerometer. The dynamic range of an accelerometer will range from a few Hertz to a fraction of the resonant frequency given by

$$f = \frac{1}{2\pi}\sqrt{\frac{k}{m}} \tag{8.80}$$

usually in the kHz range. A typical frequency response curve for a piezoelectric accelerometer is shown in Figure 8.51.

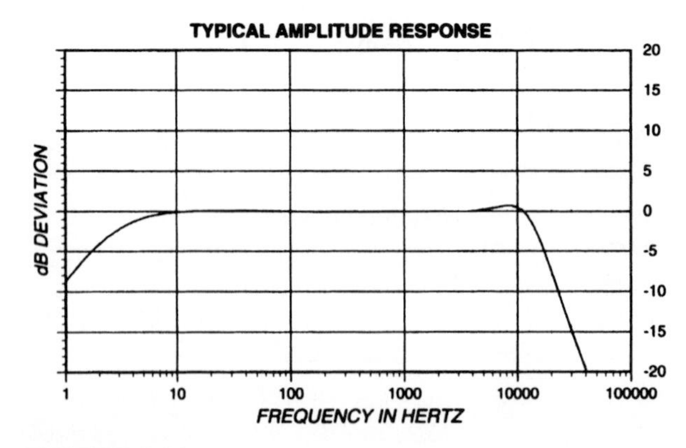


**FIGURE 8.51**
Piezoelectric accelerometer frequency response.
*(Courtesy of Endevco, San Juan Capistrano, CA)*

▼▼▼ *CLASS DISCUSSION ITEM 8.13. **Piezoelectric Sound.*** How does a piezoelectric microphone work? How about a piezoelectric buzzer?

## 8.6 PRESSURE AND FLOW MEASUREMENT

Many techniques and products have been developed to measure pressure and flow. Most techniques for pressure involve measuring a displacement or deflection and relating it to pressure through calibration or theoretical relations. One class of pressure sensor is a manometer, which measures a static pressure or pressure difference by detecting fluid displacements in a gravitational field. Another class is the elastic diaphragm, bellows, or tube where deflection of the elastic member is measured and related to pressure. Another class is the piezoelectric pressure transducer that can measure dynamic pressures as the piezoelectric crystal deforms in response to the applied pressure. References that cover these and other techniques are included in the bibliography at the end of the chapter.

There are also many techniques for measuring flow rate. A pitot tube manometer measures the difference between total and static pressure of a moving fluid. Venturi and orifice meters are based on measuring pressure drops across obstructions to flow. Turbine flow meters detect the rate of flow by measuring the rate of rotation of an impeller in the flow. Coriolis flow meters measure mass flow rate through a U-tube in rotational vibration. Hot wire anemometers sense the resistance changes in a hot current-carrying wire. The temperature and resistance of the wire depend on the amount of heat transferred to the moving fluid. The heat transfer coefficient is a function of the flow rate. Laser Doppler velocimeters (LDVs) sense the frequency shift of laser light scattered from particles suspended in a moving fluid. Most fluid mechanics textbooks present a variety of flow measurement techniques. References that cover these and other techniques are included in the bibliography at the end of the chapter.

## 8.7 SEMICONDUCTOR SENSORS AND MICRO-ELECTROMECHANICAL DEVICES

The advent of wide-scale semiconductor electronic design and production changed more than the way we process electronic signals. Many of the techniques developed for producing integrated circuits have been adapted for the design of a new class of semiconductor sensors and actuators called ***micro-electromechanical (MEM)*** devices. In 1980 the first MEM sensor was conceived using integrated circuit technology to etch silicon and produce a device that responded to mechanical action. This was a tiny silicon bar with an even tinier semiconductor strain gage placed on it, with a mass at the end so that acceleration deformed the bar and the strain gage sensed the magnitude of acceleration. The strain gage and the bar were etched from a single piece of silicon using the processing methods developed earlier to modify silicon for semiconductor electronics. MEM accelerometers are now used on automobiles to control air bag systems.

ICs are made by a series of processes consisting of photoresist lithographic layering, light exposure, controlled chemical etching, vapor deposition, and doping. The chemical etching process is important in that tiny mechanical devices can be created by a technique known as micromachining. Using carefully designed masks and timed immersion in chemical baths, microminiature versions of accelerometers, static electric motors, and hydraulic or gas-driven motors can be formed.

Semiconductor sensor designs are based on different electromagnetic properties of doped silicon and gallium arsenide and the variety of ways that semiconductors function in different physical environments. Sensors can be designed in two ways, sensors on semiconductor and sensors in semiconductor, implying a microelectronic device placed on the surface of silicon or as an integral part of the small crystal of silicon.

Semiconductor sensors can be classified by function. The following list summarizes some of these function classes along with applications:

- ***Surface acoustic wave (SAW)*** devices function by propagating stress waves between two piezoelectric regions in the semiconductor. The wave propagation properties of the silicon alter the frequency and delay the propagation of the transmitted signal. The SAW method is the basis for gas monitors and signal filters.
- The piezoresistive characteristic of doped silicon, meaning the coupling between resistance change and deformation, is the basis for semiconductor strain gages and pressure sensors.
- The magnetic characteristics of doped silicon, principally the Hall effect, are the basis of semiconductor magnetic transistors where the collector current can be modulated by an external magnetic field.
- Electromagnetic waves and nuclear particle radiation and beams impart changes in semiconductors, forming the basis of light color sensors and other radiation detectors.
- Thermal properties of semiconductors are the basis for thin film resistors, thermocouples, thermal conductivity sensors, humidity sensors, and temperature ICs.

With the development of so many different semiconductor sensors, engineers have now begun to integrate sensors and signal processing circuits together in a hybrid circuit that contains a transducer array, A/D converter, programmable memory, and microprocessor. The complete package provides a ***micromeasurement system*** (MMS) that is small, accurate, and less expensive than a discrete measuring system. Distributed micromeasurement systems will be used more and more in future mechatronic system design.

## QUESTIONS AND EXERCISES

**8.1.** You are using a poorly constructed 4-bit natural binary-coded absolute encoder and observe that as the encoder rotates through the single step from code 3 to code 4, the encoder outputs several different and erroneous values. If misalignment between the photosensors and the code disk is the problem, what possible codes could result during the 3-to-4 transition?

**8.2.** Derive Equation 8.9 for a circular conductor instead of a rectangular conductor.

**8.3.** A steel bar with modulus of elasticity 200 GPa and diameter 10 mm is loaded in tension with an axial load of 50 kN. If a strain gage of Gage Factor 2.115 and resistance 120 Ω is mounted on the bar in the axial direction, what is the change in resistance of the gage from the unloaded state to the strained state? If the strain gage is placed in one branch of a Wheatstone bridge ($R_1$) with the other three legs having the same base resistance ($R_2 = R_3 = R_4 = 120\ \Omega$), what is the output voltage of the bridge ($V_o$) in the strained state? What is the stress in the bar?

**8.4.** Draw a schematic diagram illustrating the wiring of a Wheatstone bridge circuit that takes advantage of both 3-wire lead connections and 1/2 bridge dummy gage temperature compensation.

**8.5.** A strain gage bridge used in load cell will dissipate energy. Why? Compare the power dissipated in a bridge circuit with equal resistance arms for gages of 350 Ω and 120 Ω when the excitation voltage is 10.0 V. What strategy can one employ to lower the power if heating becomes a problem? Does the strategy have any downsides?

**8.6.** Even when we use 3-wire connections to a strain gage to reduce the effects of leadwire resistance to temperature, we can have a phenomenon called leadwire desensitization. If the magnitude of the leadwire resistance exceeds 0.1% of the nominal gage resistance, significant error occurs. Assuming 22AWG leadwire (0.050 Ω/m) and a standard 120 Ω gage, how long can the leads be before there is leadwire desensitization?

**8.7.** A spring-mass-damper vibrometer is designed with a seismic mass of 1 kg, a coil spring of spring constant 2 N/m, and a dashpot with damping constant 2 Ns/m. Determine an expression for the steady state displacement of the seismic mass $x_{out}(t)$ given an object input displacement of $x_{in}(t) = 10\sin(1.25t)$ mm.

**8.8.** Refer to the standard 2-junction thermocouple configuration shown in Figure 8.40. With a 0°C reference temperature and Type J thermocouple (iron-constantan), the circuit has the behavior shown by the calibration curve on the next page. If a temperature measurement is made with the reference junction held at 100°C, resulting in a measured voltage of 30 mV, what is the actual measured temperature?

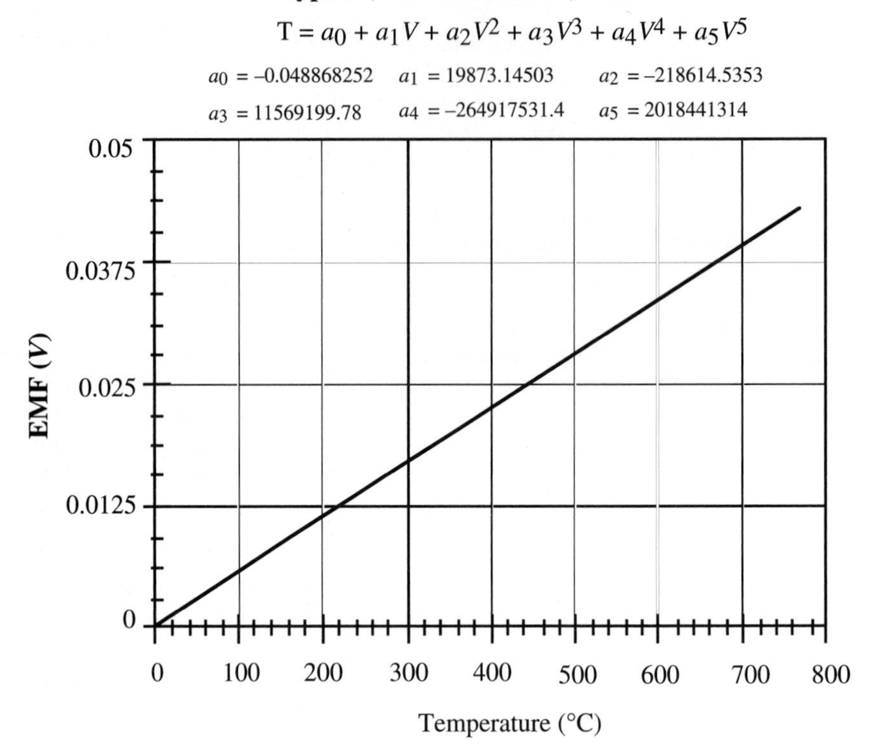


## BIBLIOGRAPHY


Beckwith, T., Marangoni, R., and Lienhard, J., *Mechanical Measurements,* 5th Edition, Addison-Wesley, Reading, MA, 1993.

Burns, G., Scroger, M., and Strouse, G., "Temperature-Electromotive Force Reference Functions and Tables for the Letter-Designated Thermocouple Types Based on the ITS-90," NIST Monograph 175, April 1993.

*Complete Temperature Measurement Handbook and Encyclopedia,* v. 28, Omega Engineering, Inc., Stamford, CT, 1992.

Dally, J. and Riley, W., *Experimental Stress Analysis*, 3rd Edition, McGraw-Hill, New York, 1991.

Doeblin, E., *Measurement Systems Application and Design,* 4th Edition, McGraw-Hill, New York, 1990.

Figliola, R. and Beasley, D., *Theory and Design for Mechanical Measurements,* 2nd Edition, John Wiley, New York, 1995.

Gardner, J., *Microsensors: Principles and Applications,* John Wiley, New York, 1994.

Hauptmann, P., *Sensors, Principles and Applications*, Carl Hanser Verlag, 1991.

Holman, J., *Experimental Methods for Engineers,* 6th Edition, McGraw-Hill, New York, 1994.

Janna, W., *Introduction to Fluid Mechanics,* Brooks/Cole Engineering Division, Monterey, CA, 1983.

Measurements Group Education Division, "Strain Gage Based Transducers: Their Design and Construction," Measurements Group, Inc., Raleigh, NC, 1988.

Measurements Group Education Division, "Student Manual for Strain Gage Technology," Measurements Group, Inc., Raleigh, NC, 1991.

Miu, D., *Mechatronics: Electromechanics and Contromechanics*, Springer-Verlag, New York, 1993.

Pallas-Areny, R. and Webster, J., *Sensors and Signal Conditioning,* John Wiley, New York, 1991.

Sze, S., *Semiconductor Sensors,* John Wiley, New York, 1994.

Walton, J., *Engineering Design: From Art to Practice,* pp. 117–119, West Publishing, St. Paul, MN, 1991.

White, F., *Fluid Mechanics,* 3rd Edition, McGraw-Hill, New York, 1994.

# 9

# ACTUATORS

**OBJECTIVES:** After you read, discuss, study, and apply ideas in this chapter, you will be able to:

- Identify the different classes of actuators, including solenoids, dc motors, ac motors, hydraulics, and pneumatics
- Understand the differences between series, shunt, compound, permanent magnet, and stepper dc motors
- Understand how to design the circuitry to control a stepper motor
- Select a motor for a mechatronics application

## 9.1 INTRODUCTION

Most mechatronic systems involve motion or action of some sort. This motion or action can be applied to anything from a single atom to a large articulated structure. It is created by a force or torque that results in acceleration and displacement. ***Actuators*** are devices used to produce this motion or action.

To this point in the book, we have focused on electronic components and sensors and associated signals and signal processing, all of which are required to produce a specific mechanical action or action sequence. Sensor input measures how well a mechatronic system produces its action, open loop or feedback control helps to regulate the specific action, and much of the electronics we have learned about is required to manipulate and communicate this information by a wide variety of methodologies. Actuators produce physical changes such as linear and angular displacement. They also modulate the rate and power associated with these changes. How well the actuator meets a design requirement is the measure of our overall creativity. This chapter will cover some of the most important actuators: solenoids, electric motors, hydraulic cylinders and rotary motors, and pneumatic cylinders. Putting it poetically, this chapter is "where the rubber meets the road."

## 9.2 ELECTROMAGNETIC PRINCIPLES

Many actuators rely on electromagnetic forces to create their action. When a current in a conductor is moved in a magnetic field, a force is produced in a direction perpendicular to the current direction and magnetic field direction. ***Lorentz's Force Law*** in vector form states that

$$\vec{F} = \vec{I} \times \vec{B} \tag{9.1}$$

where $\vec{F}$ is the force vector, $\vec{I}$ is the current vector, and $\vec{B}$ is the magnetic field vector. Figure 9.1 illustrates the relationship between these vectors and indicates the ***right-hand-rule*** analogy, which states that if your right-hand index finger points in the direction of the current and your middle finger is aligned with the field direction, then your extended thumb (perpendicular to the index and middle fingers) will point in the direction of the force.

Another important electromagnetic effect important to actuator design is field intensification within a coil. Recall that when discussing inductors in Chapter 2 we stated that the magnetic flux through a coil is proportional to the current through the coil and the number of windings. The proportionality constant is a function of the permeability of the material within the coil. The permeability of a material characterizes how easily magnetic flux will penetrate the material. Iron has a permeability a few hundred times that of air; therefore, a coil wound around an iron core will support a magnetic flux a few hundred times that of the same coil with no

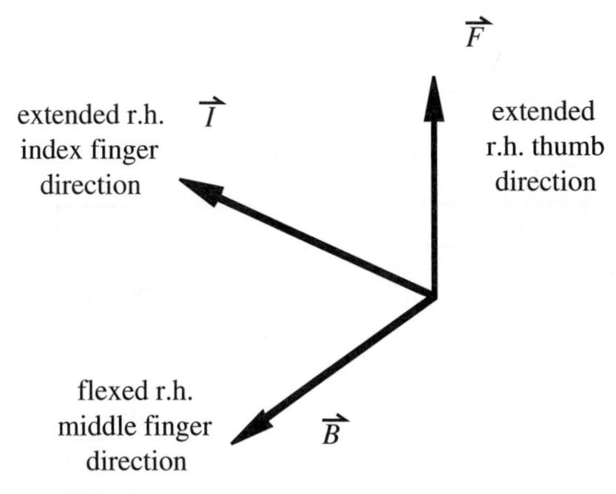


**FIGURE 9.1**
Direction of force produced by a magnetic field and a current.

core. Most electromagnetic devices we will discuss use iron cores of one form or another to enhance the magnetic flux. Cores are usually laminated (made up of insulated layers of iron stacked parallel to the coil-axis direction) to reduce eddy currents that are induced when the cores experience changing magnetic fields. Eddy currents, which are a result of Faraday's Law of Induction, result in inefficiencies and undesirable core heating.

## 9.3 SOLENOIDS AND RELAYS

As illustrated in Figure 9.2, a ***solenoid*** consists of a coil and a movable iron core called the ***armature***. When the coil is energized with current, the core moves to increase the flux linkage by closing the air gap between the cores. The movable core is usually spring-loaded to allow the core to retract when the current is switched off. The force generated is approximately proportional to the square of the current and inversely proportional to the square of the length of the air gap. Solenoids are inexpensive, and their use is primarily limited to on-off applications such as latching, locking, and triggering. They are frequently used in home appliances (e.g., washing machine valves), office equipment (e.g., copy machines), automobiles (e.g., door latches and the starter solenoid), pinball machines (e.g., plungers and bumpers), and factory automation.

An electromechanical ***relay*** is a solenoid used to make or break mechanical contact between electrical leads. A small voltage input to the solenoid controls a potentially large current through the relay contacts. Applications include power switches and electromechanical control elements. A relay performs a function sim-

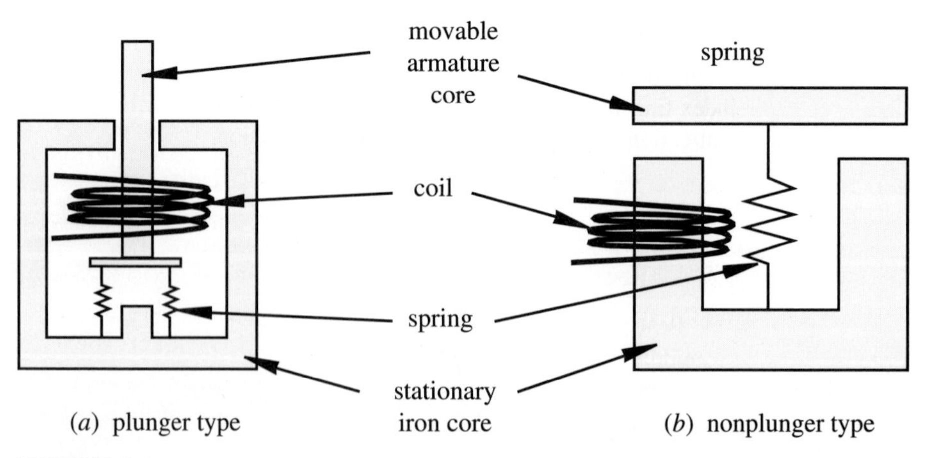


**FIGURE 9.2**
Solenoids.

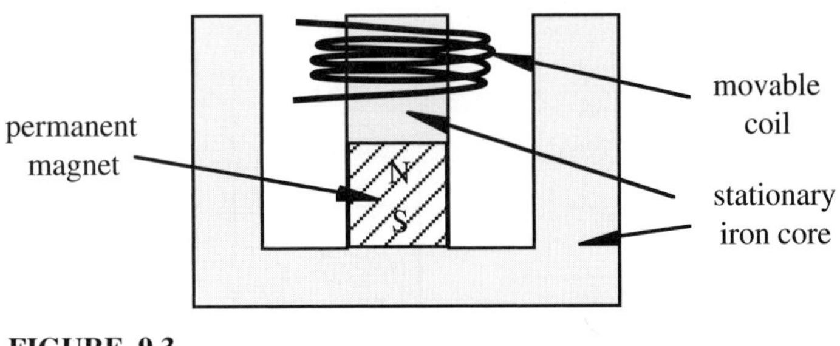


**FIGURE 9.3**
Voice coil.

ilar to a power transistor but has the capability to switch extremely large currents if necessary. However, transistors have a much shorter switching time than relays.

As illustrated in Figure 9.3, a ***voice coil*** consists of a coil that moves in a magnetic field produced by a permanent magnet and intensified by an iron core. The force on the coil is directly proportional to the current in the coil. The coil is usually attached to a movable load such as the diaphragm of an audio speaker, the spool of a hydraulic proportional valve, or the read-write head of a computer disk drive. The linear response and bidirectional capability make voice coils more attractive than solenoids for control applications.

▼▼▼ ***CLASS DISCUSSION ITEM 9.1. Examples of Solenoids, Voice Coils, and Relays.*** Make of list of common household and automobile devices that contain solenoids, voice coils, and relays. Describe the reasons why these devices were selected in the applications you cite.

## 9.4 ELECTRIC MOTORS

Electric motors are by far the most ubiquitous of the actuators, occurring in virtually all electromechanical systems. Electric motors can be classified either by function or by electrical configuration. In the functional classification, motors are given names suggesting how the motor is to be used. Examples of functional classifications include torque, gear, servo, instrument servo, and stepping. However, it is usually necessary to know something about the electrical design of the motor to make judgments about its application for delivering power and controlling position. Figure 9.4 provides a configuration classification of electrical motors found in engineering applications. The differences are due to motor winding (field) and rotor designs, resulting in a large variety of operating characteristics, which will be discussed. The price-performance ratio of electric motors continues to improve, making them important additions to all sorts of mechatronic systems from appliances to automobiles.

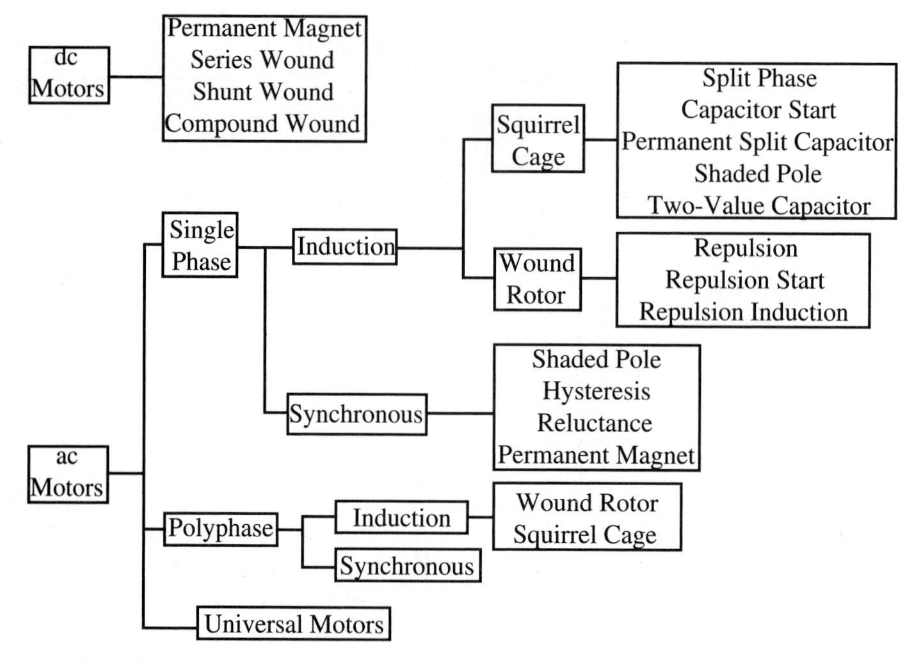


**FIGURE 9.4**
Classification of electric motors.

Figure 9.5 illustrates the construction and components of a typical electric motor. The stationary outer housing, called the ***stator***, supports radial magnetized poles. These poles consist of either permanent magnets or wire coils, called ***field coils,*** wrapped around laminated iron cores. The purpose of the stator poles is to provide radial magnetic fields. The iron core intensifies the magnetic field inside the coil by increasing the permeability. The purpose for laminating the core is to reduce the effects of eddy currents, which are induced in a conducting material (see Class Discussion Item 9.2). The ***rotor*** consists of a rotating shaft supported by bearings, an iron core into which windings are anchored, and, in ***dc motors***, a ***commutator*** to deliver current and control its direction in the rotor windings. The rotor and its windings are sometimes referred to as the ***armature***. For motors with a commutator, the ***brushes*** provide stationary electrical contact to the moving commutator conducting segments. Brushes in early motors consisted of bristles of copper wire flexed against the commutator, hence the term *brush*; but now they are usually made of conducting solid graphite that provides a larger contact area and is self-lubricating. The brushes are usually spring-loaded to ensure continual contact between them and the commutator. There is a small ***air gap*** between the rotor and the stator where the magnetic fields interact.

Figure 9.6 shows examples of commercially available assembled motors. In the top figure, the motor on the left is an ac induction motor with a gearhead speed

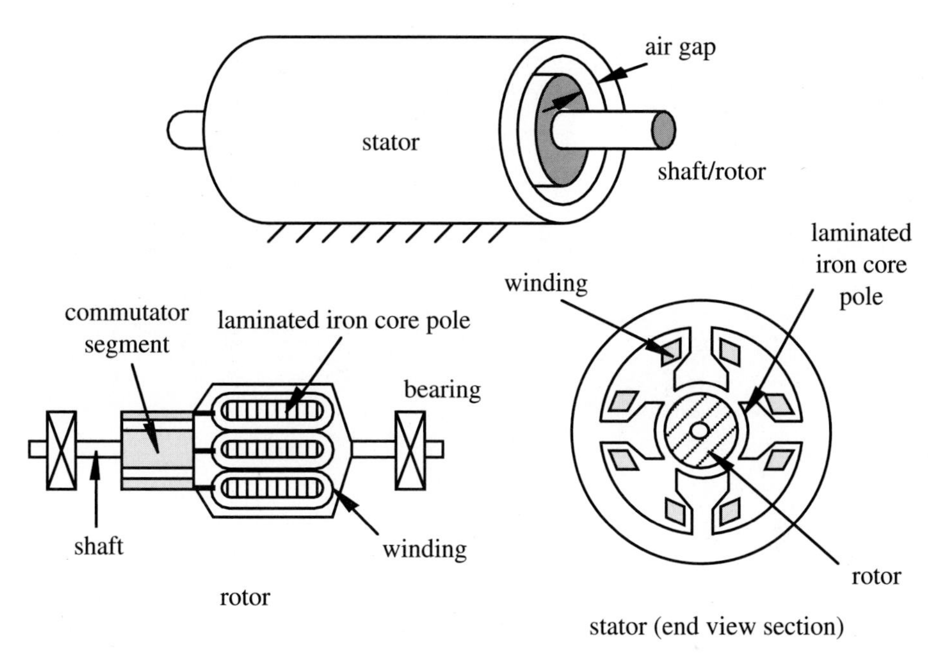


**FIGURE 9.5**
Motor construction and terminology.

reduction unit attached. The motor on the right is a two-phase stepper motor. Motors come in standard sizes with standard mounting brackets, and they usually include nameplates listing some of the motor's specifications. The bottom figure shows the internal construction of a permanent magnet rotor stepper motor.

▼▼▼ ***CLASS DISCUSSION ITEM 9.2. Eddy Currents.*** Describe the causes of eddy currents that are induced in a conducting material experiencing a changing magnetic field. The iron core in a motor rotor is usually laminated. Explain why. What is the best orientation for the laminations?

Torque is produced by an electric motor either through the interaction of stator fields and armature currents or through the interaction of stator fields and armature fields. We will illustrate both principles starting with the first. Figure 9.7 illustrates a dc motor with six armature windings. The direction of current flow in the windings is illustrated in the figure. As a result of Equation 9.1, the interaction of the fixed stator field and the currents in the armature windings produce a torque in the clockwise direction. You can verify this torque direction by applying the right-hand rule to the stator field and armature current directions. To maintain the torque as the rotor rotates, the spatial arrangement of the armature currents relative to the stator field must remain fixed. A commutator performs this task by switching the currents in the armature windings in the correct sequence.

(*a*) ac induction and stepper motor

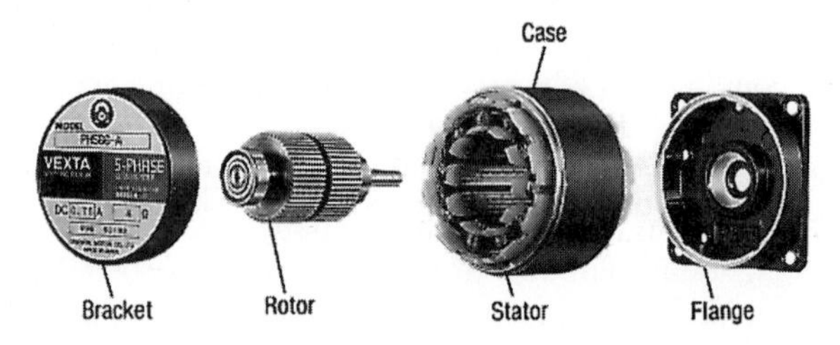


(*b*) exploded view of stepper motor with a permanent magnet rotor

**FIGURE 9.6**
Examples of commercial motors. *(Courtesy of Oriental Motor, Torrance, CA)*

Figure 9.8 illustrates an example of a commutator. It consists of a ring of alternating conductive and insulating materials connected to the rotor windings. Current is directed through the windings via the brushes, which slide on the surface of the commutator. In the position shown, current flows through windings A, B, and C in the clockwise direction and through F, E, and D in the counterclockwise direction.

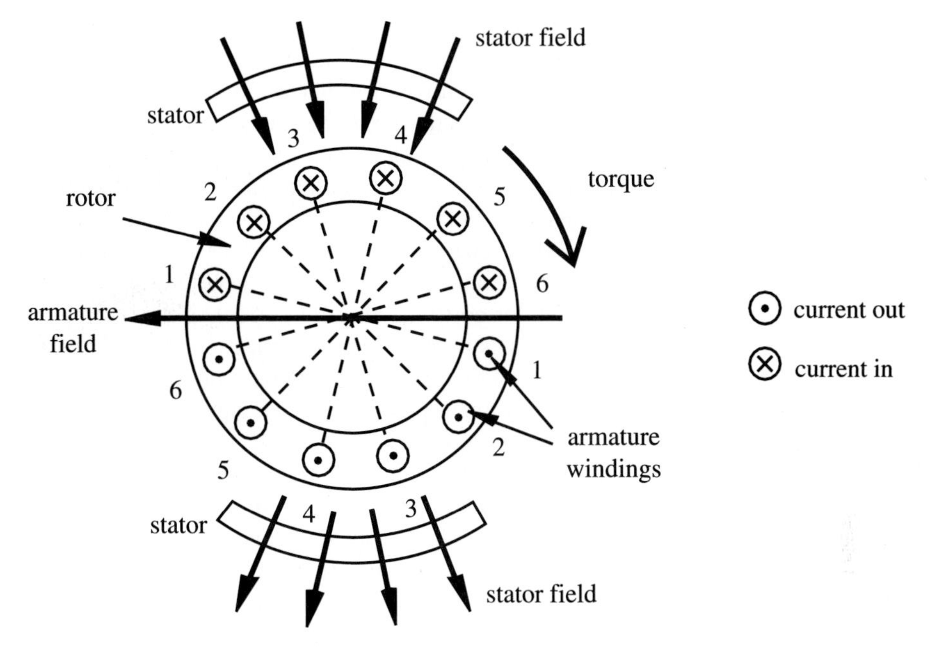


**FIGURE 9.7**
Electric motor field-current interaction.

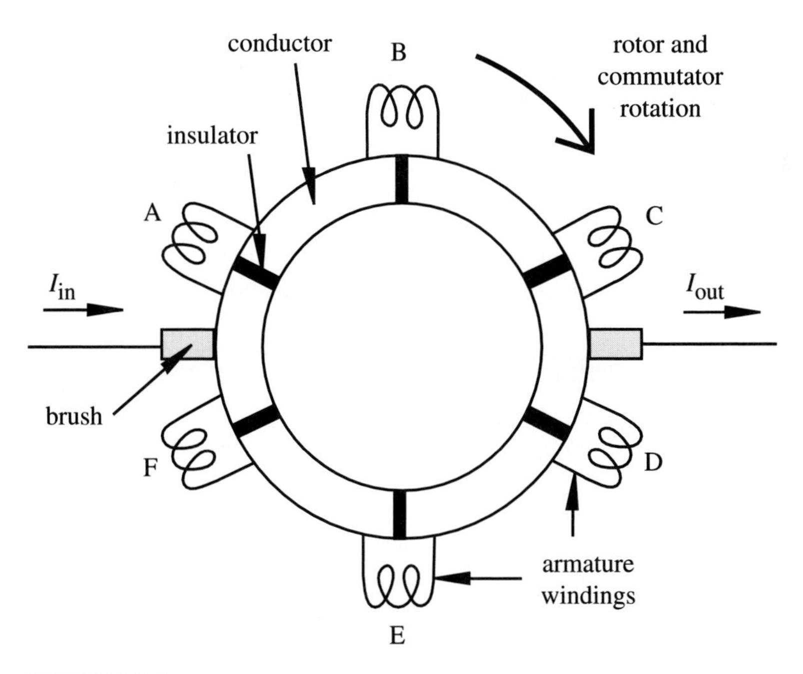


**FIGURE 9.8**
Electric motor six-winding commutator.

When the rotor turns 1/6 of a full rotation from the position shown, the currents in windings C and F switch directions. As the brushes slide over the rotating commutator, this process continues in sequence. With appropriate winding configurations, the commutator maintains a consistent spatial arrangement of the currents relative to the fixed stator fields, resulting in a constant clockwise torque.

Another method by which electric motors can create torque is through the interaction of stator and rotor magnetic fields. The torque is produced by the fact that like field poles attract and unlike poles repel. Figure 9.9 illustrates this principle of operation with a simple two-pole dc motor. The stator poles generate fixed magnetic fields with permanent magnets or coils carrying dc current. The winding in the rotor is commutated to cause changes in direction of its magnetic field. The interaction of the changing rotor field and the fixed stator fields produce a torque on the shaft, causing rotation. With the rotor in position (i), the right brush contacts commutator segment A and the left brush contacts segment B, creating a current in the rotor winding, resulting in the magnetic poles as shown. The rotor magnetic poles oppose the stator magnetic poles, creating a torque causing clockwise motion of the rotor. In position (ii), the stator poles both oppose and attract the rotor poles to enhance the clockwise rotation. Between positions (iii) and (v) the commutator contacts switch, changing the direction of the rotor current and hence the direction of the magnetic field. In position (iv), both brushes temporarily lose contact with the commutator, but the rotor continues to move due to its momentum. In position (v), the reversed magnetic field in the rotor again opposes the stator field, continuing the clockwise torque and motion.

▼▼▼ ***CLASS DISCUSSION ITEM 9.3. Field-Field Interaction in a Motor.*** Does the armature field in Figure 9.7 have any effect on the torque produced by the motor?

A problem with the simple two-pole design illustrated in Figure 9.9 is that starting would not occur if the motor happens to be in position (iv), where the brushes are located over the commutator gaps. This problem can be avoided by designing the motor with more poles and more commutation segments with overlapping switching. This allows the brushes to always contact two active segments, even while switching.

Other problems not discussed with the simple models above are back emf and induction. As the rotor windings cut through the stator magnetic field, a back emf is induced opposing the voltage applied to the rotor. Also, when the commutator switches the direction of current, a voltage is induced to oppose the change in current direction. These problems are beyond the scope of our discussion.

The principles of operation of ***ac motors*** are similar regarding interaction of the magnetic fields, but commutation is not required. This is due to the fact that the magnetic field rotates around the stator as a result of the ac voltages and the arrangement of the coils around the stator housing. The rotor windings of ***asynchronous*** ac motors have no external voltage applied; rather, voltages are induced in the rotor windings due to the rotating fields around the stator. The rotor rotates at slower speeds than the rotating stator fields (this is called ***slip***), making the induction possible, hence the term *asynchronous*. Because of this action, asynchronous

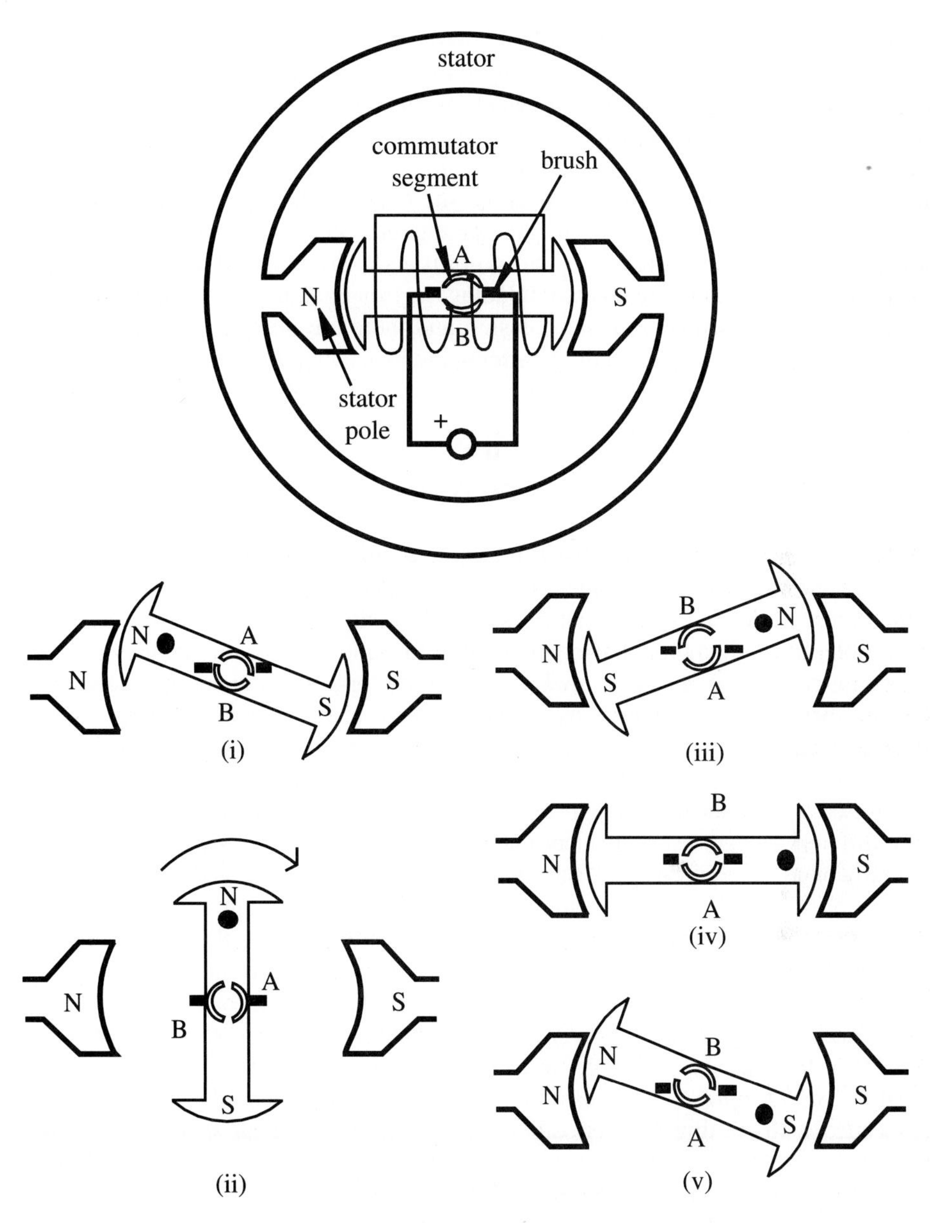


**FIGURE 9.9**
Electric motor theory of operation.

motors are sometimes referred to as ***induction machines***. With ***synchronous*** ac motors, the rotor windings are energized, but through ***slip rings*** instead of a commutator. Brushes provide constant uninterrupted contact with the slip rings, causing fields to rotate around the rotor windings at the same rate as the fields rotate around the stator. Due to the interaction of these fields, the rotor rotates at the same speed as the stator fields, hence the term *synchronous*.

As we saw above, dc motors often contain a commutator and brushes. When this is the case, they are called ***brushed motors***. Another class of dc motors is the ***brushless dc motor,*** which has permanent magnets on the rotor and a rotating field in the stator. The permanent magnets on the rotor eliminate the need for the commutator. Instead, transistors switch the dc current between the coils in response to proximity sensors that are triggered as the shaft rotates. Brushes are no longer required in the absence of a mechanical commutator. This is attractive from a maintenance point of view since there are no brushes to be replaced when they wear down. Also, since there are no rotor windings or iron core, the rotor inertia is much smaller, sometimes making control easier, and there are no rotor heat dissipation problems since there is no winding current, and hence no $I^2R$ heating.

## 9.5 dc MOTORS

Direct current (dc) motors are used in a large number of engineering designs because of the torque-speed characteristics achievable with different electrical configurations. dc Motor speeds can be smoothly controlled and in most cases are reversible. Since dc motors have a high ratio of torque-to-rotor inertia, they can respond quickly. Also, ***dynamic braking***, where motor-generated energy is fed to a resistor dissipater, or ***regenerative braking***, where motor-generated energy is fed back to the dc power supply, can be implemented in applications where quick stops and high efficiency are desired.

Based on how the stator magnetic fields are created, dc motors are classified into four different categories: permanent magnet, shunt-wound, series-wound, and compound-wound. The electrical schematics, torque-speed curves, and current-torque curves for each configuration are illustrated in Figures 9.11 through 9.14. Figure 9.10 illustrates a motor ***torque-speed curve*** that displays the torques that the motor can provide at different speeds at rated voltage. For a given torque provided by the motor, the current-torque curve can be used to determine the amount of current required when rated voltage is applied. As a general rule of thumb, motors deliver large torques at low speeds, and large torques imply large motor currents. The ***starting torque*** or ***stall torque*** $T_s$ is the maximum torque the motor can produce at zero speed associated with starting or overloading the motor. The ***no-load speed*** $\omega_{max}$ is the maximum sustained speed the motor can attain; this speed can only be reached when there is no load or torque applied to the motor (i.e., only when it is free running).

In Figures 9.11 through 9.14, $V$ is the dc voltage supply, $I_A$ is the current in the rotor (armature) windings, $I_F$ is the current in the stator (field) windings, and $I_L$ is the total load current delivered by the dc supply.

The stator fields in ***permanent magnet (PM) motors*** (see Figure 9.11) are provided by permanent magnets, which require no external power source and therefore produce no $I^2R$ heating. The case of a PM motor is lighter and smaller than other equivalent dc motors because the field strength of permanent magnets is high. The radial width of the permanent magnet is roughly one-fourth that of the width of an equivalent field winding. PM motors are easily reversed by switching

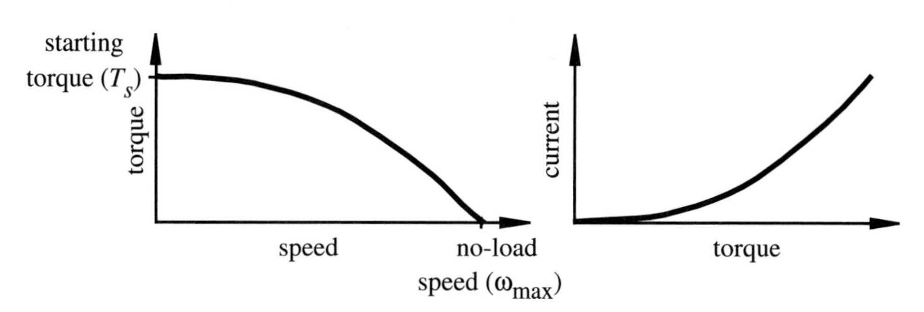


**FIGURE 9.10**
Motor torque-speed curve.

the direction of the applied voltage since the current and field change direction only in the rotor. The PM motor is ideal in computer control applications because of the linearity of its torque-speed relation. The design of a controller is always easier when the actuator is linear since the system analysis is greatly simplified. When a motor is used in a position or speed control application with sensor feedback to a controller, it is referred to as a ***servomotor***. PM motors are only used in low power applications since their rated power is usually limited to 5 hp (3728 W) or less, with fractional horsepower ratings being more common. PM dc motors can be brushed, brushless, or stepper motors.

***Shunt motors*** (see Figure 9.12) have armature and field windings connected in parallel, which are powered by the same supply. The total load current is the sum of the armature and field currents. Shunt motors exhibit nearly constant speed over a large range of loading, have starting torques (maximum torque at zero speed) about 1.5 times the rated operating torque, have the lowest starting torque of any of the dc motors, and can be economically converted to allow adjustable speed by placing a potentiometer in series with the field windings.

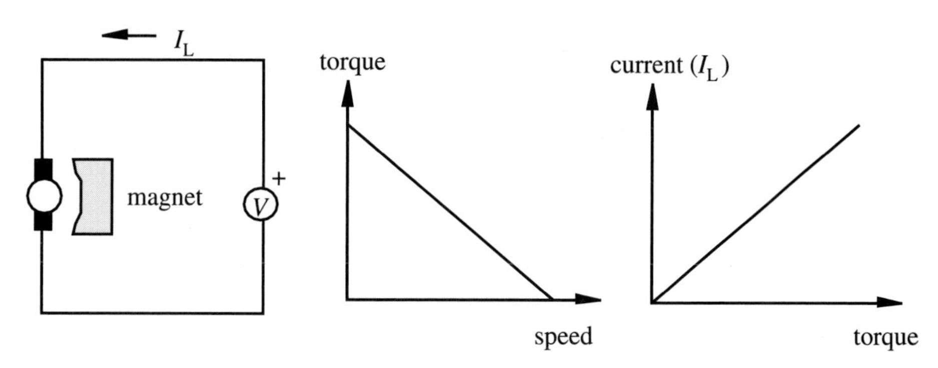


**FIGURE 9.11**
dc Permanent magnet motor schematic and torque-speed curve.

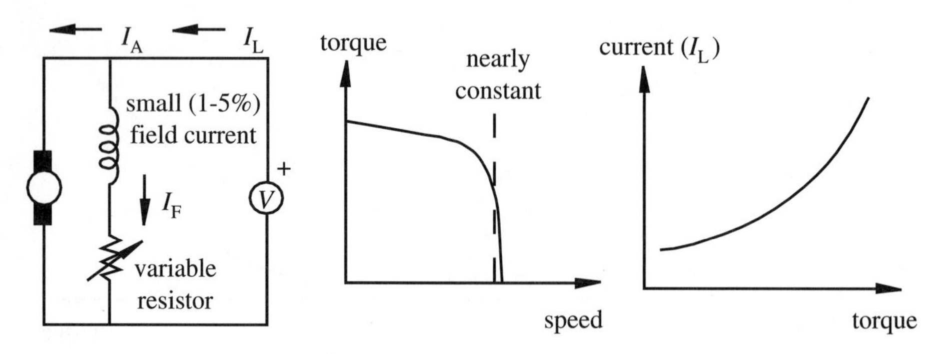


**FIGURE 9.12**
dc Shunt motor schematic and torque-speed curve.

***Series motors*** (see Figure 9.13) have armature and field windings connected in series so the armature and field currents are equal. Series motors exhibit very high starting torques, highly variable speed depending on load, and very high speed when the load is small. In fact, large series motors can fail catastrophically when they are suddenly unloaded (e.g., in a belt drive application when the belt fails) due to dynamic forces at high speeds. This is called ***run-away***. As long as the motor remains loaded, this does not pose a problem. The torque-speed curve for a series motor is hyperbolic in shape, implying an inverse relationship between torque and speed and nearly constant power over a wide range.

***Compound motors*** (see Figure 9.14) include both shunt and series field windings, resulting in combined characteristics of both shunt and series motors. Part of the load current passes through both the armature and series windings, and the remaining load current passes through the shunt windings only. The maximum speed of a compound motor is limited, unlike a series motor, but its speed regulation is not as good as with a shunt motor. The torque produced by compound motors is somewhat lower than that of series motors of similar size.

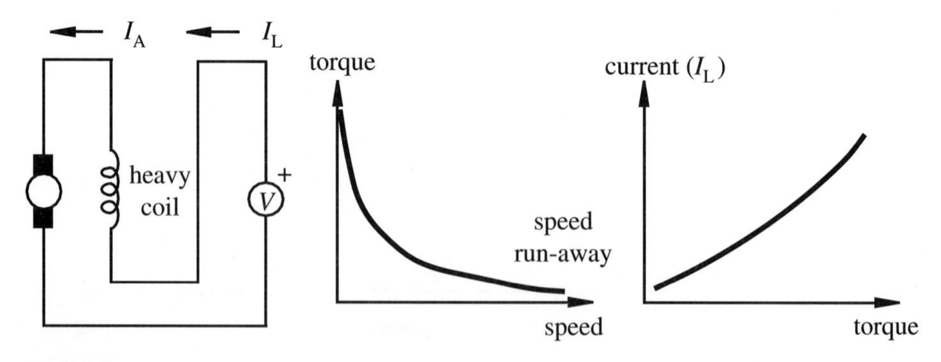


**FIGURE 9.13**
dc Series motor schematic and torque-speed curve.

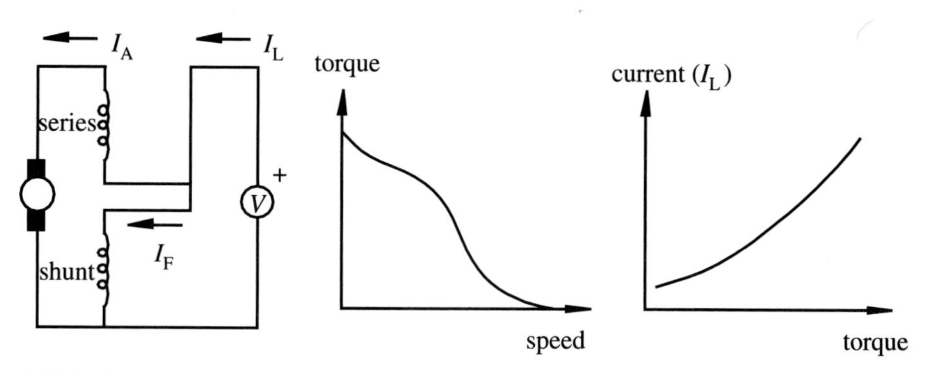


**FIGURE 9.14**
dc Compound motor schematic and torque-speed curve.

Note that unlike the permanent magnet motor, when voltage polarity for a shunt, series, or compound dc motor is changed, the direction of rotation would not change. The reason for this is that the polarity of both the stator and rotor changes since the field and armature windings are excited by the same source.

### 9.5.1 dc Motor Electrical Equations

When the armature of a motor is tested with an impedance meter with the armature locked in one position, the motor impedance appears to be equivalent to a resistance $R$ in series with the parallel combination of an inductance $L$ and a second resistance $R_L$. However, as the armature begins to rotate a voltage is induced in the armature windings called the back emf $V_{emf}$ opposing the applied voltage. Therefore, the equivalent electrical circuit for the armature is shown in Figure 9.15.

$R_L$, the resistive loss in the magnetic circuit, is roughly an order of magnitude larger than $R$, the resistance of the windings, and is usually neglected. If we assume that the voltage applied to the armature is $V_{in}$ and current through the armature is $I_{in}$, the electrical equation for the motor is

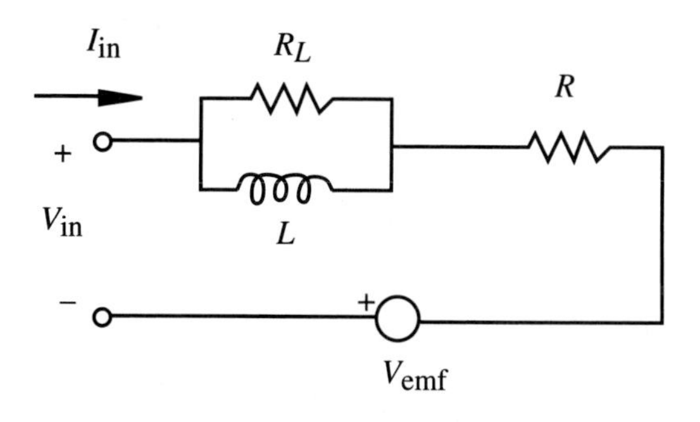


**FIGURE 9.15**
Motor armature equivalent circuit.

$$V_{in} = L\frac{dI_{in}}{dt} + RI_{in} + k_e\omega \tag{9.2}$$

where $\omega$ is the rotational speed of the motor in rad/sec and $k_e$ is the ***electrical constant*** of the motor defined as

$$k_e = \frac{V_{emf}}{\omega} \tag{9.3}$$

### 9.5.2 PM dc Motor Dynamic Equations

Since the permanent magnet (PM) motor is the easiest to understand and analyze, we will look at its governing equations in more detail. Due to the interaction between the stator field and the armature current, the torque generated by a PM dc motor is directly proportional to armature current:

$$T = k_t I_{in} \tag{9.4}$$

where $k_t$ is defined as the ***torque constant*** of the motor. A PM motor's electrical constant $k_e$ and torque constant $k_t$ are very important parameters, and they are often reported in manufacturer specifications.

When the dynamics of the system is considered, the motor torque $T$ is also given by

$$T = (J_a + J_L)\frac{d\omega}{dt} + T_f + T_L \tag{9.5}$$

where $J_a$ and $J_L$ are the polar moments of inertia of the armature and the attached load, $T_f$ is the frictional torque opposing armature rotation, and $T_L$ is the resisting torque of the load.

When the motor is connected to the power supply, the armature will accelerate until a steady state operating condition is attained. At steady state, Equation 9.2 becomes

$$V_{in} = RI_{in} + k_e\omega \tag{9.6}$$

Note that at steady state, from Equation 9.5, the motor torque balances the friction and load torques.

By solving for $I_{in}$ in Equation 9.4 and substituting into Equation 9.6, we get

$$V_{in} = \left(\frac{R}{k_t}\right)T + k_e\omega \tag{9.7}$$

and by solving for the motor torque in this equation we get

$$T = \left(\frac{k_t}{R}\right)V_{in} - \left(\frac{k_e k_t}{R}\right)\omega \tag{9.8}$$

This equation predicts the linear torque-speed relation for a PM dc motor with a fixed applied voltage.

Figure 9.16 shows the torque-speed curve and power-speed curve for a permanent magnet dc motor. Since the torque-speed relation is linear, it can be expressed in terms of the starting torque $T_s$ and the no-load speed $\omega_{max}$ as

$$T(\omega) = T_s\left(1 - \frac{\omega}{\omega_{max}}\right) \tag{9.9}$$

The power delivered by the motor at different speeds can then be expressed as

$$P(\omega) = T\omega = \omega T_s\left(1 - \frac{\omega}{\omega_{max}}\right) \tag{9.10}$$

The maximum power output of the motor occurs at the speed where

$$\frac{dP}{d\omega} = T_s\left(1 - \frac{2\omega}{\omega_{max}}\right) = 0 \tag{9.11}$$

Solving for the speed gives

$$\omega^* = \frac{1}{2}\omega_{max} \tag{9.12}$$

Therefore, the best speed to run a permanent magnet motor to achieve maximum power output is half of the no-load speed.

In addition to the electrical and torque constants, manufacturers often also specify the armature resistance $R$. This value is useful in determining the ***stall current*** $I_s$ of the motor:

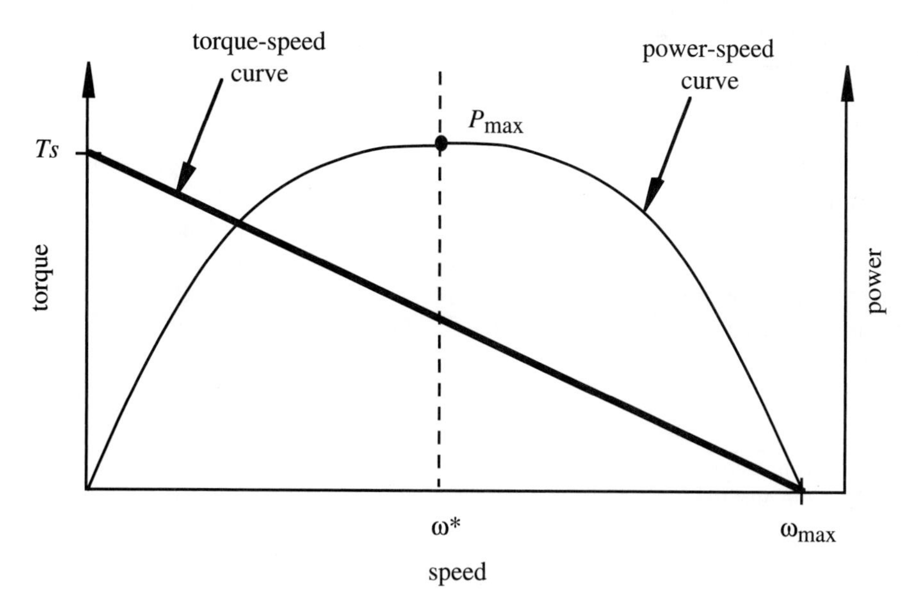


**FIGURE 9.16**
Permanent magnet dc motor characteristics.

$$I_s = \frac{V_{\text{in}}}{R} \tag{9.13}$$

This equation for current is valid only when the motor rotor is not turning; otherwise, the rotor current is affected by the back emf induced in the rotor windings. The stall current is the maximum current through the motor for a given supply voltage. Equations 9.4 and 9.13 can be used to relate the stall torque $T_s$ to the torque constant, supply voltage, and armature resistance:

$$T_s = k_t \frac{V_{\text{in}}}{R} \tag{9.14}$$

### 9.5.3 Electronic Control of a PM dc Motor

The simplest form of motor control is open loop control, where one simply sets the drive voltage value and the motor characteristics and load determine the operating speed and torque. But most of the interesting problems require some sort of automatic control where the voltage is automatically varied to produce a desired motion. This is called ***closed loop*** or ***feedback control***, and it requires an output speed and/or torque sensor to feed back output values in order to continuously compare the actual output to a desired value called the ***set point***. The controller then actively changes the motor output to move closer to the set point. Electronic speed controllers are of two types: linear amplifiers and pulse width modulators. Although both systems can be designed to function well, pulse width modulation controllers have the advantage that they drive bipolar power transistors rapidly between cutoff and saturation where operation is very efficient (power dissipation is minimized) or turn FETs on and off. Servo amplifiers using linear power amplification are satisfactory but require dissipation of a lot of heat since they function in the linear region of transistor operation. You will find commercial servo controllers using linear amplifiers, but because of lower power requirements, ease of design, smaller size, and lower cost, we will focus on the switched amplifier designs, which are generally called ***pulse width modulation (PWM)*** amplifiers.

The principle of a PWM amplifier is shown in Figure 9.17. A dc power supply voltage is rapidly switched at a fixed frequency $f$ between two values (e.g., "ON" and "OFF"). This frequency is often in excess of 1 KHz. The high value is held during a variable pulse width $t$ during the fixed period $T$ where

$$T = \frac{1}{f} \tag{9.15}$$

The resulting asymmetric waveform has a ***duty cycle,*** defined as the ratio between the ON time and the period of the waveform, usually specified as a percentage:

$$\text{duty cycle} = \frac{t}{T}100\% \tag{9.16}$$

As the duty cycle is changed (by the controller), the average current through the motor will change, causing changes in speed and torque at the output. It is pri-

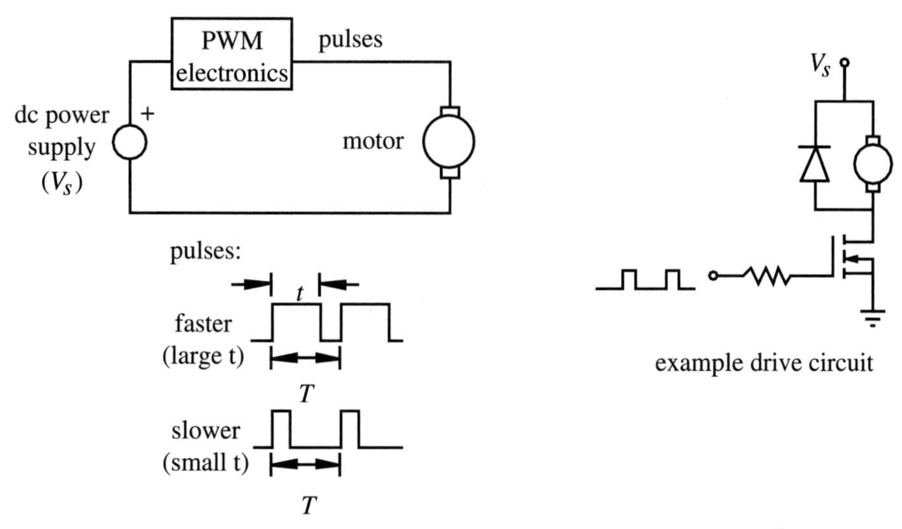


**FIGURE 9.17**
Pulse width modulation of a dc motor.

marily the change in the duty cycle and not the value of the power supply voltage alone that determines the output characteristics of the motor.

The block diagram of a PWM speed feedback control system for a dc motor is shown in Figure 9.18. A voltage tachometer produces an output linearly related to the motor speed. This is compared to the desired speed set point (another voltage that can be manually set or computer controlled). The error and the motor current are sensed by a pulse width modulation regulator that produces a width-modulated square wave as an output. This signal is amplified to a level appropriate for the voltage drive for the motor.

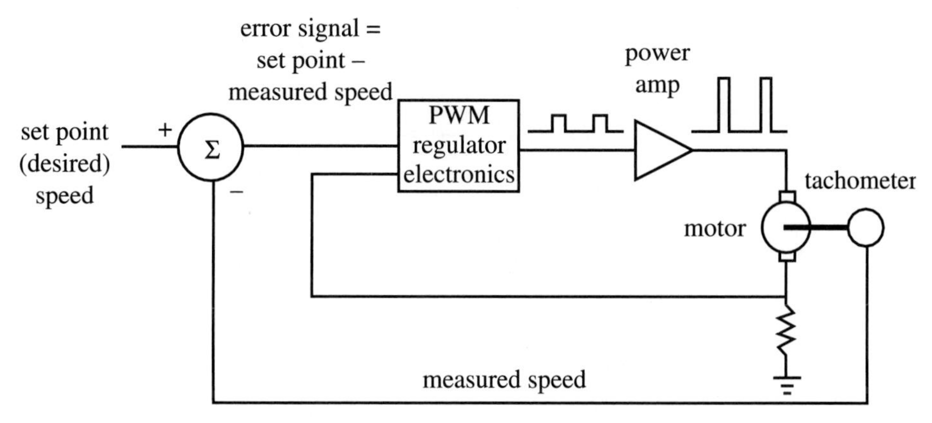


**FIGURE 9.18**
PWM velocity feedback control.

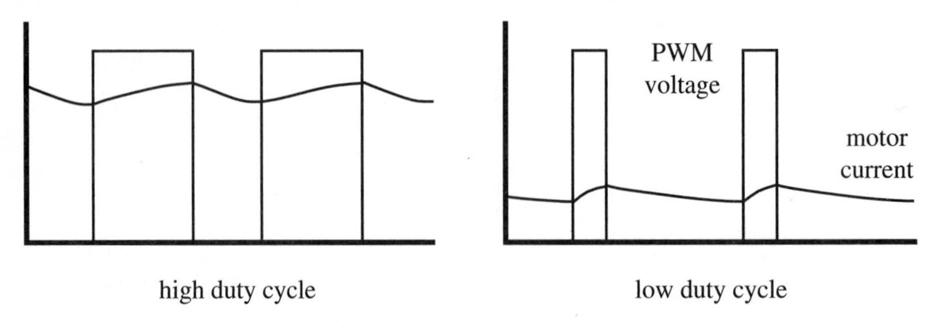


**FIGURE 9.19**
PWM voltage and motor current.

In PWM motor control the voltage is switching rapidly across the armature, and the current through the motor is affected by the motor inductance and resistance. Since the switching speed is high, the resulting current through the motor has a small fluctuation around an average value, as illustrated in Figure 9.19. As the duty cycle gets larger, the average current gets larger and the motor speed increases.

---

▼▼ ***DESIGN EXAMPLE 9.1. H-Bridge Drive for a dc Motor.*** Although it is possible to design and build a drive circuit for a servomotor with discrete control and power components, there are several integrated circuit designs available that save enormously on time and money in mechatronic design. Consider the very ordinary problem of controlling a dc motor. Your ultimate intent may be to control speed, direction of rotation, angle, and/or torque.

In order to control the speed of a dc motor, we must be able to change the current supplied to the motor. To control the direction of rotation, the direction of current supplied to the motor must be reversible. This requires a current amplifier and some means to switch the current direction. The concept of an ***H-bridge*** meets these requirements. We use four power transistors arranged in an H configuration around the dc motor (see the figure below) and alternately turn on one pair at a time for the desired direction of motion.

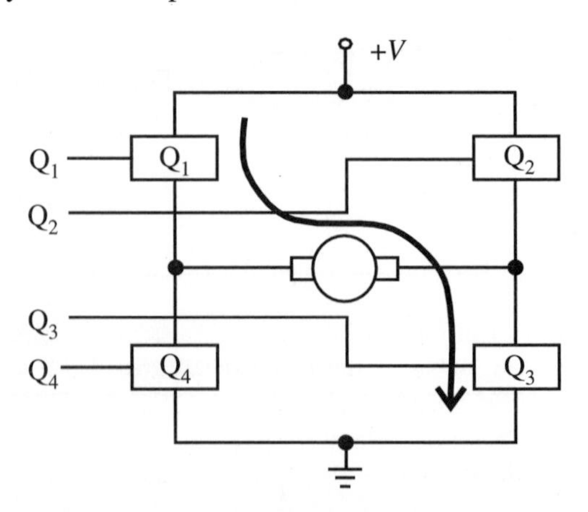

If transistors $Q_1$ and $Q_3$ are turned on and $Q_2$ and $Q_4$ are off, current will flow through the motor in the direction shown and the motor will rotate in one direction. Alternatively, if transistors $Q_2$ and $Q_4$ are turned on and $Q_1$ and $Q_3$ are off, the motor will rotate in the other direction. One can build a discrete H-bridge with power BJTs or MOSFETs, but it may be difficult to properly choose and bias the transistors. Therefore, we will utilize a monolithic solution using National Semiconductor's line of motion control ICs, which can be conveniently adapted for driving dc motors. Consider the LMD18200, a 3 A, 55 V H-bridge specifically designed to drive dc and stepper motors. In addition to simple control strategies, the user can employ overcurrent and overtemperature safety features, can pulse width modulate the system, and can dynamically brake the motor. The functional diagram is shown below.

**Functional Diagram**

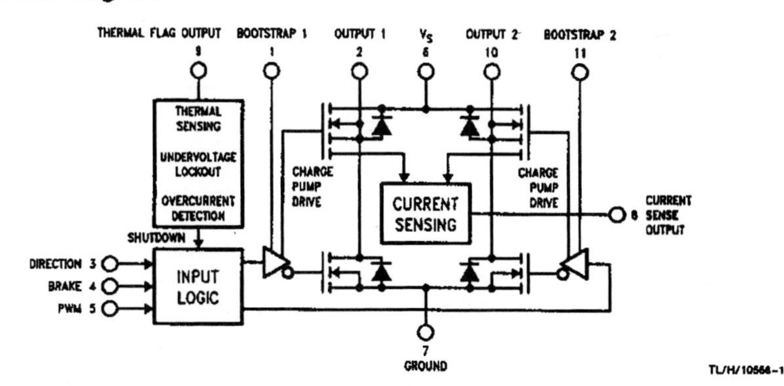


**Function Diagram and Ordering Information**

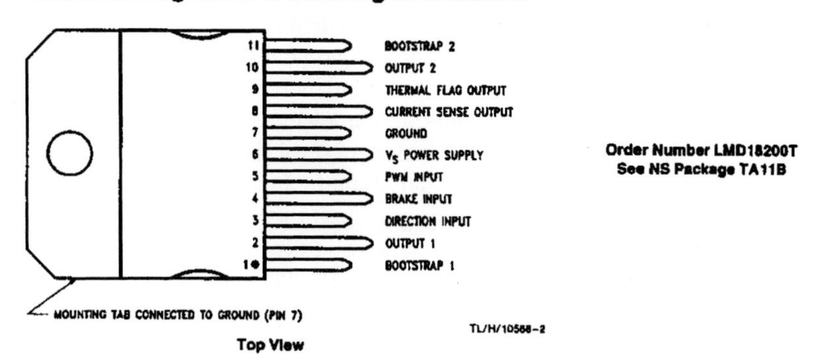


*(Courtesy of National Semiconductor Inc., Santa Clara, CA)*

This design uses power MOSFETs with flyback protection diodes that suppress large reverse transients across the transistors. The motor poles are connected between Output 1 and Output 2. The voltage supply can be up to 55 V. External digital signals control direction, braking, and pulse width modulation. There is a thermal sensor that will shut down the outputs when 170°C is exceeded. Thermal warming is indicated when the temperature of the device exceeds 145°C.

The complete block diagram for our speed controller design is shown in the figure below. The speed set point is fixed with a potentiometer or input voltage value. A quadrature tachometer is added to the motor as the sensor to provide a measure of the motor speed. The National Semiconductor pulse width modulation IC LM 3525A is conveniently used here to drive the H-bridge motor controller input.

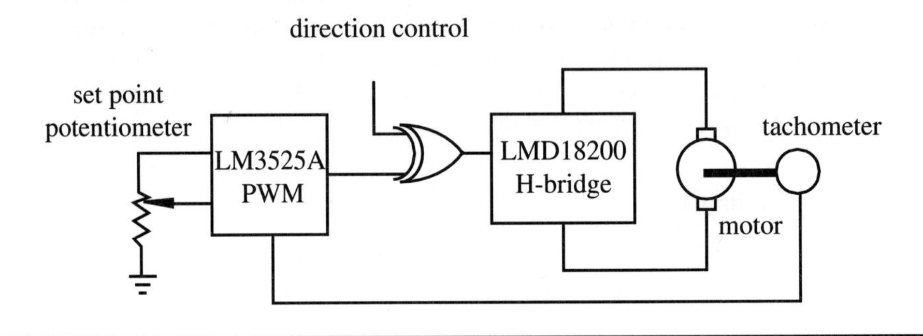


## 9.6 STEPPER MOTORS

A special type of dc motor known as a ***stepper motor*** is a permanent magnet or variable reluctance dc motor that has the following performance characteristics: rotation in both directions, precision angular incremental changes, repetition of accurate motion or velocity profiles, a holding torque at zero speed, and capability for digital control. It can move in accurate angular increments known as steps in response to the application of digital pulses to an electric drive circuit from a digital controller. The number and rate of the pulses control the position and speed of the motor shaft. Generally, stepper motors are manufactured with steps per revolution of 12, 24, 72, 144, 180, and 200, resulting in shaft increments of 30°, 15°, 5°, 2.5°, 2°, and 1.8° per step. Special ***micro-stepping*** circuitry is also sometimes provided to allow many more steps per revolution, often 10,000 steps/rev or more.

Stepper motors are either ***bipolar***, requiring two power sources or a switchable polarity power source, or ***unipolar***, requiring only one power source. They are powered by dc current sources and require digital circuitry to produce the coil energizing sequences for rotation of the motor. Feedback is not always required for control, but the use of an encoder or other position sensor can ensure accuracy when it is essential. The advantage of operating without feedback (i.e., in open loop mode) is that a closed loop control system is not required. Generally, stepper motors produce less than 1 horsepower (746 W) and are therefore frequently used in low-power position control applications.

A commercial stepper motor has a large number of poles that define a large number of equilibrium positions of the rotor. In the case of a permanent magnet stepper motor, the stator consists of wound poles, and the rotor poles are permanent magnets. Exciting different stator winding combinations moves and holds the rotor in different positions. The ***variable reluctance*** stepper motor has a ferromagnetic rotor rather than a permanent magnet rotor. Motion and holding are results of minimization of the magnetic reluctance between the stator and rotor poles. A variable reluctance motor has the advantage of a lower rotor inertia and therefore a faster dynamic response. The permanent magnet motor has the advantage of a small residual holding torque, called the ***detent torque***, even when the stator is not energized.

To understand how the rotor moves in an incremental fashion, consider a simple design consisting of four stator poles and a permanent magnet rotor as illustrated in Figure 9.20. In step 0, the rotor is in equilibrium since opposite poles on the stator and rotor are adjacent to and attract each other. Unless the magnetization of the stator poles is changed, the rotor remains in this position and can withstand an opposing torque up to a value called the ***holding torque***. When the pole magnetization is changed as shown (step 0 to step 1), a torque is applied to the rotor, causing it to move 90° in the clockwise direction to a new equilibrium position shown as step 1. When the magnetization of the poles is again changed as shown (step 1 to step 2), the rotor experiences a torque driving it to step 2. By successively changing the magnetization of the poles in this manner, the motor can rotate to successive equilibrium positions in the clockwise direction. We can see that the sequencing of the pole excitation is the means by which the direction of rotation occurs. Counterclockwise motion can be achieved by applying the polarization sequence in the opposite direction. The motor torque is directly related to the magnetic field strength of the poles and the rotor.

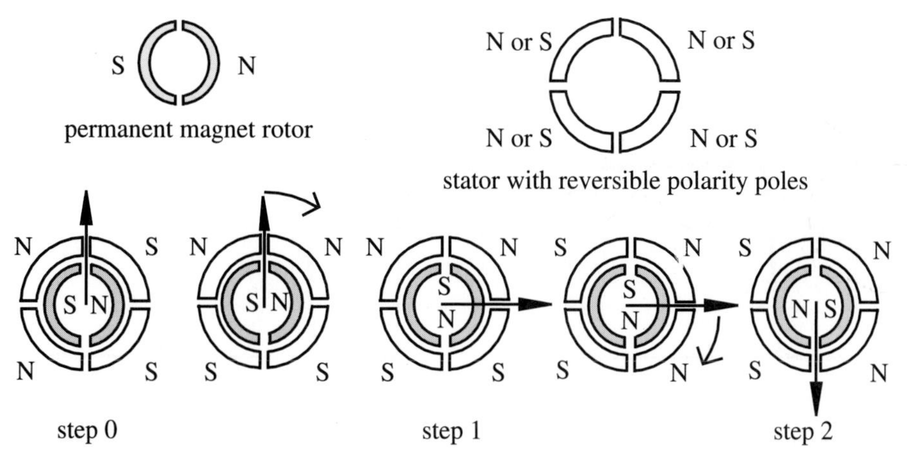


**FIGURE 9.20**
Stepper motor step sequence.

The dynamic response of the rotor and attached load must be carefully considered in applications that involve starting or stopping quickly, changing or ramping speeds quickly, or driving large or changing loads. Due to the inertia of the rotor and attached load, rotation can exceed the desired number of steps. Also, a stepper motor driving a typical mechanical system through one step will exhibit an underdamped response, as illustrated in Figure 9.21, as it increments. If damping is increased in the system, for example, with mechanical, frictional, or viscous damping, the response can be modified to reduce oscillation, as shown in Figure 9.21. Note, however, that even with an ideal choice for damping, the motor does require time to totally settle into a given position, and this settling time varies with the step size and the amount of damping. It is also important to note that the torque required from the motor increases with added damping.

The torque-speed characteristics for a stepper motor are usually divided into two regions as illustrated in Figure 9.22. In the ***locked step mode***, the rotor decelerates and may even come to rest between each step. Within this region, the motor can be instantaneously started, stopped, or reversed without losing step integrity. In the ***slewing mode***, the speed is too fast to allow instantaneous starting, stopping, or reversing. The rotor must be gradually accelerated to enter this mode and gradually decelerated to leave the mode. While in slewing mode, the rotor is in synch with the stator field rotation and doesn't settle between steps. The curve between the regions in the figure indicates the maximum torques that the stepper can provide at different speeds without slewing. The curve bordering the slewing mode region represents the absolute maximum torques the stepper can provide at different speeds.

A unipolar stepper motor field coil schematic is illustrated in Figure 9.23 with external power transistors that must be switched on and off to produce the controlled sequence of pole magnetization to cause rotation. For full-step motion, only two transistors are on at any one time. The wire colors indicated in the figure are standard for most manufacturers.

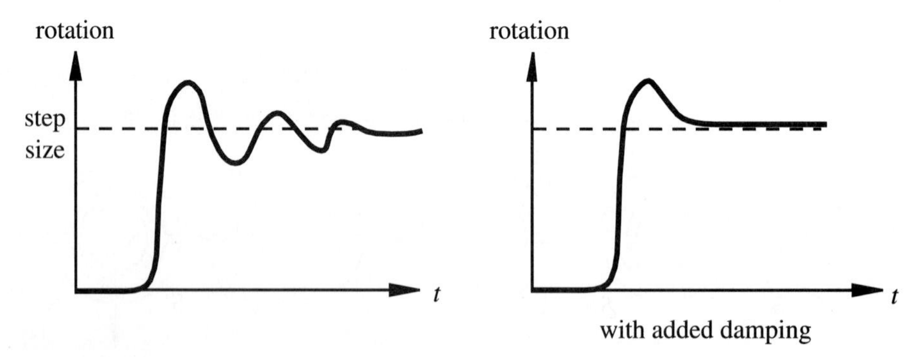


**FIGURE 9.21**
Dynamic response of a single step.

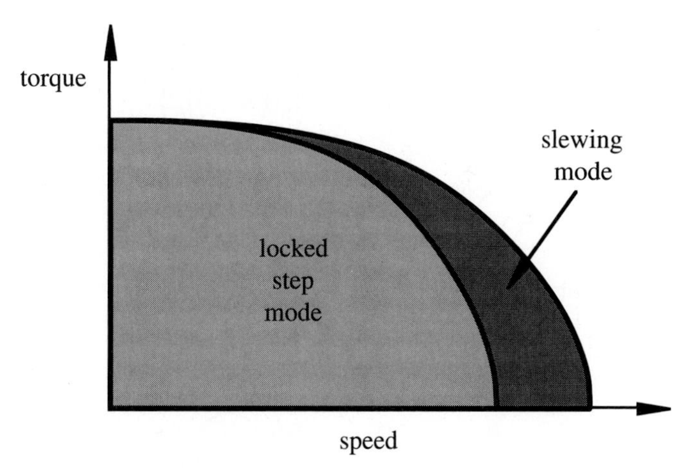


**FIGURE 9.22**
Stepper motor torque-speed curves.

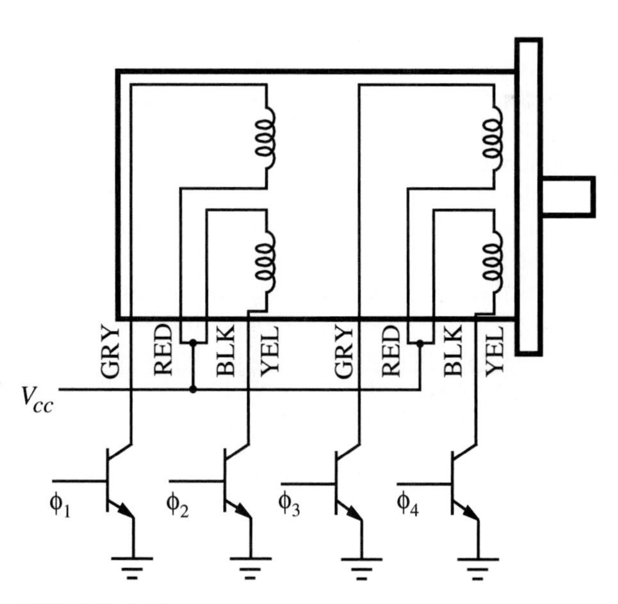


**FIGURE 9.23**
Standard unipolar stepper motor field coil schematic.

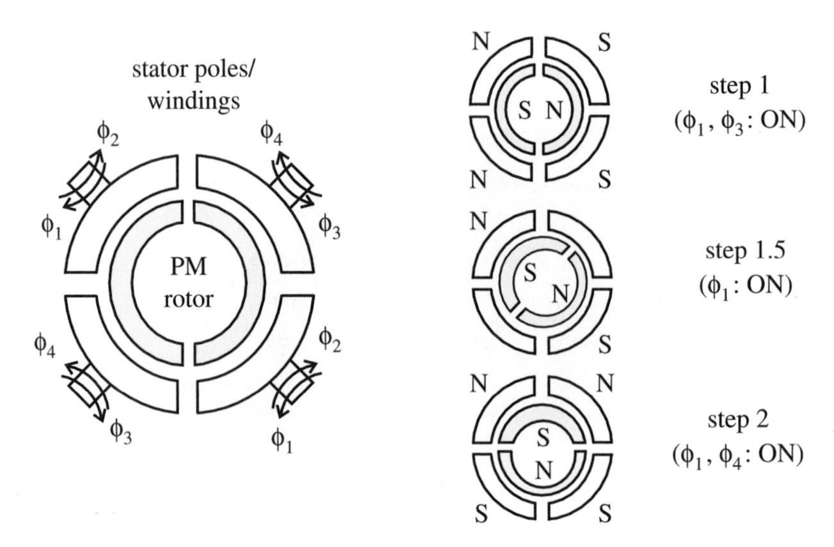


**FIGURE 9.24**
Example of a unipolar stepper motor.

Figure 9.24 illustrates the construction of and stepping sequence for a four-phase unipolar stepper motor. It consists of a two-pole permanent magnet rotor and a four-pole stator with each pole wound by two complementary windings (e.g., $\phi_1$ and $\phi_2$ wound in opposite directions on the top left pole). Table 9.1 lists the phase sequence to step the motor in full steps, where two of the four phases are energized (ON) and each stator pole is energized. Table 9.2 lists the phase sequence for half-stepping, where between each full step only one phase is on and only two stator poles are energized. The resolution or number of steps of the motor is twice as large in the half-step mode (8 steps/rev at 45°) than in the full-step mode (4 steps/rev at 90°), but the holding torque and drive torque change between two values on alternate cycles. Another technique for increasing the number of steps is called micro-stepping, where the phase currents are controlled by fractional amounts (rather than just ON/OFF), resulting in more magnetic equilibrium positions between the poles. In effect, discretized sine waves are applied to the phases instead of square waves. The most common commercially available stepper motors have 200 steps/rev in full-step mode and are sometimes referred to as 1.8° (360°/200) steppers. In micro-stepping mode, 10,000 or more steps per revolution can be achieved.

**TABLE 9.1**
**Unipolar full-step phase sequence**

| | Step | $\phi_1$ | $\phi_2$ | $\phi_3$ | $\phi_4$ |
|---|---|---|---|---|---|
| CW ↓ | 1 | ON | OFF | ON | OFF |
| | 2 | ON | OFF | OFF | ON |
| CCW ↑ | 3 | OFF | ON | OFF | ON |
| | 4 | OFF | ON | ON | OFF |

TABLE 9.2
**Unipolar half-step phase sequence**

| | Step | $\phi_1$ | $\phi_2$ | $\phi_3$ | $\phi_4$ |
|---|---|---|---|---|---|
| CW ↓ | 1 | ON | OFF | ON | OFF |
| | 1.5 | ON | OFF | OFF | OFF |
| | 2 | ON | OFF | OFF | ON |
| | 2.5 | OFF | OFF | OFF | ON |
| CCW ↑ | 3 | OFF | ON | OFF | ON |
| | 3.5 | OFF | ON | OFF | OFF |
| | 4 | OFF | ON | ON | OFF |
| | 4.5 | OFF | OFF | ON | OFF |

Figure 9.25 illustrates the structure, pole geometry, and coil connections of an actual stepper motor in more detail. This particular stepper motor can be wired as a four-phase unipolar motor or a two-phase bipolar motor. Figure 9.26 shows the 50-tooth split rotor with one side magnetized north and the other south.

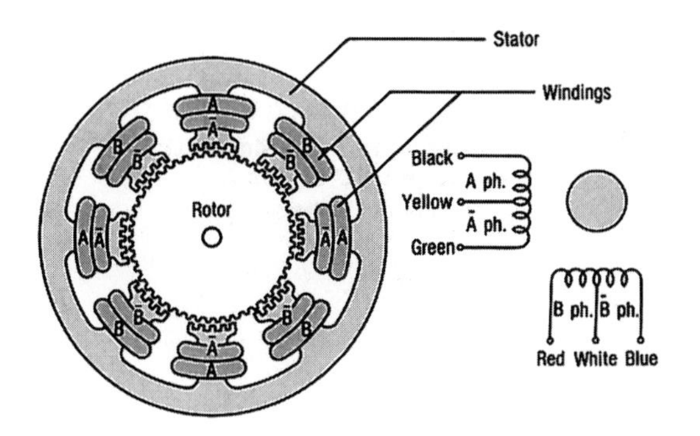


**FIGURE 9.25**
Typical stepper motor rotor and stator configuration. *(Courtesy of Oriental Motor, Torrance, CA)*

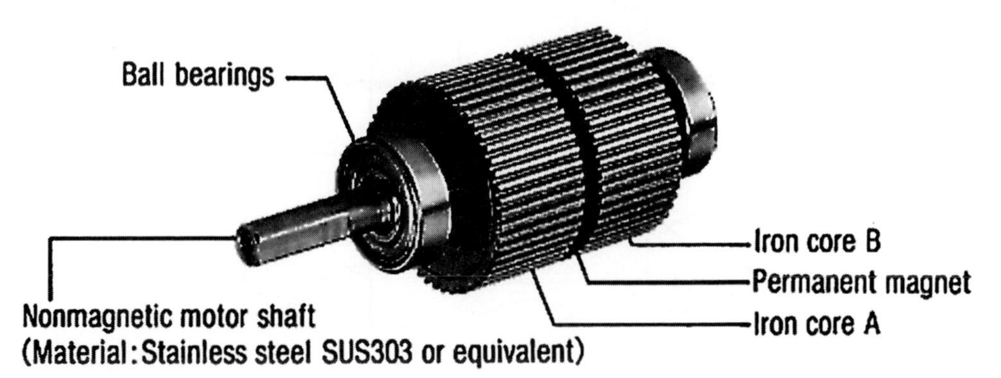


**FIGURE 9.26**
Actual stepper motor rotor. *(Courtesy of Oriental Motor, Torrance, CA)*

## 9.7 STEPPER MOTOR DRIVE CIRCUITS

A drive circuit for properly phasing the signals applied to the poles of the unipolar stepper motor for rotation in full-step mode is easily and economically produced using the components illustrated in Figure 9.27. A similar drive circuit can be purchased as a single monolithic IC (e.g., Signetic SAA1027 or Allegro Microsystems' UCBN 5804B). The discrete circuit includes 7414 Schmitt Trigger Buffers, a 74191 up-down counter, and 7486 exclusive OR gates. As shown in Chapter 6 (see Figure 6.32), the Schmitt triggers produce well-defined control signals with sharp rise and fall times in the presence of noise or fluctuations. The hysteresis of the Schmitt triggers provides sharp square wave signals for the direction (CW/CCW), initialization (RESET), and single step (STEP) inputs. The up-down counter and the XOR gates in turn create four properly phased motor drive signals. These four digital signals ($\phi_1$, $\phi_2$, $\phi_3$, $\phi_4$) are coupled to the bases of power transistors that sequentially energize the respective motor coils connected to the dc motor supply, resulting in its rotation. Each square wave pulse received at the STEP input causes the motor to rotate a full step in the direction determined by the CW/CCW input.

The timing diagram for the two least significant output bits $B_0$ and $B_1$ of the counter and the phase control signals is shown in Figure 9.28. Compare the signals to the sequence in Table 9.1. They are in agreement.

Boolean expressions that produce the four desired phased outputs from the two counter bits can represented in both AND-OR-NOT and XOR forms:

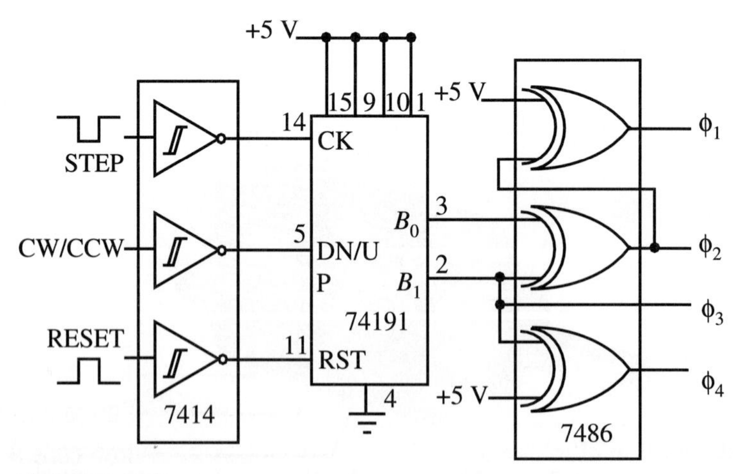


**FIGURE 9.27**
Unipolar stepper motor full-step drive circuit.

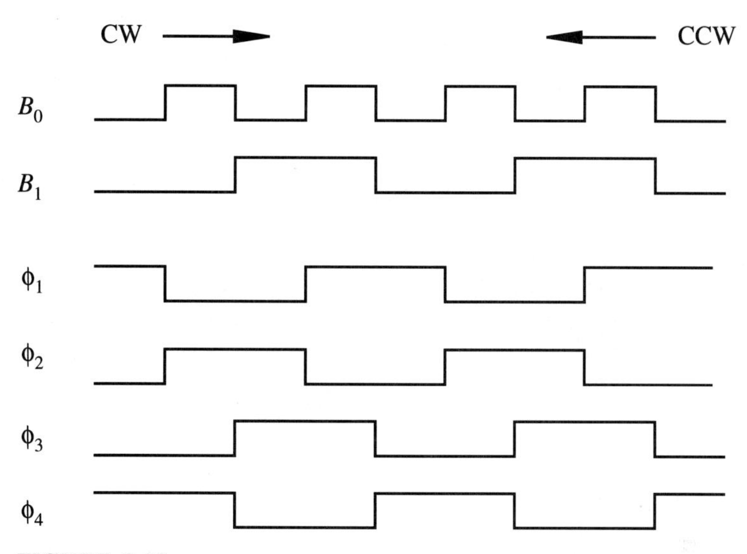


**FIGURE 9.28**
Timing diagram for full-step stepper motor drive circuit.

$$\begin{aligned} \phi_1 &= \overline{\phi_2} = \phi_2 \oplus 1 \\ \phi_2 &= (B_0 \cdot \overline{B_1}) + (\overline{B_0} \cdot B_1) = B_0 \oplus B_1 \\ \phi_3 &= B_1 \\ \phi_4 &= \overline{B_1} = B_1 \oplus 1 \end{aligned} \tag{9.17}$$

These expressions can be verified by checking the signal values at different times in the timing diagram shown in Figure 9.28. The purpose for representing the Boolean expressions in XOR form is to allow the logic to be executed using a single IC (the quad XOR 7486); otherwise three ICs would be required for the AND, OR, INV representation.

## 9.8 SELECTING A MOTOR

When selecting a motor for a specific engineering application, the designer must consider many factors and specifications, including speed range, torque-speed variations, reversibility, duty cycle, starting torque, and power required. These and other factors are described below in a list of questions that a designer must consider when selecting and sizing a motor in consultation with a motor manufacturer. As we will see, the torque-speed curve provides important information, helping to

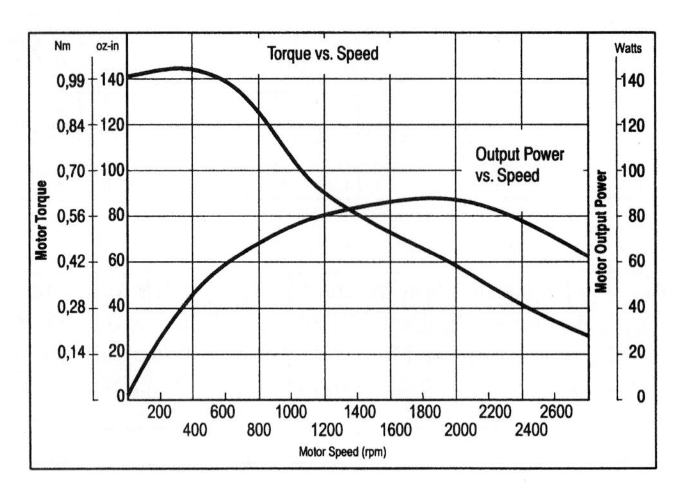


**FIGURE 9.29**
Typical stepper motor performance curves. *(Courtesy of Aerotech, Pittsburgh, PA)*

answer many questions about a motor's performance. Recall that the torque-speed curve displays the torques the motor can deliver at different speeds at rated voltage. Figure 9.29 shows an example of a torque-speed curve for a stepper motor, and Figure 9.30 shows an example of a torque-speed curve for a servo motor.

Some of the salient questions a designer may need to consider when choosing a motor for an application include the following:

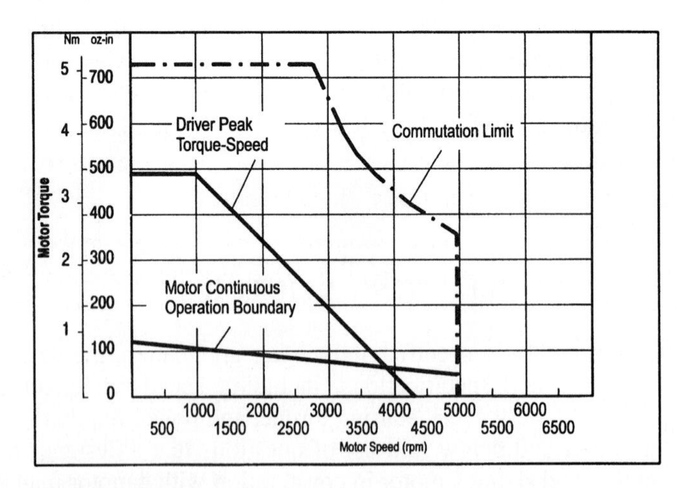


**FIGURE 9.30**
Typical servomotor performance curves. *(Courtesy of Aerotech, Pittsburgh, PA)*

- *Will the motor start and will it accelerate fast enough?* The torque at zero speed, called the ***starting torque***, is the torque that the motor can deliver when rotation begins. For the system to be self-starting, the motor must generate torque sufficient to overcome friction and any load torques. The series configuration provides the greatest torque at low speeds for dc motors.

  The acceleration of the motor and load at any instant is given by

  $$\alpha = (T_{\text{motor}} - T_{\text{load}})/J \tag{9.18}$$

  where $\alpha$ is the angular acceleration in rad/sec$^2$, $T_{\text{motor}}$ is the torque produced by the motor, $T_{\text{load}}$ is the torque dissipated by the load, and $J$ is the total polar moment inertia of the motor rotor and the load. The difference between motor and load torques determines the acceleration of the system. When the motor torque is equal to the load torque, the system has reached a steady state operating speed.

- *What is the maximum speed the motor can produce?* The zero load point on the torque-speed curve determines the maximum speed a motor can reach. Note that the motor cannot deliver any torque at this speed. When the motor is loaded, the maximum ***no-load speed*** cannot be achieved.

- *What is the duty cycle?* When a motor is not operated continuously, one must consider the operating cycle of the system. The ***duty cycle*** is defined as the ratio of the time the motor is on with respect to the total elapsed time. If a load has a low duty cycle, a lower power motor may be selected that can operate above rated levels but will still perform adequately without overheating during repeated on-off cycles.

- *How much power does the load require?* The power rating is a very important specification for a motor. Knowing the power requirements of the load, a designer should choose a motor with adequate power based on the duty cycle.

- *What is the load inertia?* As Equation 9.18 implies, for fast dynamic response, it is desirable to have low motor rotor and load inertia $J$. When the load inertia is large, the only way to achieve high acceleration is to size the motor so it can produce much larger torques than the load requires. Printed circuit rotor pancake motors have very low rotor inertia allowing for extremely fast response.

- *Is the load to be driven at constant speed?* The simplest method to achieve constant speed is to select an ac synchronous motor or a dc shunt motor that runs at constant speed over a significant range of load torques.

- *Is accurate position or speed control required?* In the cases of angular positioning at discrete locations and incremental motion, a stepper motor is a good choice. A stepper motor is easily rotated to and held at discrete positions. It also can rotate at a wide range of speeds by controlling the step rate. The stepper motor can be operated with open loop control, where no sensor feedback is required. However, if you attempt to drive a stepper motor at too fast a step rate or if the load torque is too large, the stepper motor may slip and not execute the number of steps expected. Therefore, a feedback sensor such as an encoder might be included with a stepper motor to check if the motor has achieved the desired motion.

  For some complex motion requirements where precise position or speed profiles are required (e.g., in automation applications where machines need to perform prescribed programmed motion), servomotors may be used. A ***servomotor*** is a dc, ac, or brushless dc motor combined with a position sensing device (e.g., a digital encoder). The servomotor is driven by a programmable controller that processes the sensor input and generates amplified voltages and currents to the motor to achieve specified motion profiles. This is called closed loop control since it includes sensor feedback. A servomotor is typically more expensive than a stepper motor, but it has a much faster response.

- *Is a transmission or gearbox required?* Often loads operate at low speeds and require large torque. Since motors function at high speed and low torque, a speed reducing transmission (gear box or belt drive) is needed to match the motor output to the load requirements. The term ***gear motor*** is used to refer to a motor-gearbox assembly.

  When a transmission is used, the effective inertia of the load as seen by the motor is also changed according to

  $$J_{\text{eff}} = J_{\text{load}}\left(\frac{\omega_{\text{load}}}{\omega_{\text{motor}}}\right)^2 \tag{9.19}$$

  where $J_{\text{eff}}$ is the effective polar moment of inertia of the load as seen by the motor. The sum of this inertia and the motor rotor inertia can be used in Equation 9.18 to calculate acceleration.

- *Is the motor torque-speed curve well matched to the load?* If the load has a well-defined torque-speed relation, called a ***load line***, it is wise to select a motor with a similar torque-speed characteristic. If this is the case, the motor torque can match the load torque over a large range of speeds, and the speed can be controlled easily by making small changes in voltage to the motor.

- *For a given motor torque-speed curve and load line, what will the operating speed be?* As the figure below illustrates, for a given motor curve and for a well-defined load line, the system will settle at a fixed speed operating point. The operating point is self-regulating. At lower speeds the motor torque exceeds the load torque and the system accelerates toward the operating point, but at higher speeds the load torque exceeds the motor torque, bringing the speed down toward the operating point.

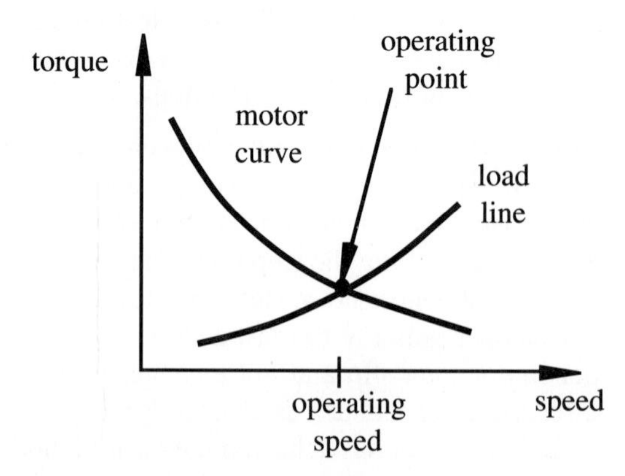


The operating speed can be actively changed by adjusting the voltage supplied to the motor, which in turn changes the torque-speed characteristic of the motor.

- *Is it necessary to reverse the motor?* Some motors are not reversible due to their construction and control electronics, and care must be exercised in selecting a motor for an application requiring rotation in two directions.
- *Are there any size and weight restrictions?* Motors can be large and heavy, and designers need to be aware of this early in the design phase.

▼▼▼ ***CLASS DISCUSSION ITEM 9.4. Motor Sizing.*** Why is it important not to oversize a motor for a particular application?

▼▼▼ ***CLASS DISCUSSION ITEM 9.5. Examples of Electric Motors.*** Make of list of the different types of electric motors found in household devices and automobiles. Describe the reasons why you think the type of motor is used in the examples you cite.

## 9.9 HYDRAULICS

Hydraulic systems are designed to move large loads by controlling a high-pressure fluid in distribution lines and pistons with mechanical or electromechanical valves. A hydraulic system, illustrated in Figure 9.31, consists of a pump to deliver high-pressure fluid, a pressure regulator to limit the pressure in the system, valves to control flow rates and pressures, a distribution system composed of hoses or pipes, and linear or rotary actuators. The infrastructure, which consists of the elements contained in the dashed box in the figure, is typically used to power many hydraulic valve-actuator subsystems.

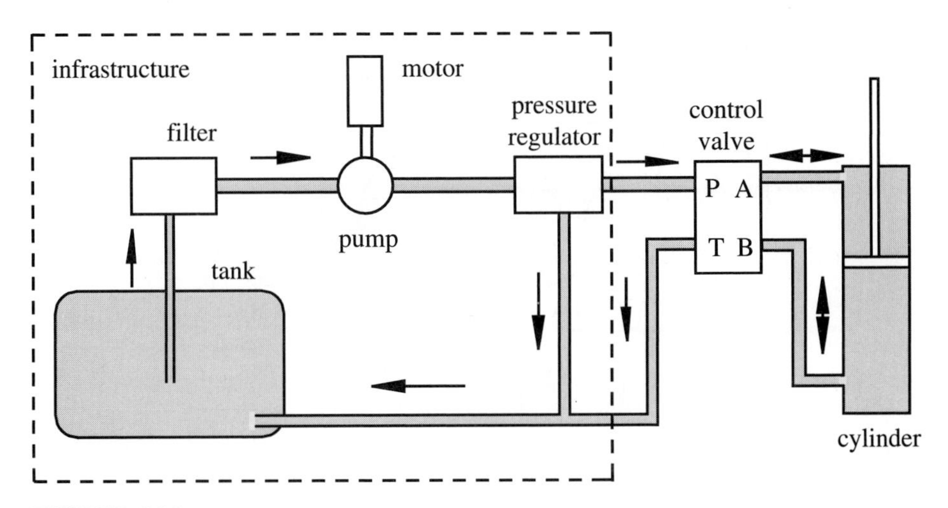


**FIGURE 9.31**
Hydraulic system components.

A hydraulic ***pump*** is usually driven by an electric motor (e.g., a large ac induction motor) or an internal combustion engine. Typical fluid pressures generated by pumps used in heavy equipment (e.g., construction equipment and large industrial machines) are in the 1000 psi (6.89 MPa) to 3000 psi (20.7 MPa) range. The hydraulic fluid is selected to have the following characteristics: good lubrication to prevent wear in moving components (e.g., between pistons and cylinders), corrosion resistance, and incompressibility to provide rapid response. Most hydraulic pumps are ***positive displacement,*** which means they deliver a fixed volume of fluid with each cycle or rotation of the pump. The three main types of positive displacement pumps used in hydraulic systems are gear pumps, vane pumps, and piston pumps. An example of a ***gear pump,*** which displaces the fluid around a housing between teeth of meshing gears, is shown in Figure 9.32. Note that the meshing teeth provide a seal, and the fluid is displaced from the inlet to the outlet along the nonmeshing side of the gears.

Figure 9.33 illustrates an example of a ***vane pump,*** which displaces the fluid between vanes guided in rotor slots riding against the housing and vane guide. The vane guide supports the vanes from one side of the housing to the next and is constructed to the allow the fluid to pass. The output displacement can be varied (with a constant motor speed) by moving the shaft vertically relative to the housing.

Figure 9.34 illustrates an example of a ***piston pump***. The cylinder block is rotated by the input shaft, and the piston ends are driven in and out as they ride in the fixed ***swash plate*** slot that is angled with respect to the axis of the shaft. A piston draws fluid from an inlet manifold over half the swash plate and expels fluid into the outlet manifold during the other half. The displacement of the pump can be changed simply by changing the angle of the fixed swash plate. Table 9.3 lists and compares the general characteristics of the different pump types.

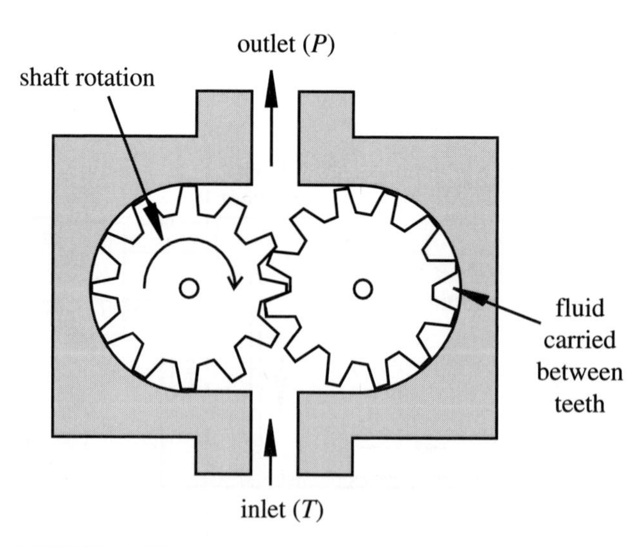


**FIGURE 9.32**
Gear pump.

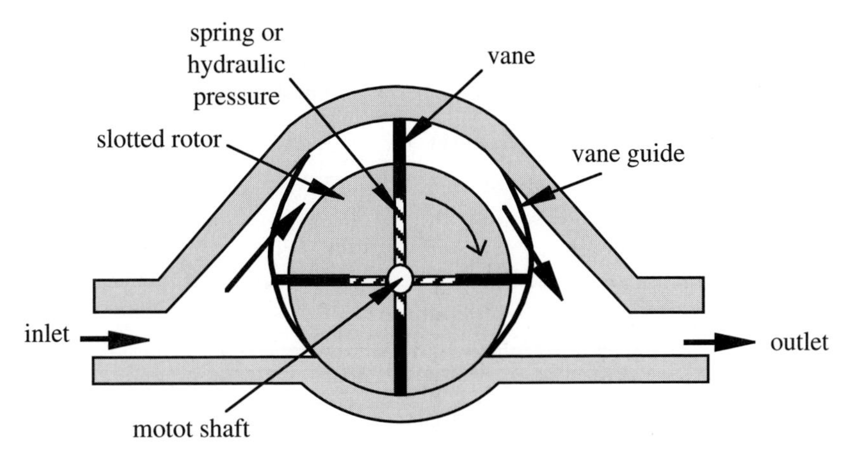


**FIGURE 9.33**
Vane pump.

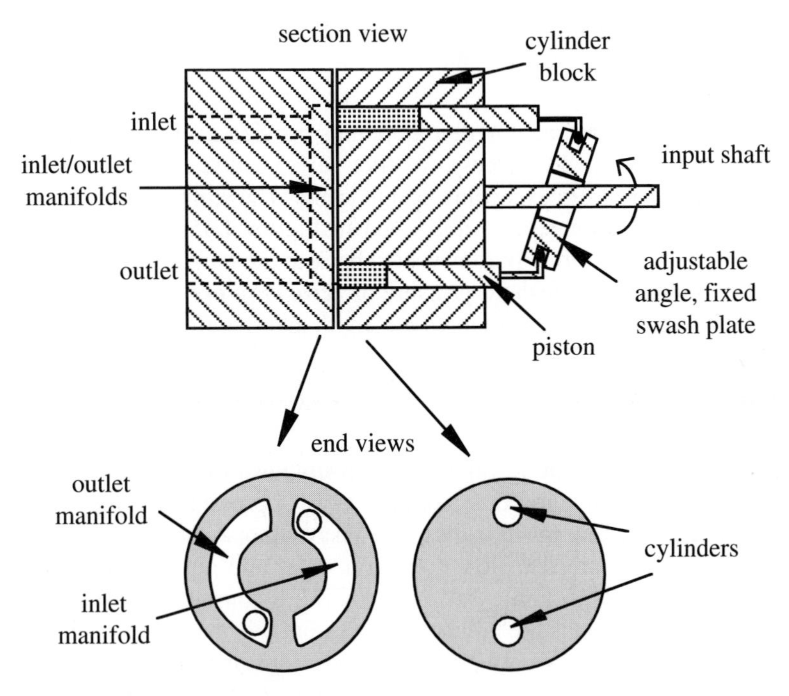


**FIGURE 9.34**
Swash plate piston pump.

**TABLE 9.3**
**Comparison of pump characteristics**

| Pump type | Displacement | Typical pressure (psi) | Cost |
|---|---|---|---|
| Gear | Fixed | 2000 | Low |
| Vane | Variable | 3000 | Medium |
| Piston | Variable | 6000 | High |

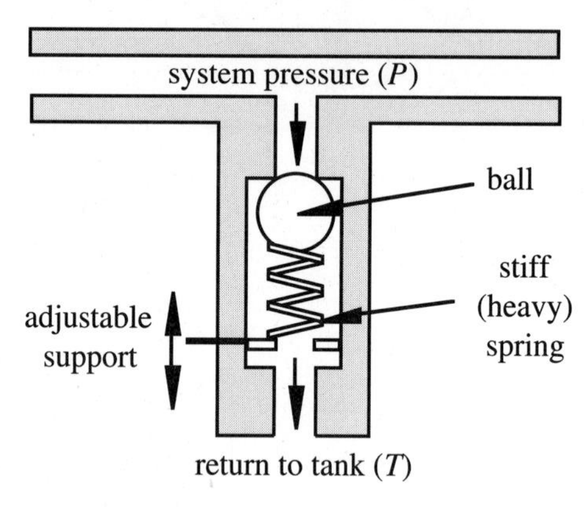


**FIGURE 9.35**
Pressure regulator.

Since positive displacement hydraulic pumps provide a fixed volumetric flow rate, it is necessary to include a pressure relief valve, called a ***pressure regulator***, to prevent the pressure from exceeding design limits. The simplest pressure regulator is the spring-ball arrangement illustrated in Figure 9.35. When the pressure force exceeds the spring force, fluid is vented back to the tank, preventing a further increase in pressure. The threshold pressure or ***"cracking pressure"*** is usually adjusted by changing the spring's compressed length and therefore its resisting force.

### 9.9.1 Hydraulic Valves

There are two types of hydraulic valves: the ***infinite position valve*** that allows any position between open and closed to modulate flow or pressure and the ***finite position valve*** that has discrete positions, usually just two, open and closed, each providing a different pressure and flow condition. Inlet and outlet connections to a valve are also called ***ports***. Finite position valves are commonly described by an "$x/y$" designation, where $x$ is the number of ports and $y$ is the number of positions. As an example, a 4/3 valve, with 4 ports and 3 positions, is illustrated in schematic form in Figure 9.36. In position 1, system pressure is vented to tank; in position 2, output port A is pressurized and port B is vented to tank; and in position 3, output port B is pressurized and port A is vented to tank. As illustrated in Figure 9.37, this particular valve is useful in controlling a double-acting hydraulic cylinder where ports A and B connect to opposite ends of the cylinder applying or venting pressure on opposite sides of the piston. In position 1, the cylinder does not move since the pressure is vented to the tank. In position 2, the cylinder moves to the right since pressure is applied to the left side of the piston. In position 3, the cylinder moves to the left since pressure is applied to the right side of the piston.

Common types of fixed position valves are: check valves, poppet valves, spool valves, and rotary valves. Figure 9.38 illustrates the check and poppet valves. The ***check valve*** allows flow in one direction only. The ***poppet valve*** is a check valve that can be forced open to allow reverse flow.

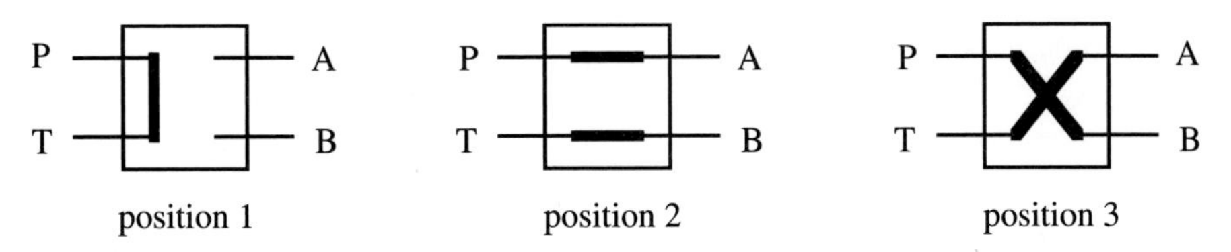


**FIGURE 9.36**
4/3 Valve schematic.

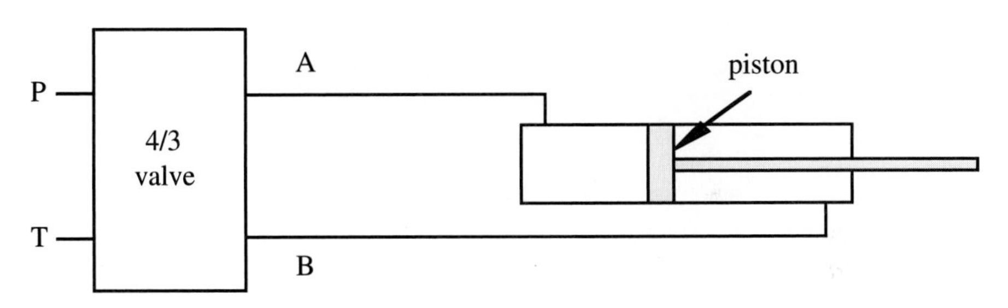


**FIGURE 9.37**
Double-acting hydraulic cylinder.

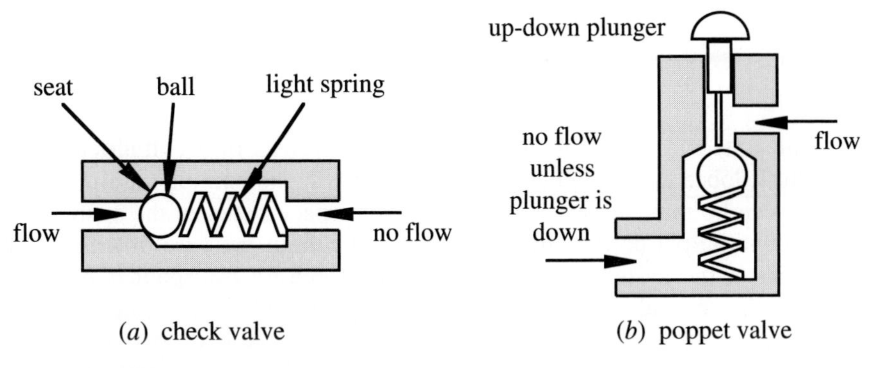


**FIGURE 9.38**
Check and poppet valves.

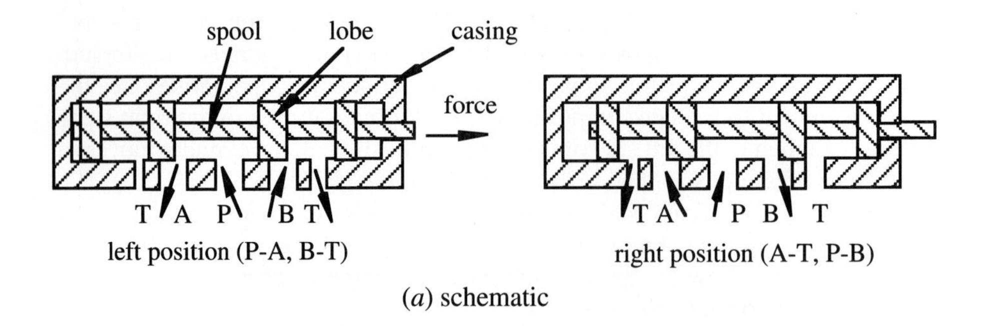


(*a*) schematic

SPOOL "U" GROOVES
Improves contamination tolerance of spools over conventional "V" grooves. Significantly reduces spool hang-up resulting in loss of production.

LOW PRESSURE DROPS
Reduces heat loss and increases efficiency.

SEALED WET ARMATURE SOLENOIDS
Maximum protection against moisture, corrosion and dirt.

OPERATOR PROTECTION
High temperature elements are isolated from direct contact.

STANDARD MOUNTINGS
Conforms to NFPA and ANSI/ISO standards.

2 WIRE LEAD
50/60 Hz coils standard for increased application flexibility.

INTERCHANGEABLE SPOOLS
Provide easy field maintenance. No matching of parts.

(*b*) actual
*(Courtesy of Continental Hydraulics, Savage, MN)*

**FIGURE 9.39**
Spool valve.

As illustrated in Figure 9.39, a ***spool valve*** consists of a cylindrical spool with multiple lobes moving within a cylindrical casing containing multiple ports. The spool can be moved back and forth to align space between the spool lobes with input and output ports in the housing to direct high-pressure flow to different conduits in the system. The static pressure force on the spool is balanced since the pressure is always applied to opposing internal faces of the lobes. To move the spool, an axial force is required to overcome the hydrodynamic forces associated with changing the momentum of the flow. In the left position, port A

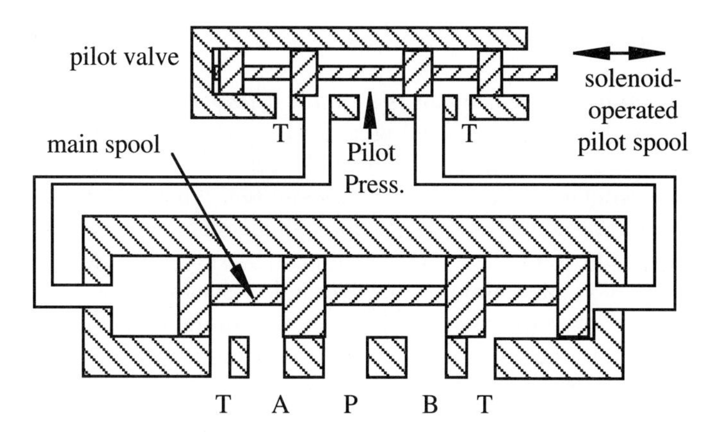


**FIGURE 9.40**
Pilot-operated spool valve.

is pressurized and port B is vented to the tank. A force is required to move the spool from this position to the right position, where port B is pressurized and port A is vented.

In the design of a spool valve where large hydrodynamic forces occur, a ***pilot valve*** is added, as shown in Figure 9.40. The pilot valve operates at a lower pressure, called ***pilot pressure***, and at much lower flow rates and therefore requires lower force to actuate. The pilot valve directs pilot pressure to one side of the main spool, and the force generated by the pressure acting over the main spool lobe face is large enough to actuate the main valve. The effect of the pilot valve is to amplify the force provided by the solenoid acting on the pilot spool. In the figure, the pilot spool is in the full left position, causing pilot pressure to be applied to the left side of the main spool and venting fluid from the right side of the main spool to tank, thus driving the main spool to the full right position. This will apply main pressure to port B and vent port A.

The discussion of spool valves so far has been limited to operation between two positions only: on and off. Continuous operation can be achieved by using a ***proportional valve***, one whose spool moves a distance proportional to a mechanical or electrical input (e.g., a lever or a solenoid), thus changing the rate of flow and varying the speed and force of the actuator. When the spool position is controlled by electrical solenoids, the proportional valve is called an ***electrohydraulic valve***. These valves may be used in open loop control situations with no feedback, but they often include sensors to monitor spool position or actuator output. Proportional valves equipped with sensor and control circuitry are often called ***servovalves***. Electrohydraulic valves are often pilot-operated where the solenoids drive the pilot spool, which in turn controls the position of the primary spool. The pilot spool can be driven by a single solenoid with spring return or a set of opposing solenoids. The solenoid currents can be controlled by amplifiers linked to analog or digital controllers.

### 9.9.2 Hydraulic Actuators

The most common hydraulic actuator is a simple ***cylinder*** with a piston driven by the pressurized fluid. As illustrated in Figure 9.41, a cylinder can be ***single-acting,*** where it is held in one position by pressure and returned to the other position by a spring or by the weight of the load, or ***double-acting,*** where pressure is used to drive the piston in both directions. Hydraulic cylinders are usually double-acting. As illustrated in Figure 9.42, the linear actuator can be very versatile in achieving a variety of motions. Cylinder motion in the hydraulic elevator drives the elevator directly. The scissor jack converts small linear motion in the horizontal direction to larger linear motion in the vertical direction. Linear motion of the cylinder in the crane results in rotary motion of its pivoted boom.

Rotary motion can also be achieved with hydraulic systems directly with a rotary actuator. One type of rotary actuator, called a ***gear motor***, is simply a gear pump (see Figure 9.32) driven in reverse where pressure is applied, resulting in rotation of a shaft.

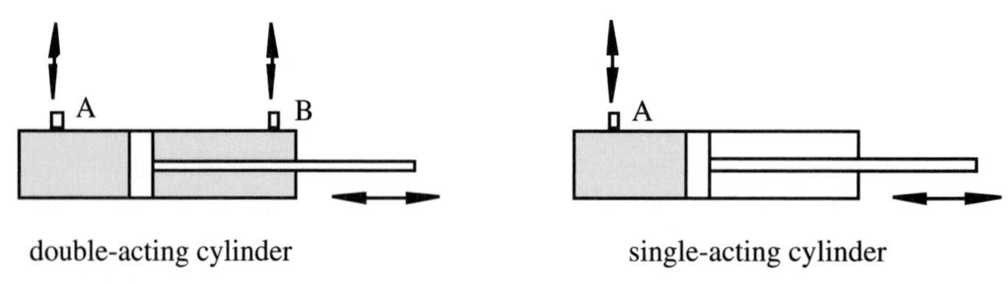


**FIGURE 9.41**
Single-acting and double-acting cylinders.

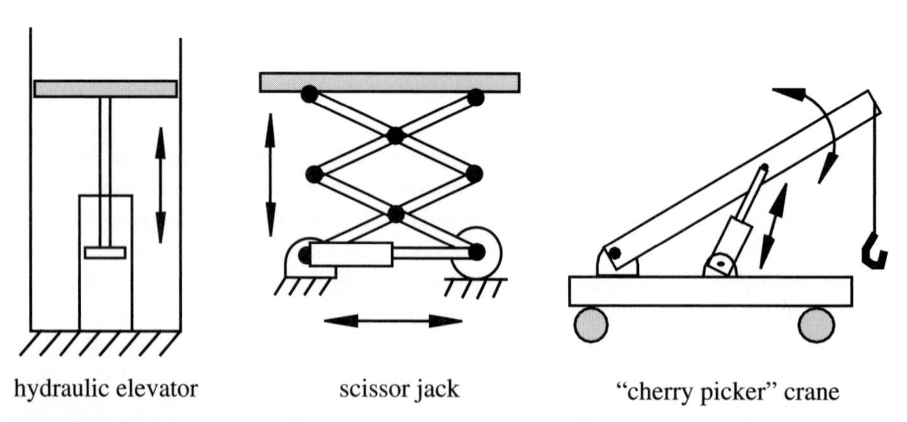


**FIGURE 9.42**
Example mechanisms driven by hydraulic cylinder/piston.

▼▼▼ ***CLASS DISCUSSION ITEM 9.6. Force Generated by a Double-Acting Cylinder.*** For a given system pressure, is the force generated by a double-acting cylinder different depending on the direction of actuation? How is the force determined for each direction of motion?

Hydraulic systems have the advantage of generating extremely large forces from very compact actuators. They also can provide precise control at low speeds and have built-in travel limits defined by the cylinder stroke. Drawbacks of hydraulic systems include the need for a large infrastructure (high-pressure pump, tank, distribution lines); potential for fluid leaks, which are undesirable in a clean environment; possible hazards associated with high pressures (e.g., a ruptured line); noisy operation; vibration; and maintenance requirements. Because of the disadvantages, electric motor drives are often the preferable choice. However, in large systems, which require extremely large forces, hydraulics often provide the only alternative.

## 9.10 PNEUMATICS

Pneumatic systems are similar to hydraulic systems, but they use compressed air as the working medium rather than hydraulic fluid. The components in a pneumatic system are illustrated in Figure 9.43. A compressor is used to provide pressurized air, usually on the order of 70 to 150 psi (482 kPA to 1.03 MPa), which is much lower than hydraulic system pressures. As a result of the lower operating pressures, pneumatic systems generate much lower forces than hydraulic systems.

After the inlet air is compressed (see Figure 9.43), excess moisture and heat are removed from the air with an air treatment unit. Unlike hydraulic pumps, which provide positive displacement of fluid at high pressure on demand, compressors cannot provide high volume of pressurized air responsively; therefore, a large volume of compressed air is stored in a reservoir or tank. The reservoir is equipped with a pressure-sensitive switch that activates the compressor when the pressure starts to fall below the desired level. Control valves and actuators act much in the same way as in hydraulic systems, but instead of returning fluid to a tank, the pneumatic air is simply returned (exhausted) to the atmosphere. Pneumatic systems are open systems, always processing new air, and hydraulic systems are closed systems, always recirculating the same oil. This eliminates the need for a network of return lines in pneumatic systems. Another advantage of pneumatic systems is that air is "cleaner" than oil, although air does not have the self-lubricating features that hydraulic oil does.

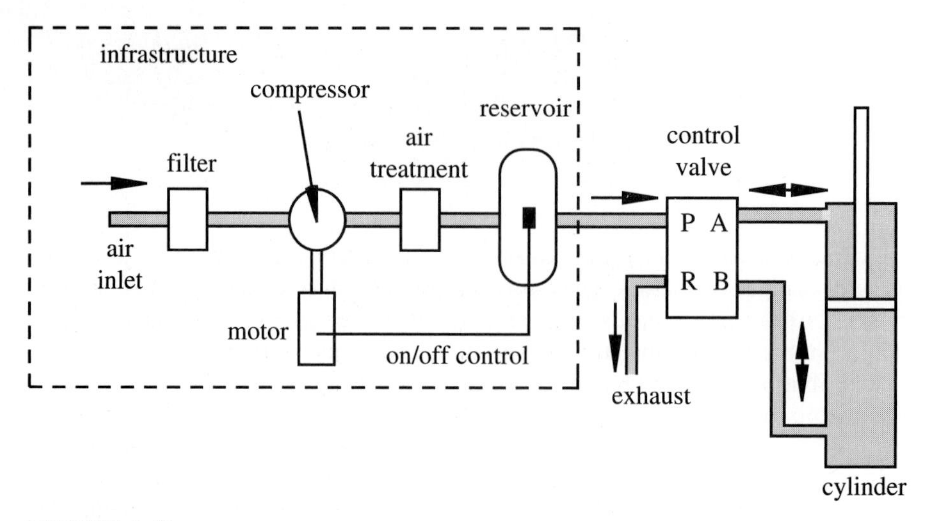


**FIGURE 9.43**
Pneumatic system components.

If sources of compressed air are readily available, as they often are in engineering-related facilities, then pneumatic actuators may be a good choice. Double-acting or single-acting pneumatic cylinders are ideal for providing low-force linear motion between two well-defined endpoints. Since air is compressible, pneumatic cylinders are not typically used for applications requiring accurate motion between the endpoints, especially in the presence of a varying load.

Another advantage of pneumatic systems is the possibility of replacing the infrastructure with a high-pressure storage tank. The tank serves an analogous function to a battery in an electrical system, making light, mobile pneumatic systems possible (e.g., a pneumatically actuated walking robot). In these applications, the capacity of the tanks limits the range of the system.

## QUESTIONS AND EXERCISES

**9.1.** A solenoid can be modeled as an inductor in series with a resistor. Design a system to use a digital output to control a 24 V solenoid.

**9.2.** A permanent magnet motor is coupled to a load through a gearbox. If the polar moment of inertia of the rotor and load are $J_r$ and $J_l$, the gearbox has a $N$:$M$ reduction from the motor to the load, the motor has a starting torque $T_s$ and a no-load speed $\omega_{max}$, and the load torque is proportional to its speed ($T_l = k\omega$),

(*a*) What is the maximum acceleration that the motor can produce in the load?

(*b*) What is the steady state speed of the motor and the load?

(*c*) How long will it take for the system to reach a steady state speed?

**9.3.** If a manufacturer's specifications for a PM dc motor are as listed below, what are the motor's no-load speed, stall current, starting torque, and maximum power for an applied voltage of 10 V?

- Torque constant = 0.12 Nm/A
- Electrical constant = 12 V/1000 RPM
- Armature resistance = 1.5 Ω

**9.4.** Recognizing that the H-bridge IC in Design Example 9.1 includes a current sense output, draw the block diagram for torque control of a brushed dc motor using the IC discussed. (Hint: The current sense output pin from the H-bridge yields 377 mA of current for each amp that is output to the motor. Convert this output to a voltage and use it as the input to a PWM chip like the LM3525. Include a torque adjust potentiometer on the LM3525.)

**9.5.** A unipolar stepper motor full-step drive circuit is available as a single component from some manufacturers. Explore the Internet to find a stepper motor driver, draw its schematic, and indicate how it will be connected to a stepper motor.

**9.6.** A designer wishes to use a stepper motor coupled to a gearbox to drive an indexed conveyor belt to achieve a linear resolution of 1 mm and a maximum speed of 10 cm/s. The gearbox is a speed reducer with a gear ratio of 3 to 1, and the conveyor is driven by a 10 cm drum attached to the output shaft of the gearbox. What is the minimum resolution required for the stepper motor? Also, what step rate would be required to achieve the maximum speed at this resolution?

**9.7.** Why does the presence of electric motors or solenoids affect the function of nearby electronic circuits?

**9.8.** For each of the following applications, what is a good choice for the type of electric motor used? Justify your choice.

(*a*) Robot arm joint

(*b*) Ceiling fan

(*c*) Electric trolley

(*d*) Circular saw

(*e*) NC milling machine

(*f*) Electric crane

(*g*) Disk drive head actuator

(*h*) Disk drive motor

(*i*) Windshield wiper motor

(*j*) Industrial conveyor motor

(*k*) Washing machine

(*l*) Clothes dryer

**9.9.** In designing a pneumatic system, you will use a pneumatic valve. Find the specifications for a specific pneumatic valve capable of handling 100 psi and draw a schematic showing how you would interface it to a digital system.

**9.10.** You are designing a pneumatic actuator using a dual acting cylinder to produce 100 lb of force using a 2000 psi scuba tank as a pressure source. Specify the elements required in the system in block diagram form. Be as specific as possible about component specifications.

## BIBLIOGRAPHY

Kenjo, T., *Electric Motors and Their Controls,* Oxford Science Publications, Oxford, England, 1994.

Khol, R., editor, "Electrical & Electronics Reference Issue," *Machine Design*, v. 57, n. 12, May 30, 1985.

McPherson, G., *An Introduction to Electrical Machines and Transformers*, John Wiley, New York, 1981.

National Semiconductor Corporation, "National Power ICs Databook," 2900 Semiconductor Drive, P.O. Box 58090, Santa Clara, CA 95052.

Norton, R. L., *Design of Machinery,* McGraw-Hill, New York, 1992.

Shultz, G. P., *Transformers and Motors,* Macmillan, Carmel, IN, 1989.

Westbrook, M. H. and Turner, J. D., *Automotive Sensors,* Institute of Physics Publishing, Philadelphia, PA, 1994.

Williamson, L., "What You Always Wanted to Know about Solenoids," *Hydraulics and Pneumatics*, September, 1980.

# 10

# MECHATRONIC SYSTEMS—CONTROL ARCHITECTURES AND CASE STUDIES

**OBJECTIVES:** After you read, discuss, study, and apply ideas in this chapter, you will be able to:

- Identify why many engineering designs today can be classified as mechatronic systems
- Understand the integration of control in mechatronic systems via digital electronics and computers
- Appreciate the variety of designs that can result from limited specifications of a mechatronic system
- Recognize that many aerospace, automotive, and consumer products are mechatronic systems

## 10.1 INTRODUCTION

In the previous chapters of this book, we have developed the foundations for the integration of mechanical devices, sensors, and signal and power electronics as mechatronic systems. As we progressed through the chapters, we included increasingly challenging mechatronic design examples to help you connect the theory and analysis with actual applications in order to expand your knowledge and design skills beyond the traditional bounds of so-called nuts-and-bolts engineering. To obtain completeness in the integration of mechanical devices, sensors, and signal and power electronics into the most advanced mechatronic systems, we must also include embedded microprocessors. While this book is not designed as a control systems or microprocessor text, some of the rudiments of control approaches and architectures are included here. We start with a broad-brush treatment of the varieties of control architectures that exist in contemporary mechatronic systems and follow with two very detailed case studies of complete mechatronic systems: an articulated walking machine for movement in unusual environments and a coin counter. These projects have been used as mechatronic projects in coursework at Colorado State University. To provide a final perspective, we list cursory examples of mechatronic systems in many industries.

## 10.2 CONTROL ARCHITECTURES

Since many mechatronic systems have an array of inputs and outputs with some deterministic relationship required among them, the systems require some form of control. There is a wide spectrum of control architectures from which a designer can choose, ranging from simple open loop control to complex feedback control. Implementation of the control can be as simple as using a single op amp or as complicated as programming massively parallel microprocessors. Below, we describe a hierarchy of basic control approaches you may consider in the design of a mechatronic system.

### 10.2.1 Analog Circuits Alone

Many simple mechatronic operations require specific action of an actuator based on an analog input signal. Simple analog signal processing circuits consisting of op amps and transistors singly or in combination can be employed to effect the desired control. Op amps can be used to perform comparisons and mathematical operations such as analog addition, subtraction, integration, and differentiation. They can also be used in amplifiers for linear control of actuators. Analog controllers are often simple to design and easy to implement. See Chapter 5 for more details.

### 10.2.2 Digital Circuits Alone

If the input signals are digital or can be converted to a finite set of states, then combinational or sequential logic controllers may be easy to implement. The simplest designs use a few digital chips to create a digital controller. To generate complex binary or Boolean functions on single chips, there are a variety of specialized digital devices easing the design task, such as multiplexers, program array logic controllers (PAL), and program logic arrays (PLA). See Chapter 6 for more details.

### 10.2.3 Programmable Logic Controller

Specialized industrial devices for interfacing to analog and digital devices have been designed with restricted instruction sets for industrial control applications. ***Programmable logic controllers (PLCs)*** offer more flexibility in developing complex control algorithms and are best adapted to industrial monitoring and control in production environments. They are usually programmed with ***ladder logic,*** which is a graphical method of laying out the connectivity and logic between system inputs and outputs. They are designed with industrial control and industrial environments specifically in mind. Therefore, besides being flexible and easy to program, they are robust and relatively immune to external interference.

### 10.2.4 Microcontroller

The microcontroller, which is a microcomputer on a single IC, provides a small and flexible control platform that can be easily embedded in a mechatronic system. The microcontroller can be programmed to perform a wide range of control tasks. Designing with microcontrollers usually requires knowledge of a procedural programming language, often assembly language, and requires experience in interfacing digital and analog devices. See Section 6.19 in Chapter 6 for more information.

### 10.2.5 Single Board Computer

When an application requires more features or resources than can be found on a typical microcontroller and when size is not a major concern, a single board computer offers a good alternative (see Section 6.19 in Chapter 6 for more information). Most single board computers have enough RAM and have compilers available to support programming in a high-level language such as C. Single board computers are also easily interfaced to a personal computer. This is useful in the testing and debugging stages of development and for downloading software into the single board computer's memory.

### 10.2.6 Personal Computer

In the case of large sophisticated mechatronic systems, a desktop or laptop personal computer (PC) may serve as an appropriate control platform. Also, for those not experienced with microcontrollers and single board computers, the PC may be an attractive alternative. The PC can be easily interfaced to sensors and actuators using commercially available plug-in data acquisition cards (e.g., the one shown in Figure 7.5 in Chapter 7). These cards typically include software drivers that allow easy programming within a standard high-level language compiler or development environment. Due to the ease and convenience of this approach, PC-controlled mechatronic systems are especially common in R&D testing and product development laboratories, where fast prototyping is required but where large quantity production and miniaturization are not concerns.

## 10.3 CASE STUDY 1—MECHATRONIC DESIGN OF A COIN COUNTER

Imagine the following problem is presented to you: Design a coin counter that includes an electromechanical device to accept a handful of mixed denomination US coins and that will align and present the coins in serial order to a sensor array capable of acquiring data that can be used to determine the denominations of individual coins. The sensor output is to be interfaced to electronics that can compute both the number of coins presented and the total value of the coins, and display those two values in some sort of multiplexed fashion on a single display visible to the user.

When we presented this design challenge to a class of 80 students broken down into groups of 4, we obtained complete, successful solutions from more than half the groups given a design period of 6 weeks. All groups were able to present the coins sequentially and display a count of the number of coins, but some groups required relaxing the constraints such as dropping the coins one-by-one. Multiplexing the value of the coins with the number on a single display was a more demanding problem requiring creative use of digital logic, and not all groups were successful in implementing a design for this part. Given more time to redesign, most groups would have a successful experience. As you will see, there is no single correct solution to the problem; the designs are as varied as the people who developed them.

Therefore, let us continue with the general design discussion. This problem has two significant parts: the design of an electromechanical coin presentation system to sequentially align coins in some fashion so they may be presented to an array of sensors, and the design of an electronic calculator to use the sensor data to display the number and value of the coins. We will consider each part in succession, realizing that the student design groups often subdivided their efforts between the mechanical design of the coin presenter and the electronic design of the calculator. Students with significant machine shop experience tackled the electromechanical coin presentation design while students comfortable with the electronic experience they had gained in the course focused on the sensor and counter design. As in all design projects, certain people will have affinities for certain parts of the overall design, and important tasks are assignment of responsibilities, communication of progress, documentation of work, and assurance of compatibility among the various pieces. We don't want to digress too much on the design process itself here, other than to mention that human elements in the design process are as critical as the engineering design itself.

The mechanical component of the design requires a chamber to accept a handful of mixed denomination coins and a mechanism to select coins individually and present them to a sensor array that can read their attributes and output digital signals for computation. By observing existing mechanical coin sorters, many students designed inclined rotating disk mechanisms that had circular holes cut in them to accept individual coins. The inclined disk was necessary to entrain all coins and prevent jamming. Examples of the mechanical systems designed to present the group of coins in series are shown in Figure 10.1.

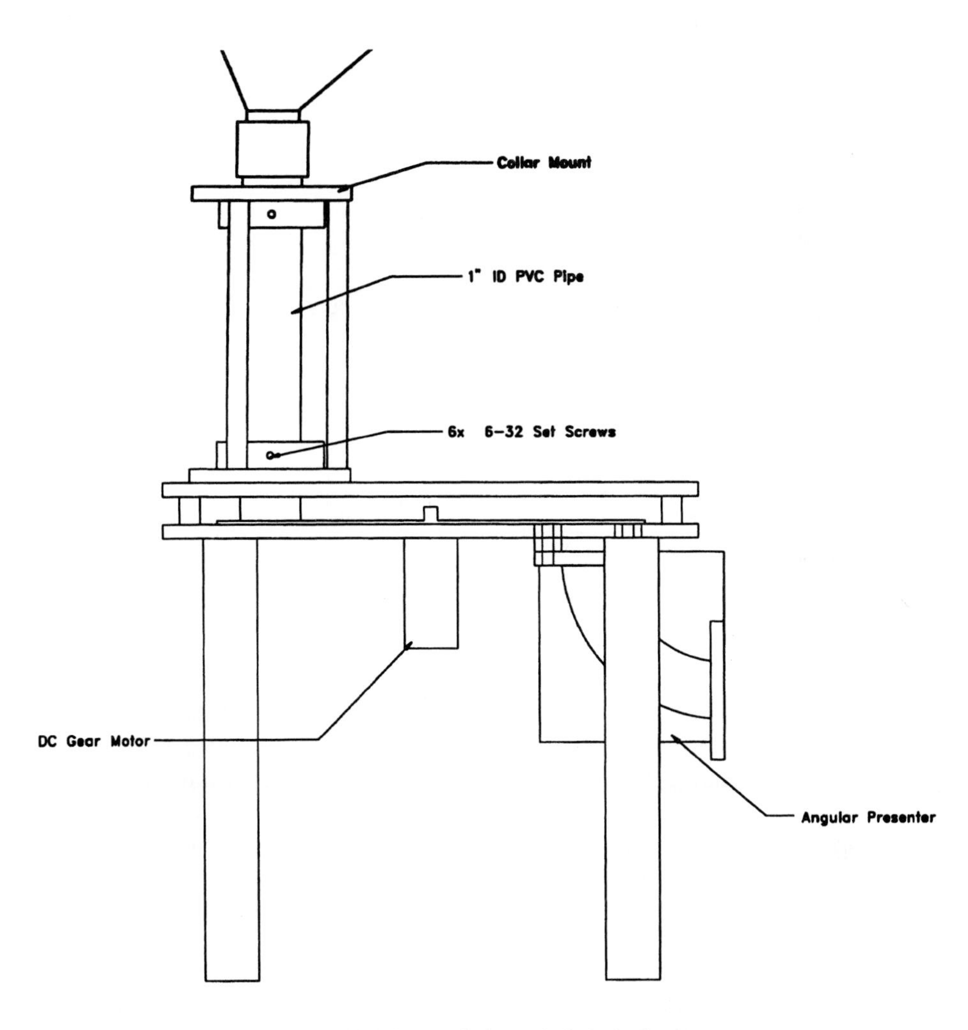


(*a*) horizontal slotted disk design

**FIGURE 10.1**
Example of coin counter presentation mechanisms.

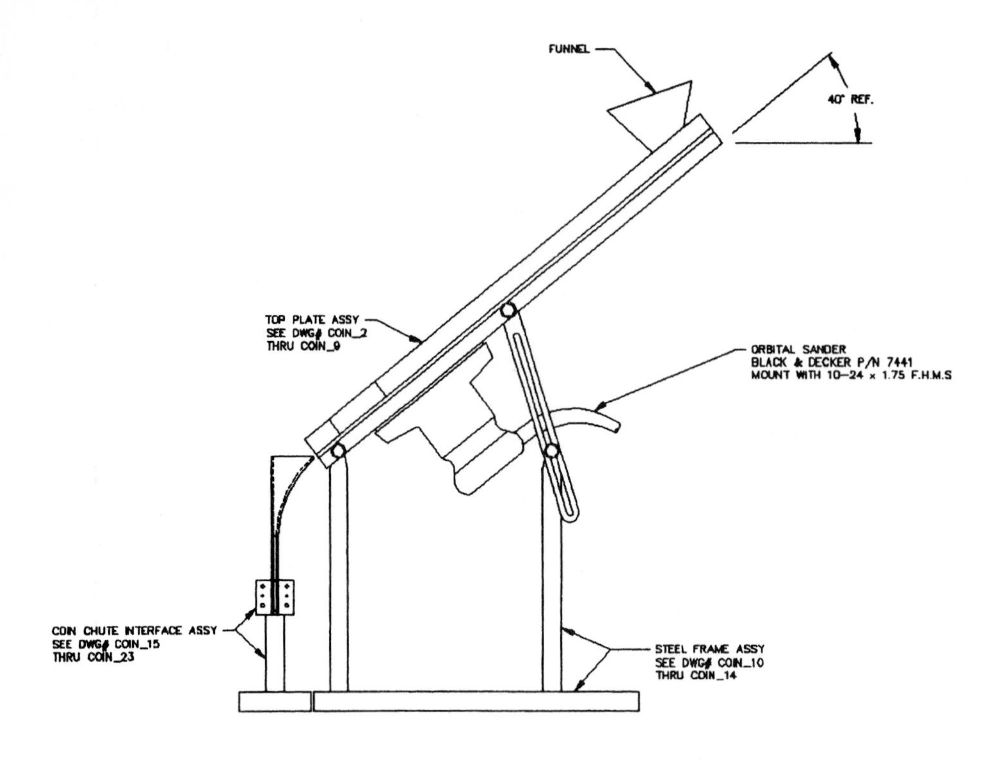


(*b*) "pachinko" design

**FIGURE 10.1**
*(continued)*

Two of the systems were manufactured from Plexiglas and the other from aluminum. Both used a dc motor to rotate a perforated disk that entrained a single coin. As the disk rotated, the coin would drop into the slot containing the sensor array. Most designs included a dc motor to rotate the disk continuously. Other less successful designs included vibrating chutes and pachinko-like mechanical arrays; in these designs, jamming often occurred. As individual coins were separated, they entered a chute and rolled down a rectangular slot containing the sensor array.

Although people apprehend the value of coins by sight and feel, automated methods require a well-designed sensing system. The size, weight, and thickness of coins are also unique for different denominations. For the sake of simplicity, all groups elected to design a height sensor using a set of phototransistor-photodiode pairs.

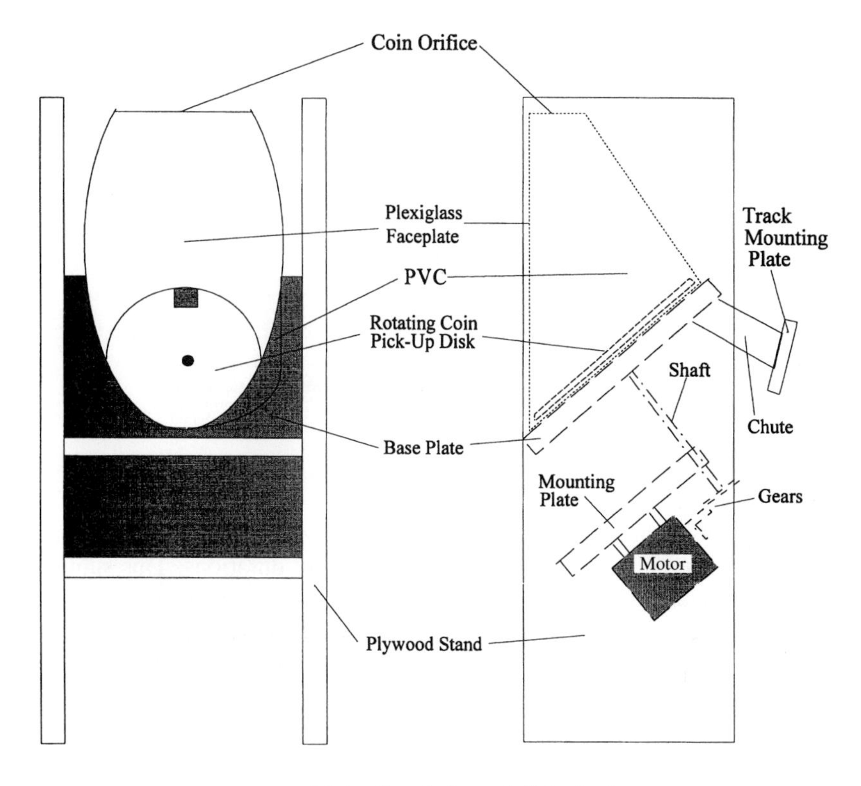


(*c*) inclined slotted disk design

**FIGURE 10.1**
*(continued)*

Each sensor was carefully positioned so that the combination of signals from the sensors would be unique for each different denomination of coin (see Figure 10.2). For this project, the coin denominations were limited to the penny, nickel, and quarter to provide significant size differences. The height of the coins (diameter) was the primary attribute for assessing each coin's value, assuming that they rolled down the slot smoothly. This meant that the largest coin, the quarter, activated all sensors and the smallest, the penny, only one. Furthermore, the signals from the sensors would be rough square waves of different pulse widths affected by the size and speed of the coins. The square waves would not begin or end at the same time, complicating the synchronizing of logic in the design of the evaluator and counter (see Figure 10.3). The lowest sensor pair would serve as the basic coin counter. The sensor outputs were converted to TTL signals using 7404 Schmitt triggers, making the outputs compatible with the computational part of the mechatronic design.

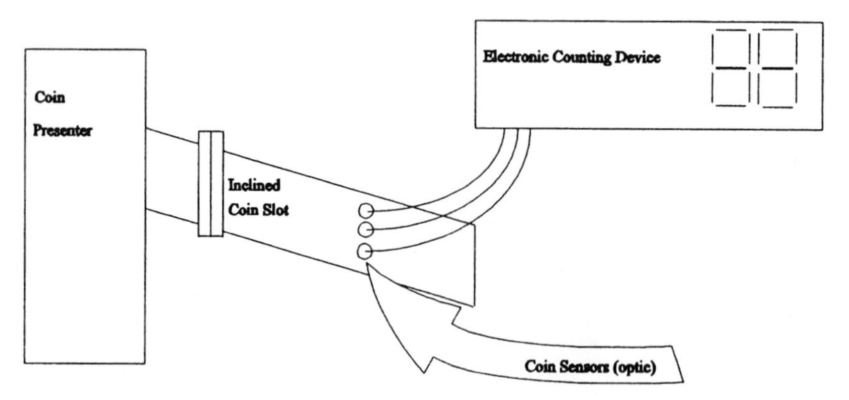


**FIGURE 10.2**
Sensor array and chute design.

As stated above, the sensor outputs are not well synchronized. The sensors produce pulses at different starting times and for different durations. This subtlety requires careful sequential logic design to ensure that the correct coin denomination is detected. If the TTL signals from the sensors were synchronized, then the Boolean expressions for the different denomination coins could be easily written as follows:

$$X = A \cdot B \cdot C \tag{10.1}$$

$$Y = A \cdot B \cdot \overline{C} \tag{10.2}$$

$$Z = A \cdot \overline{B} \cdot \overline{C} \tag{10.3}$$

where $C$ corresponds to the output of the top sensor, $B$ to the output of the middle sensor, and $A$ to the output of the bottom sensor. $X$ corresponds to the passage of a quarter, $Y$ to the passage of a nickel, and $Z$ to that of a penny.

Combinational and sequential logic is required to determine the denomination of the coin and to increment the displayed output by a value corresponding to the denomination. There were as many solutions for this problem as there were design groups. Figures 10.4 and 10.5 (pages 358–361) display two interesting and successful solutions for the coin counter processing and display circuit.

The outputs were transmitted to a digital display driver to multiplex the real-time number of coins with the accumulated value. The values remaining at the end represented the accumulated value and total number of coins.

## 10.4 CASE STUDY 2—MECHATRONIC DESIGN OF A ROBOTIC WALKING MACHINE

As a culmination of the design examples for this book, we close with a detailed case study of the mechatronic design of an articulated walking machine. It was executed by undergraduate engineering students in 1994, following the completion of our introductory mechatronics course for which this book is used. In 1987 the Society of

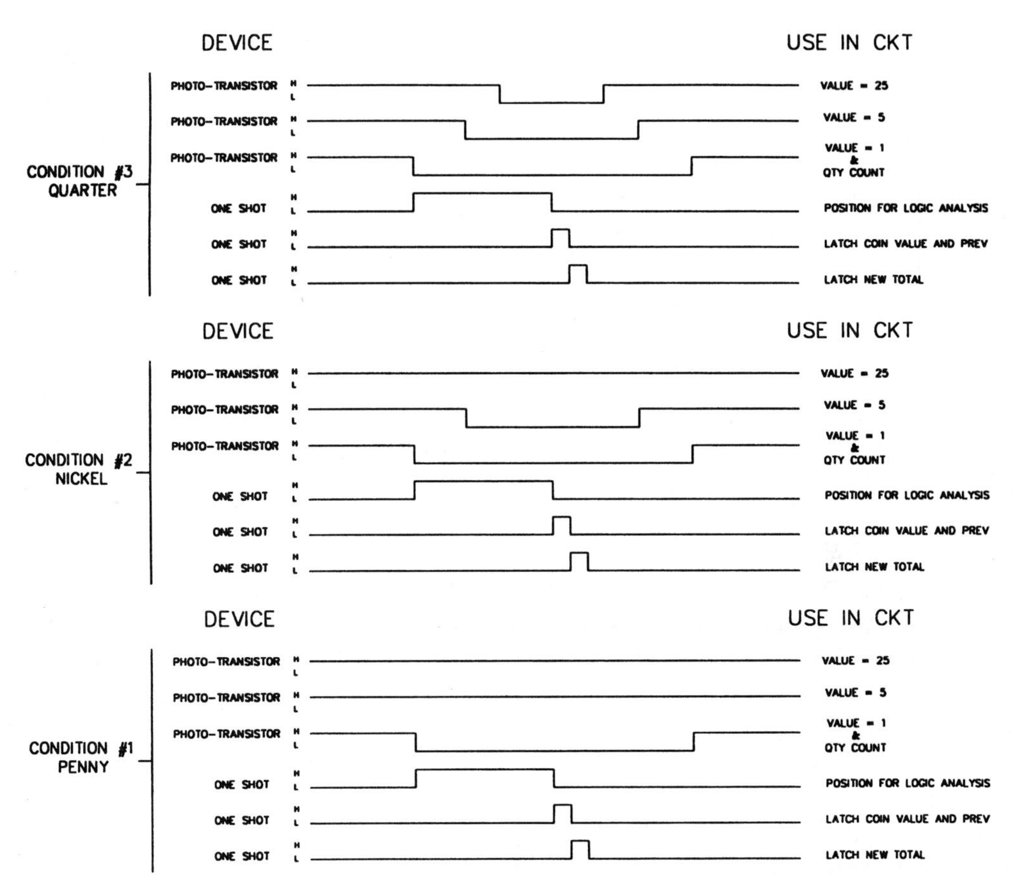


**FIGURE 10.3**
TTL outputs corresponding to different coins.

Automotive Engineers began sponsoring an annual Robotic Walking Machine Decathlon, pitting teams from different universities in a challenge to design articulated walking machines that could execute 10 different performance events, such as a dash, a slalom, obstacle avoidance, and crossing a crevasse. Half of the events included walking motion and obstacle avoidance under tether control, i.e., control from a human-operated switch box with an electrical umbilical to the machine; the other half of the events required autonomous action, i.e., by control effected by onboard, preprogrammed systems on the machine itself with no human intervention. Over the last 10 years we have seen over 100 different walking machine designs, some very simple and capable of completing just a few events to machines of great sophistication and creativity capable of completing all events. The excitement and fun of such competitions is in seeing the fruit of design concepts actually functioning to specifications. Our intent here is not to examine the design varieties of walking machines but to focus on a specific design example to illustrate the mechatronic aspects. We begin by displaying three different walking machines, all of which won the national SAE contest (see Figure 10.6, pages 362–363).

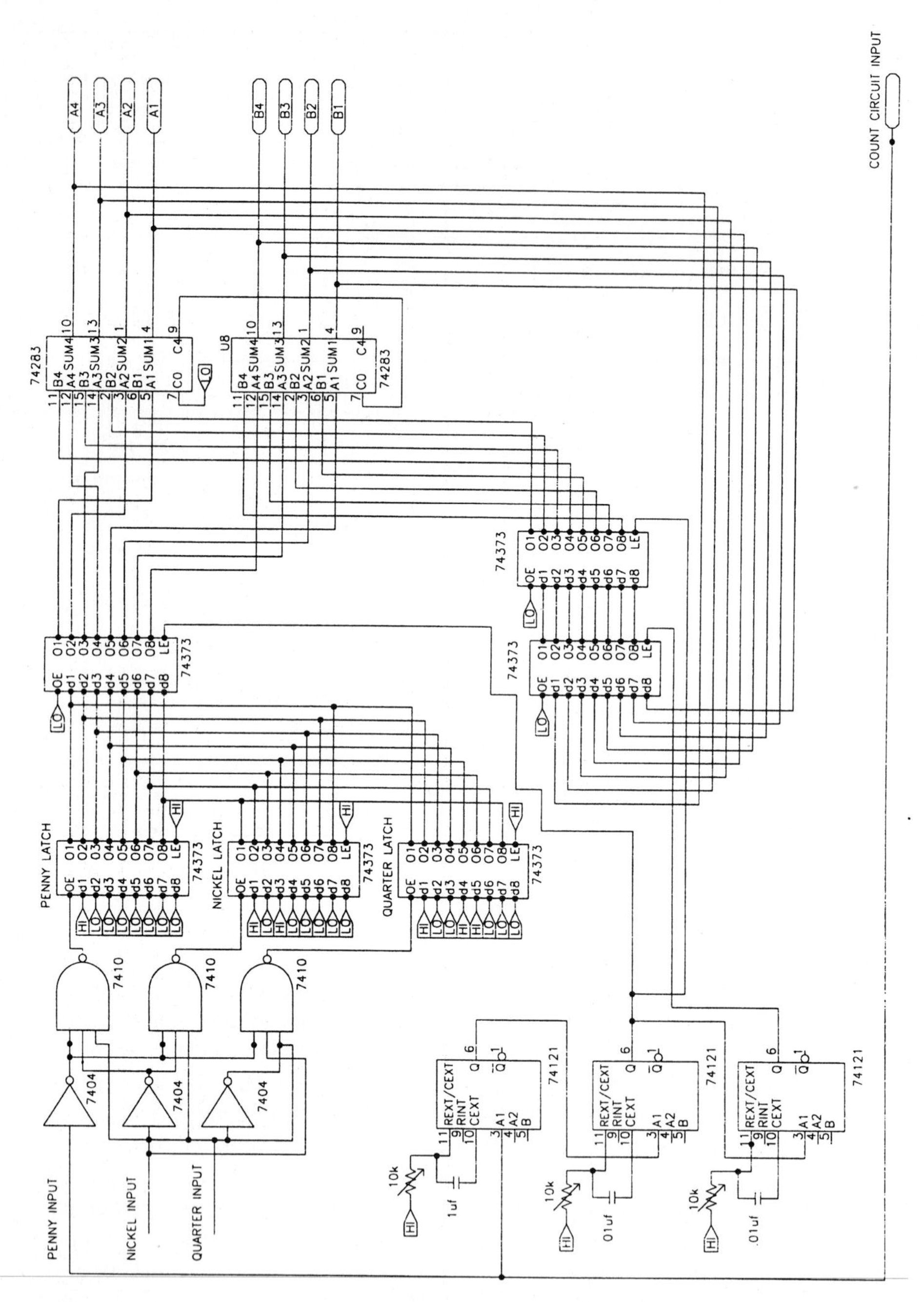

(*a*) first half of circuit

**FIGURE 10.4**
Counter design 1.

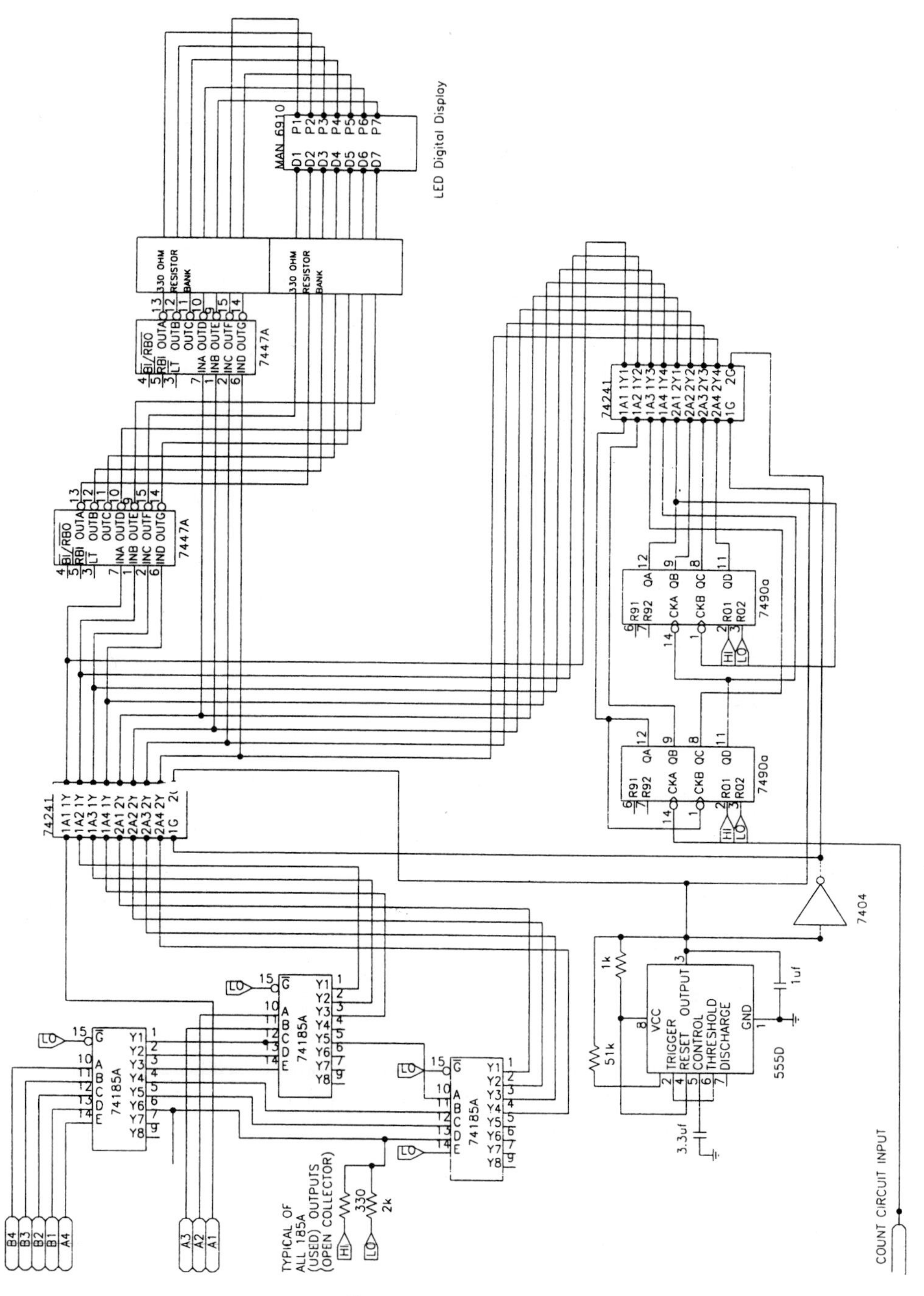


(*b*) second half of circuit

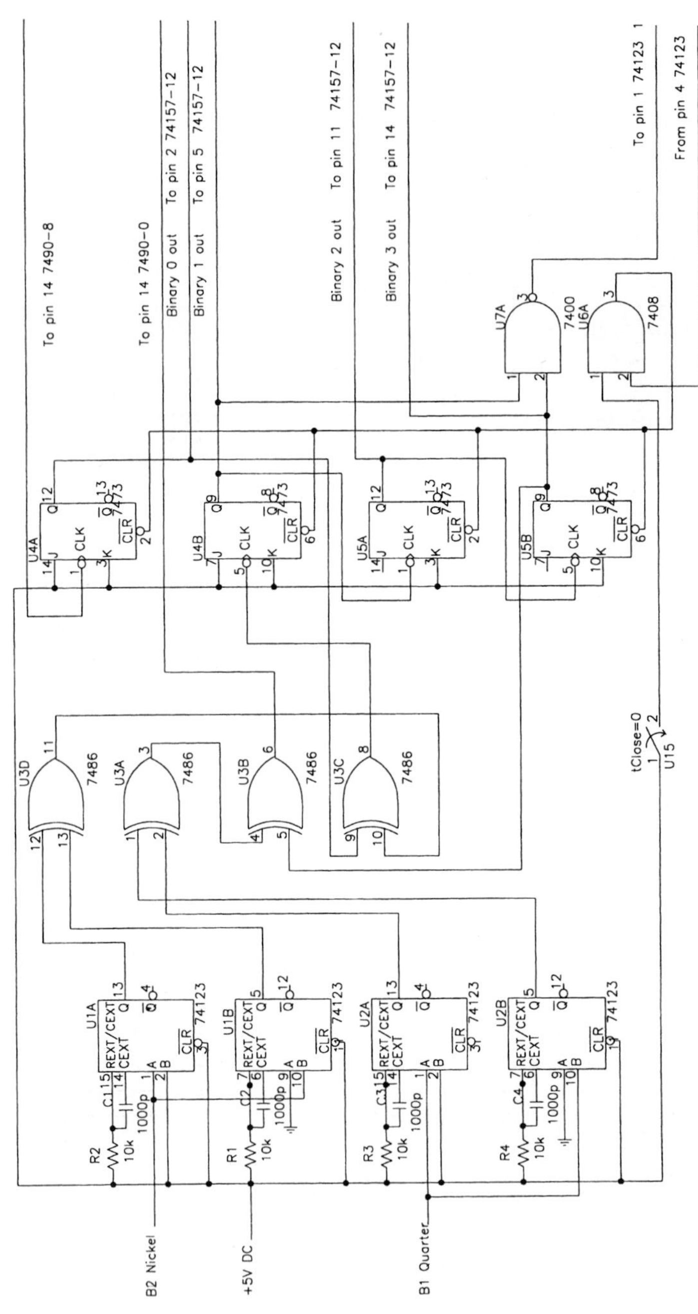


(*a*) first half of circuit

**FIGURE 10.5**
Counter design 2.

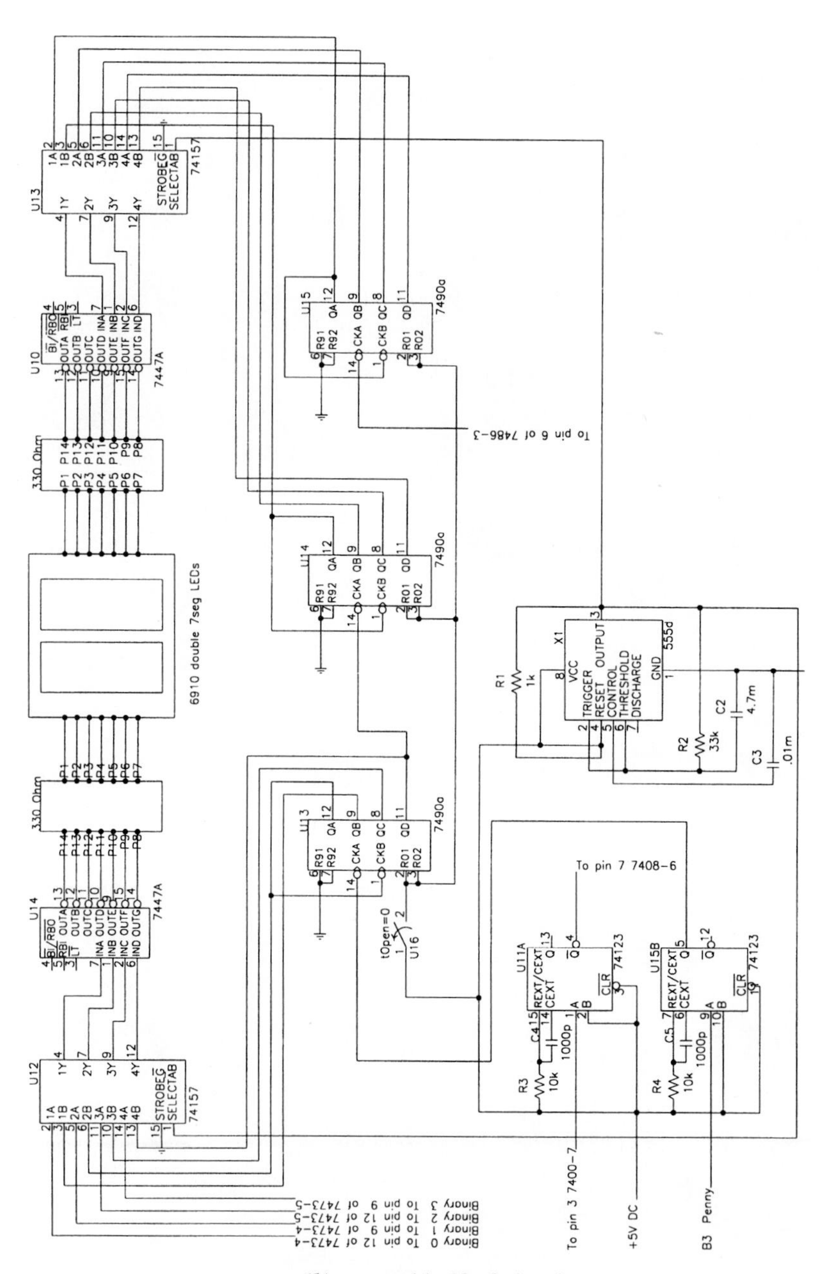

(*b*) second half of circuit

(*a*) “Lurch”—Scotch yoke mechanism leg design (1989)

(*b*) “Airachnid”—First pneumatic design (1992)

**FIGURE 10.6**
Student-designed walking machines from Colorado State University.

(*c*) "Airratic"—Refined pneumatic design (1994)

**FIGURE 10.6**
*(continued)*

We will now present as a case study the design of the 1994 Colorado State walking machine that the students, applying their ever-present wit, named "Airratic." This design was a refinement of the first pneumatic design from 1992, which was a dramatic break from the evolving electromechanical designs of the previous seven years. With the air-powered designs, the students had to contend with a whole new set of design constraints: providing an on-board source of stored, pressurized air; controlling the mechanical motion of the articulated legs with pneumatic cylinders; distributing and controlling the air; controlling the pressure; reducing coordinated walking motion to sets of computer commands; interfacing a computer to the pneumatic control system; minimizing the weight and size of the machine; ensuring the safety of the system, including sensors on the machine for obstacle avoidance; and making the design changeable and adaptable during test trials on the competition floor.

Since the designers elected to power the walking machine pneumatically, an on-board scuba tank was selected as the energy source for the strategically placed pneumatic pistons that controlled the position and movements of the legs. The scuba tank was quickly replaced by a fiber wound pressure tank to reduce weight. As shown in Figure 10.7, the skeletal structure of the walking machine consisted of welded aluminum polygons with 16 double-acting pneumatic cylinders as actuators placed in the configuration shown. The four corner legs had three degrees of freedom, and the front and rear-center legs had two degrees of freedom. The mechanical design of the six legs was such that control algorithms could easily produce forward, reverse, sidewise, and diagonal motion, as well as full-up and full-down motion of the main frame, by simple coordinated control of the 16 cylinders. Each leg

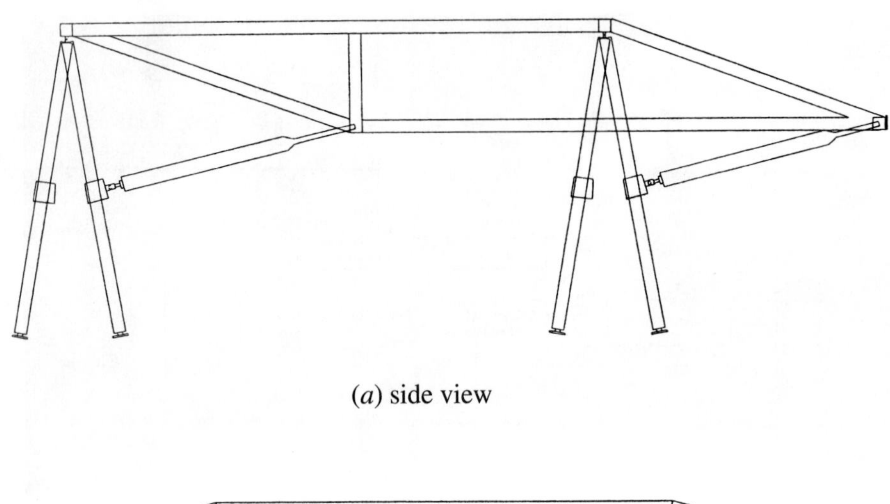

(*a*) side view

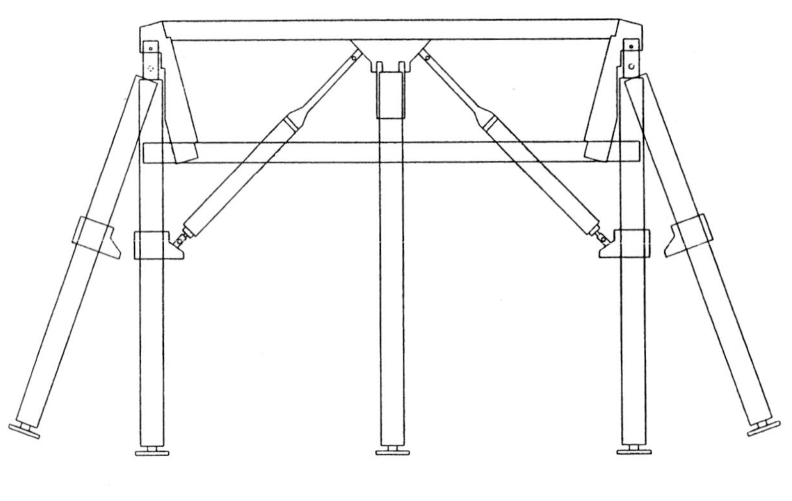

(*b*) front view

**FIGURE 10.7**
Aluminum frame and telescoping pneumatic legs.

included an axial cylinder that could be extended or retracted. Ten other cylinders were arranged to position the legs in different patterns in the horizontal plane.

For static stability the machine must be supported by a minimum of three legs at all times, requiring a total of six legs for locomotion. Most walking motions could be divided into the coordinated movement of two sets of three legs each, called Group 1 Legs and Group 2 Legs. Figure 10.8 displays the flowchart used to develop the control code for one class of coordinated motion: forward motion of the machine. For this class of motion, more subtle aspects such as the choice of the primary leg group were included.

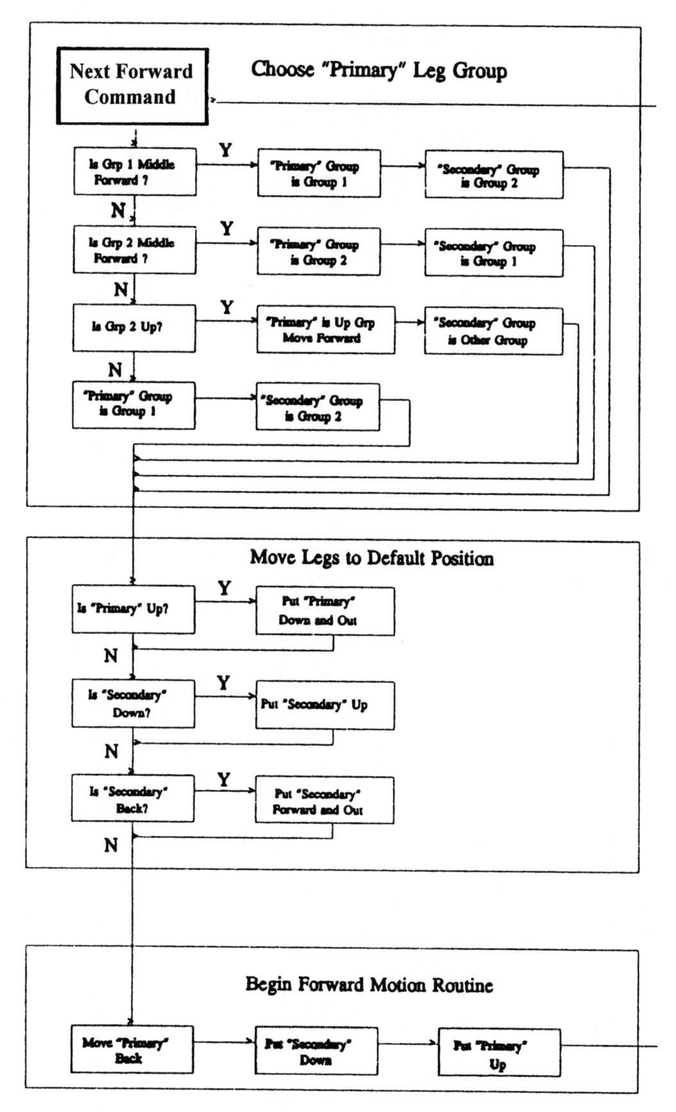


**FIGURE 10.8**
Flowchart for forward motion routine.

The essence of the control system involves the coordination of 16 pneumatic cylinders using computer-controlled solenoid valves. Each solenoid valve switches air from the main cylinder to one side of the pneumatic cylinder and exhausts pressurized air from the other side, as shown in Figure 10.9. Thus one bit of information is required to control one degree of freedom. Needle valves were manually adjusted on each solenoid valve to govern cylinder actuating speed.

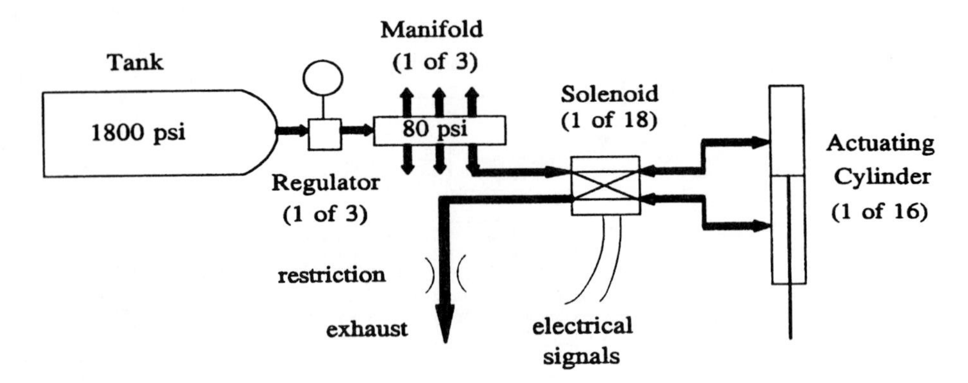


**FIGURE 10.9**
Pneumatic system.

As you can see, the flowchart in Figure 10.8 exhaustively reviews the piston positions in a logical manner complex enough to require microprocessor control. This flowchart helped in the design of the computer code necessary to control and coordinate the motion. The on-board computer controls the solenoid valves via the I/O bus to which are connected 74373 tristate inverting buffer interface chips receiving signals from the output ports of the computer. As shown in Figure 10.10, eight ports are used to provide eighteen bits of information for coordinated control of the cylinders, including a slow motion mode that we will not discuss here.

The controller is a Motorola 68HC811E2 EVM 8-bit microprocessor-based single board computer. It has 128K of RAM, and 64K acts as pseudo-ROM for downloading control programs from an external PC or laptop computer. Thirty I/O bits are provided and expanded through a multiplexer to the 48 bits required for the control of all of the components of the machine. Computer code is created on a laptop computer in the high-level language C, then complied and downloaded to the Motorola 68HC811E2 EVM for testing. This is a very convenient method to allow modifications to the control strategy and the evaluation of the interplay of sensors and actuators.

In addition to various walking motions, the machine included sensor feedback using two optical retroreflective sensors. The sensors were mounted on a rotating platform to allow scanning a portion of the region in front of the machine. A digital encoder provides position information along with the sense state of the two sensors. The sensor data is read from the I/O bus of the microprocessor, and software control routines use this data to avoid obstacles. The sensor connections to the bus are also shown in Figure 10.10.

In summary, we have examined an example of a student-designed mechatronic system that received first place in the 1994 SAE Walking Machine Decathlon. It illustrated the use of pneumatic actuators for motion, optical sensors for environmental feedback, and an on-board microcomputer for controlling the machine in both operator-controlled and autonomous modes. Projects such as this require a significant amount of time to complete. This project involved eight months of work. The

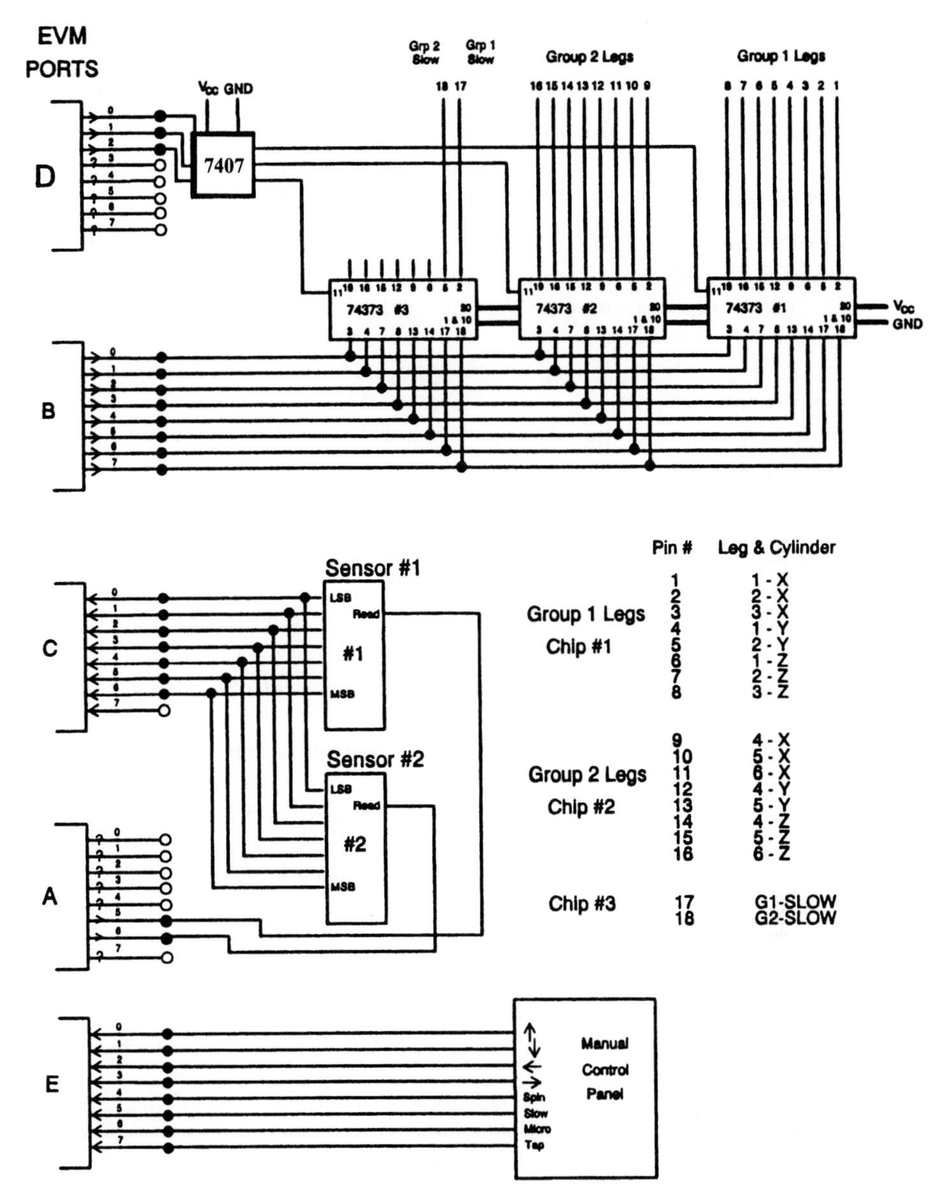


**FIGURE 10.10**
Computer ports and I/O board.

experience gained in designing a complex mechatronic system is very valuable, particularly if the design is actually built, debugged, and challenged to achieve its design goals. In this project, the students were so excited about their success that they even programmed the machine to dance. This was so well received by the audience that they all joined in a country-style line dance at the finale.

## 10.5 LIST OF VARIOUS MECHATRONIC SYSTEMS

We conclude with a partial list of mechatronic systems that you can identify in your everyday environment. You will see that a very large proportion of engineering systems designed today can be declared mechatronic systems.

- Air bag safety systems, antilock brake systems, remote automatic door locks, cruise control, and other automobile systems
- NC mills, lathes, rapid prototyping systems, and other automated manufacturing equipment
- Copy machines, FAX machines, electric typewriters, and other semiautomatic office equipment
- MRI equipment, arthroscopic instruments, ultrasonic probes, and other medical diagnostic equipment
- Autofocus 35 mm cameras, VCRs, video and CD players, camcorders, and other sophisticated consumer electronic products
- Laser printers, hard drive head positioning systems, tape drive auto load/reject cartridges, and other computer peripherals
- Welding robots, automatic guided vehicles (AGVs), NASA Mars rover, and other robots
- Flight control actuators, landing gear systems, cockpit controls and instrumentation, and other subsystems on airplanes
- Garage door openers, security systems, HVAC controls, and other home support systems
- Washing machines, dishwashers, automatic ice makers and freezers, and other home appliances
- Variable speed drills, digital torque wrenches, and other modern hand tools
- Materials testing machines, auto test dummies, and other laboratory support equipment
- PLC-controlled bar-code driver packages, conveyor systems, and other factory automation systems
- Manual and semiautomatic controllers for hydraulic cranes and other construction equipment
- Automatic labeling systems and CCD camera inspections in an IC manufacturing operation
- Video game and virtual reality input control systems

## QUESTIONS AND EXERCISES

**10.1.** For selected mechatronic systems mentioned in the list above, list the required electronics, sensors, and actuators.

**10.2.** For selected mechatronic systems mentioned in the list above, recommend a control architecture and support your choice.

# Appendix 1

# MEASUREMENT FUNDAMENTALS

**OBJECTIVES:** After you read, discuss, study, and apply ideas in this appendix, you will be able to:

- Define SI units and use them in calculations
- Use basic notions of statistics to characterize measured data
- Compute the error associated with a measurement

## A1.1
## SYSTEMS OF UNITS

Fundamental to the design, analysis, and use of any measurement system is a complete understanding of a consistent ***system of units*** used to quantify the physical parameters being measured. To define a system of units, we must select units of measure for fundamental quantities to serve as a basis for the definition of other physical parameters. Units for mass, length, time, temperature, electric current, amount of substance, and luminous intensity form one possible combination that serves this purpose. Other units used to measure physical quantities in mechatronic systems can be defined in terms of these seven ***base units.***

▼▼▼ ***CLASS DISCUSSION ITEM A1.1. Definition of Base Units.*** Although everyone has intuitive knowledge of the physical quantities associated with the base units, it is difficult to define them in everyday terms. Try to define length. You will probably use some synonym for length in your definition. Also try to define the others. All are equally difficult to put into simple terms.

The seven base units we will use to define mass, length, time, temperature, electric current, amount of substance, and luminous intensity are the kilogram, meter, second, Kelvin, ampere, mole, and candela. These units form the basis for the International System of Units, abbreviated ***SI***, from the French *Le Systeme **In**ternational d'Unites.*

The ***kilogram*** is the only unit that is defined in terms of a material standard. It is established by a platinum-iridium prototype in the laboratory of the Bureau des Poids et Mesures in Paris. Unfortunately, the name "kilogram" is confusing since it contains the prefix "kilo" which conflicts with the SI prefixing conventions described in Section A1.1.1.

The ***meter*** is defined as 1,650,763.73 wavelengths of an isotope of krypton ($^{86}Kr$) corresponding to the transition between the $2p_{10}$ and $5d_5$ electron energy levels of the krypton 86 atom. This atomic standard for the meter was proposed long ago by Maxwell (1873) but not implemented until 1960. The earlier meter definition, the distance between two scribed lines on a platinum-iridium bar, like the kilogram, required a prototype for the definition. Now the practical measurement of the unit is deliberately separated from the definition, making the definition independent of a unique prototype. An alternative standard meter was defined in 1983 as the length of the path traveled by light in vacuum during a time interval of 1/299,792,458 sec.

The ***second*** is defined as the duration of 9,192,631,770 periods of the radiation, corresponding to the transition between the two hyperfine levels of the ground state of the cesium 133 ($^{133}Cs$) atom. The definition of the second was previously based on the mean solar second, which was defined as a fraction (1/86,400) of the earth's daily rotation. Irregularities in the earth's rotation amounting to one or two seconds per year limited accuracy.

The unit of absolute thermodynamic temperature is the ***Kelvin.*** The Kelvin scale has an absolute zero of 0 K, and no temperatures exist below this level. There is a misconception that all molecular motion ceases at this value. Actually, the molecular energy is at a minimum. A standard fixed calibration point on the temperature scale is the triple point of water, which is set at a value of 273.16 K to maintain consistency with the Celsius scale. Although the Kelvin scale is established using only the absolute zero and triple points, additional fixed points have been defined based on the boiling and melting points of other materials. These points are useful when calibrating temperature measurement devices. Temperature in Kelvin and temperature in degrees ***Celsius*** are related by the following equation:

$$T_C = T_K - 273.15 \tag{A1.1}$$

where $T_K$ is the Kelvin temperature and $T_C$ is the Celsius temperature expressed in degrees Celsius (°C). Note that the triple point of water is 0.01°C, corresponding to 273.16 K. The Celsius temperature scale is sometimes referred to as the centigrade scale since it is calibrated to the 100°C temperature interval between the freezing point of water (0°C) and the boiling point of water (100°C). An interval or difference of temperature ($\Delta T$) has the same value in both the Celsius and Kelvin scales ($\Delta T_C = \Delta T_K$).

The ***ampere*** is defined as the constant current which, if maintained in two straight parallel conductors of infinite length and negligible circular cross section, and placed 1 meter apart in a vacuum, would produce a force between the conductors equal to 2 x $10^{-7}$ newtons per meter of length. Unfortunately, this creates a difficult measurement problem since the definition is in terms of other base units. Therefore, any errors in the other base units compound the errors in the measurement of the ampere.

The ***mole*** is defined as the amount of substance in a system that contains as many elementary entities as there are atoms in 0.012 kg of carbon 12 ($^{12}C$).

The ***candela*** is defined as the luminous intensity, in the perpendicular direction, of a surface area of 1/600,000 $m^2$ of a black body at the temperature of freezing platinum under a pressure of 101,325 $N/m^2$.

### A1.1.1 The Three Classes of SI Units

SI units are divided into three classes: base units, derived units, and supplementary units. The complete set of SI ***base units*** and their symbols are listed in Table A1.1.

***Derived units*** are expressed as algebraic combinations of the base units. Any known physical parameter can be quantified using a derived unit. Some examples of derived units are listed in Table A1.2. Several derived units have been given special names and symbols, which may be used themselves to express other derived units in a simpler way than in terms of base units. Some examples of these ***supplemental units*** are listed in Table A1.3.

Often the base, derived, and supplemental units are modified with prefixes to enable convenient representation of large numerical ranges. The ***prefixes*** express orders of magnitude (powers of ten) of the unit, providing an alternative to scientific notation. The prefix names, symbols, and values are listed in Table A1.4.

**TABLE A1.1**
**SI base units**

| Quantity | Name | Symbol |
|---|---|---|
| Length | Meter | m |
| Mass | Kilogram | kg |
| Time | Second | s |
| Electric current | Ampere | A |
| Thermodynamic temperature | Kelvin | K |
| Amount of matter | Mole | mol |
| Luminous intensity | Candela | cd |

**TABLE A1.2**
**Examples of SI-derived units expressed in terms of base units**

| Quantity | Name | Expression |
|---|---|---|
| Area | Square meter | $m^2$ |
| Volume | Cubic meter | $m^3$ |
| Speed | Meter per second | m/s |
| Acceleration | Meter per second squared | $m/s^2$ |
| Mass density | Kilogram per cubic meter | $kg/m^3$ |
| Current density | Ampere per square meter | $A/m^2$ |

**TABLE A1.3**
**SI-derived units with special names (supplemental units)**

| Quantity | Name | Symbol | Expression |
|---|---|---|---|
| Frequency | Hertz | Hz | 1/s |
| Force | Newton | N | $kg{\cdot}m/s^2$ |
| Pressure, stress | Pascal | Pa | $N/m^2 = kg/m{\cdot}s^2$ |
| Energy, work | Joule | J | $N{\cdot}m = kg{\cdot}m^2/s^2$ |
| Power, radiant flux | Watt | W | $J/s = kg{\cdot}m^2/s^3$ |
| Electric charge | Coulomb | C | $A{\cdot}s$ |
| Voltage, electric potential | Volt | V | $W/A = kg{\cdot}m^2/A{\cdot}s^3$ |
| Capacitance | Farad | F | $C/V = s^4A^2/m^2kg$ |
| Electric resistance | Ohm | Ω | $V/A = m^2kg/s^3A^2$ |
| Conductance | Siemens or mho | S or ℧ | $1/\Omega = s^3A^2/m^2kg$ |
| Magnetic field | Tesla | T | $N/A{\cdot}m = kg/s^2A$ |
| Magnetic flux | Weber | Wb | $T{\cdot}m^2 = m^2kg/s^2A$ |
| Inductance | Henry | H | $V{\cdot}s/A = m^2kg/s^2A^2$ |

▼ ***EXAMPLE A1.1. Unit Prefixes.*** The output of a 125 million watt power station can be expressed as

125,000,000 W or 125 MW

An example of a tiny interval of time, common in high performance electronics, can be expressed as

$5.27 \times 10^{-13}$ s or 0.527 ps

▼▼▼ ***CLASS DISCUSSION ITEM A1.2. Common Use of SI Prefixes.*** For each of the prefixes listed in Table A1.4, think of an example of a measurable physical quantity for which the prefix is commonly used to express the value.

**TABLE A1.4**
**Unit prefixes**

| Name | Symbol | Quantity |
|---|---|---|
| Yotta | Y | $10^{24}$ |
| Zetta | Z | $10^{21}$ |
| Exa | E | $10^{18}$ |
| Tera | T | $10^{12}$ |
| Giga | G | $10^{9}$ |
| Mega | M | $10^{6}$ |
| Kilo | k | $10^{3}$ |
| Hecto | h | $10^{2}$ |
| Deca | da | 10 |
| Deci | d | $10^{-1}$ |
| Centi | c | $10^{-2}$ |
| Milli | m | $10^{-3}$ |
| Micro | μ | $10^{-6}$ |
| Nano | n | $10^{-9}$ |
| Pico | p | $10^{-12}$ |
| Fempto | f | $10^{-15}$ |
| Atto | a | $10^{-18}$ |
| Zepto | z | $10^{-21}$ |
| Yocto | y | $10^{-24}$ |

## A1.1.2 Conversion Factors

There is still an extremely diverse mix of unit systems used in practice. ***English units*** are still common in engineering practice in the United States. Table A1.5 lists several conversion factors that help when interchanging between English and SI units.

**TABLE A1.5**
**Useful English to SI conversion factors**

| Physical quantity | English unit | SI unit |
|---|---|---|
| Length | 1 in. | 2.540 cm |
| | 1 ft | 0.3048 m |
| | 1 mi (mile) | 1.609 km |
| Mass | 1 lbm (pound mass) | 0.4536 kg |
| Force | 1 lbf (pound force) | 4.448 N |

*(continued)*

| Physical quantity | English unit | SI unit |
|---|---|---|
| Temperature | Fahrenheit temperature ($T_F$) | $T_K = 5/9 \cdot (T_F - 32) + 273.15$ |
| Pressure | 1 lb/in.$^2$ (psi) | 6.895 x $10^3$ Pa |
| | 1 atm | 1.013 x $10^5$ Pa |
| Power | 1 Btu/h | 0.2929 W |
| | 1 hp | 745.7 W |
| Magnetic field | 1 gauss | 1.000 x $10^{-4}$ tesla |

▼▼▼ ***CLASS DISCUSSION ITEM A1.3. Physical Feel for SI Units.*** To help gain a physical feel for SI units, it is helpful to consider and remember concrete examples for each of the units. For each of the following common physical items, list the appropriate SI unit and approximate value:

- Length of a typical human foot
- Length of a city block
- Mass of a 2-liter bottle of soda pop
- Mass of an average adult human body
- Force required to pick up a 2-liter bottle of soda pop
- Force exerted by an average adult human body on a scale
- Human body temperature
- Comfortable room temperature
- Atmospheric pressure
- Typical air pressure in a building's pneumatic system
- Power dissipated by a typical incandescent lightbulb
- Typical maximum power generated by an automobile

## A1.2 SIGNIFICANT FIGURES

Whenever we handle numerical data, we need to be aware of precision, accuracy, and different ways to present the data. Also, in establishing a rational approach to making numerical calculations with measured values, we must present decimal numbers with the appropriate number of digits.

The significant digits or ***significant figures*** in a number are those that are known with certainty. A measured value represented by $N$ digits consists of $N$ - 1 significant digits that are certain and 1 digit that is estimated. For example, when reading a dial on a pressure gage one might record 4.85 Pa. Here the 4 and 8 are certain, but the 5 may be an interpolated value. Hence, an observer is involved in the estimation of significance. If a number is reported with leading zeros, the leading zeros are not significant since they are only used to fix the decimal place (see Example A1.2).

Today, with the common use of digital computers for data processing, one must be aware of the fact that a 12-digit number resulting from a computer calculation may have only 3 significant digits! The rest may be meaningless.

---

▼ ***EXAMPLE A1.2. Significant Figures.*** Note the number of significant figures and the corresponding significant digits for each of the numbers below:

| Number | Number of significant figures | List of significant digits |
|---|---|---|
| 50.1 | 3 | 5, 0, 1 |
| 0.0501 | 3 | 5, 0, 1 |
| 5.010 | 4 | 5, 0, 1, 0 |

---

▼ ***EXAMPLE A1.3. Scientific Notation.*** The following numbers illustrate the use of scientific notation in clearly representing the number of significant figures:

| Number using scientific notation | Number of significant figures |
|---|---|
| $5.01 \times 10^{1}$ | 3 |
| $5.01 \times 10^{-2}$ | 3 |
| $5.010 \times 10^{0}$ | 4 |

---

Mathematical computations present another difficulty when values with different numbers of significant figures are combined. We must be careful in rounding off and retaining the appropriate number of significant figures in computed results.

When you are required to round a number to $N$ significant figures, dispose of all digits to the right of the $N$th place. If the discarded part exceeds one half of the $N$th digit, increase the $N$th digit by 1. If it is less than one half of the $N$th digit, leave it as it is. If it is exactly one half of the $N$th digit, leave the $N$th digit unchanged if it is even or increase it by 1 if it is odd. It is important to be consistent when you apply these rules.

When adding quantities, determine the number of significant figures to the right of the decimal place in the least accurate number. Then retain only one more decimal place in the remaining numbers by truncation. Now add the numbers and round off to the same number of decimal places as in the least accurate number. This process is demonstrated in Example A1.4.

▼ ***EXAMPLE A1.4. Addition and Significant Figures.*** We wish to add the following numbers:

$$\begin{array}{r} 5.0365 \\ +\ 1.04 \\ +\ 6.09314 \end{array}$$

Since the least number of significant figures to the right of the decimal place in the least accurate number (1.04) is two, we truncate the other numbers to three decimal places and add:

$$\begin{array}{r} 5.036 \\ +\ 1.04 \\ +\ 6.093 \end{array}$$

The result of this addition is

$$12.169$$

Since the number decimal places in the least accurate number is two, we round off the result to two decimal places, giving

$$12.17$$

When subtracting two quantities, round off the more accurate number to the number of decimal places in the least accurate number. Subtract and provide a result with the same number of decimal places as in the least accurate number, as demonstrated in Example A1.5.

▼ ***EXAMPLE A1.5. Subtraction and Significant Figures.*** We wish to subtract the following numbers:

$$\begin{array}{r} 8.59320 \\ -\ 1.04 \end{array}$$

Since the least accurate number (1.04) has two decimal places, we round off the other number also to two decimal places and subtract

$$\begin{array}{r} 8.59 \\ -\ 1.04 \end{array}$$

The result of this subtraction is

$$7.55$$

When multiplying and dividing numbers, round off the more accurate numbers to one more significant figure than the least accurate number. Compute and round the result to the same number of significant figures as in the least accurate number as demonstrated in Example A1.6.

▼ ***EXAMPLE A1.6. Multiplication and Division and Significant Figures.*** Given the multiplication and division problem:

(1.03)(51.7946)(3.01)/(695.01)(7001.59)

We round off the more accurate numbers to four significant figures (one more than the three in 1.03 and 3.01):

(1.03)(51.79)(3.01)/(695.0)(7002.)

The result of this multiplication after retaining three significant figures is

$$0.0000330 = 3.30 \times 10^{-5}$$

## A1.3 STATISTICS

When we process sets of data obtained from experimental measurements, we must handle the data in a rational, systematic, and organized fashion. The field of ***statistics*** provides models and rules for doing this properly.

Often, we seek a number or a small set of numbers to represent or characterize a large set of data. The first step is to assess the range of the data by noting the minimum and maximum values known as the ***extreme values.*** We then look at how the data points are distributed between these extreme values. This distribution can be graphically represented using a ***histogram,*** which results from sorting the values into sub-ranges and displaying the number of data points in each sub-range. The histogram for the set of experimental data given in Table A1.6 is illustrated in Figure A1.1.

**TABLE A1.6**
**Set of experimental data**

| Index | Value |
|---|---|
| 1 | 25.5 |
| 2 | 42.1 |
| 3 | 36.4 |
| 4 | 32.1 |
| 5 | 15.6 |
| 6 | 38.6 |
| 7 | 55.3 |
| 8 | 29.1 |
| 9 | 32.1 |
| 10 | 34.0 |
| 11 | 35.0 |

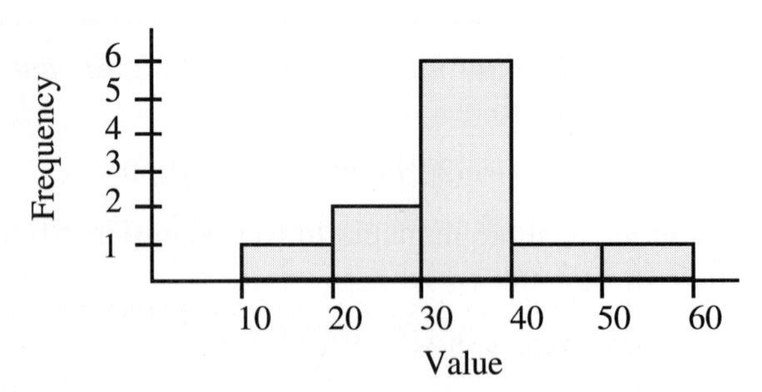


**FIGURE A1.1**
Histogram of experimental data.

The number of data points falling into each sub-range of the histogram is called the ***frequency***. The histogram may approximate a specific shape such as a normal, skewed, bimodal, or uniform distribution as illustrated in Figure A1.2. The normal distribution represents a typical statistical spread of data around an average (mean) value. The skewed distribution represents some weighting in the statistics to one side of the mean. The bimodal distribution represents a case where there may be two merged populations with two different means. The uniform distribution represents completely random data.

Many data points are required to create a histogram. Statistics provides rules for distilling the histogram down to just a few numbers representing the characteristics of the data set. The most important statistical measure is the ***arithmetic mean***, which is also called the ***average*** or simply the mean. Denoted by $\bar{x}$, it is the sum of each of the data values $x_i$ divided by the number of data points $N$:

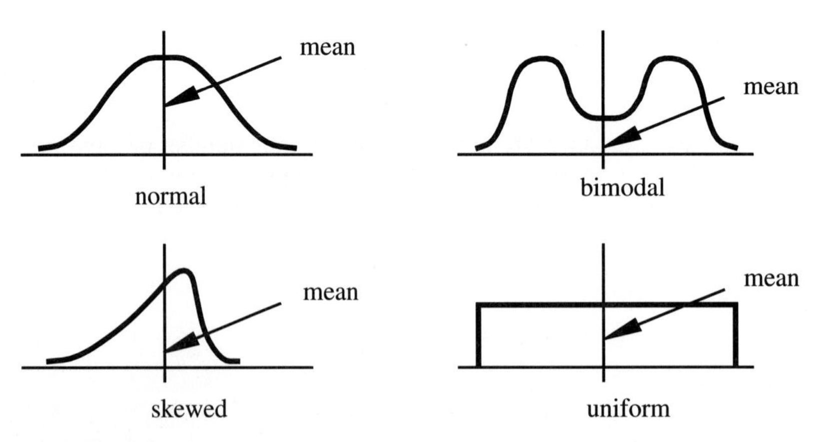


**FIGURE A1.2**
Distributions of data.

$$\bar{x} = \frac{\sum_{i=1}^{N} x_i}{N} \tag{A1.2}$$

Other statistical measures that characterize a set of data are the ***median,*** which is the data point that has an equal number of data points on both sides; the ***mode,*** which is the value that occurs most frequently; and the ***geometric mean,*** defined as the $N$th root of the product of the values:

$$\text{GM} = \sqrt[N]{x_1 x_2 \ldots x_N} \tag{A1.3}$$

The geometric mean is more desirable than the arithmetic mean for averaging ratios since the reciprocal of the GM is equal to the GM of the reciprocals.

For the set of data in Table A1.6, the mean is 34.2, the median is 34.0, the mode is 32.1, and the geometric mean is 32.7.

▼▼▼ ***CLASS DISCUSSION ITEM A1.4. Statistical Calculations.*** Verify the calculations for the mean, median, mode, and geometric mean for the data in Table A1.6.

▼▼▼ ***CLASS DISCUSSION ITEM A1.5. Your Class Age Histogram.*** In one of your class lectures, form a single line in order of birth date with the youngest being first. Now form a birth year histogram by assembling into rows according to birth year. Store the histogram data (year, frequency) for use in Question A1.2 at the end of the chapter.

The spread or dispersion of a data set over its range is characterized by another statistical measure known as the ***variance,*** defined by

$$v = \sigma^2 = \sum_{i=1}^{N} \frac{(x_i - \bar{x})^2}{N-1} \tag{A1.4}$$

where $x_i$ is an individual measurement and $N$ is the total number of measurements called the ***sample size*** of the experiment. The ***standard deviation*** also describes this distribution, but in the units of the individual measurements. It is defined as the square root of the variance:

$$\sigma = \sqrt{v} = \sqrt{\sum_{i=1}^{N} \frac{(x_i - \bar{x})^2}{N-1}} \tag{A1.5}$$

The standard deviation estimates the magnitude of the spread of the experimental data around the mean value. A small standard deviation indicates that the data set has a narrow spread.

▼▼▼ ***CLASS DISCUSSION ITEM A1.6. Relationship between Standard Deviation and Sample Size.*** The denominator in Equation A1.5 is often confusing since one might assume it to be $N$ providing a value known as the ***root mean square (rms).*** Why is the denominator $(N - 1)$ and not $N$? Consider the situation where there is only one sample $(N = 1)$. Also, consider how many data values $x_i$ must be specified if the mean is known.

## A1.4 ERROR ANALYSIS

The process of making measurements is imperfect, and there will always be uncertainty associated with measured values. It is important to recognize sources of error and estimate the magnitude of error when one makes a measurement. Usually a manufacturer will define the accuracy of an instrument in published specifications. There are three types of errors: systematic errors, random errors, and blunders. A ***systematic error*** is one that reoccurs in the same way each time a measurement is made. The method used to minimize the magnitude of systematic error is ***calibration***, where the measurement instrument is used to record values from a standard input and adjusted to compensate for any discrepancy. ***Random errors*** occur due to the stochastic variations in a measurement process. Some of the statistical tools presented in the previous section enable us to reduce the effects of these errors. ***Blunders*** occur when the engineer or scientist makes a mistake. Blunders can be avoided by careful design and review.

Figure A1.3 illustrates systematic and random errors. The center of the target represents the desired value, and the shot pattern represents measured data. The systematic error, called inaccuracy, is associated with the shift of the shot pattern from the center of the target and could be corrected by improved sighting, known as calibration. The random error, called imprecision, is the size of the shot pattern and cannot be improved by adjusting the sighting. ***Accuracy*** is the closeness to the true value, and ***precision*** is the repeatability or consistency of the measurements.

Statistical calculations help us estimate a more precise value when a sample of imprecise measurements is taken in the presence of random errors. The average, or mean, provides this estimate.

### A1.4.1 Rules for Estimating Errors

When designing a measurement protocol to compute a parameter that is defined in terms of measured variables, it is necessary to estimate the error in the parameter due to the combined errors in the variables. A procedure for computing this overall error follows:

1. Prepare a table of data including the ± error estimate for each variable. Generally, the estimated error contains no more than two significant figures.

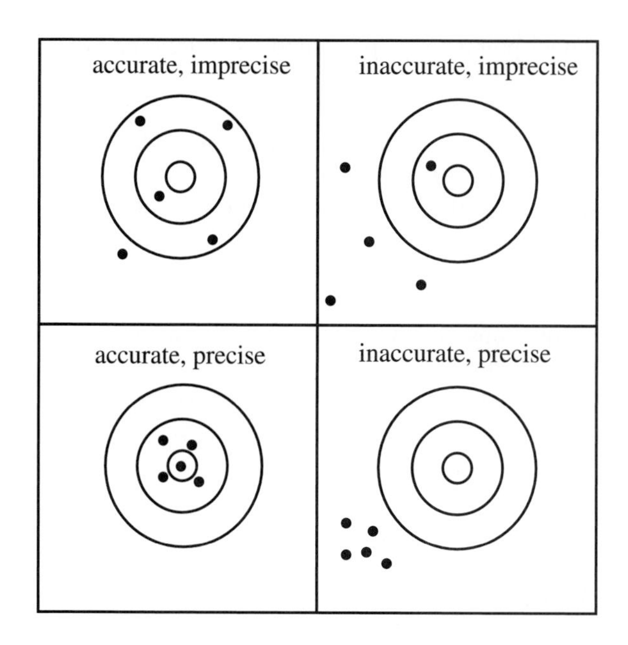


**FIGURE A1.3**
Accuracy and precision.

2. If the parameter to be computed is $X$, where $X$ is a function of measured variables ($v_i$),

$$X = X(v_1, v_2, \ldots, v_n) \tag{A1.6}$$

compute the partial derivatives $(\partial X)/(\partial v_1)$, $(\partial X)/(\partial v_2)$, . . . , $(\partial X)/(\partial v_n)$ and evaluate each to three significant figures using the recorded data ($v_1, v_2, \ldots, v_n$).

3. Compute the total absolute error $E$ using

$$E = \Delta X = \left|\frac{\partial X}{\partial v_1}\Delta v_1\right| + \left|\frac{\partial X}{\partial v_2}\Delta v_2\right| + \cdots + \left|\frac{\partial X}{\partial v_n}\Delta v_n\right| \tag{A1.7}$$

where $\Delta v_i$ is the error in the recorded value $v_i$. Round $E$ to two significant figures.

A more conventional error measure is the ***root-mean-square (rms)*** error given by the square root of the sum of the squares of the individual error terms:

$$E_{\text{rms}} = \sqrt{\left(\frac{\partial X}{\partial v_1}\Delta v_1\right)^2 + \left(\frac{\partial X}{\partial v_2}\Delta v_2\right)^2 + \cdots + \left(\frac{\partial X}{\partial v_n}\Delta v_n\right)^2} \tag{A1.8}$$

The rms error measure yields a closer approximation to the actual error.

4. Compute $X$ using Equation A1.6 to one more decimal place than the rounded error $E$. For example, if $E = \pm\ 0.039$, and $X$ is computed to be 8.9234, then this value is rounded to the same number of decimal places as $E$, giving 8.923. When computing $X$, treat $v_1, v_2, \ldots, v_n$ as exact numbers.

5. The result is

$$X = 8.923 \pm 0.039$$

Below is a summary of important points to keep in mind when treating numbers and calculating errors in laboratory data analysis:

1. Note the number of significant figures that a particular instrument displays.
2. Record all values with the correct number of significant digits. If the instrument display does not provide a digital output, this may require that the observer estimate the accuracy of the measurement.
3. Different instruments in a system may have different numbers of significant digits.
4. When computing results using equations, retain the appropriate number of significant digits when performing calculations.
5. If events can be repeated, better estimates of values can be obtained by averaging. The standard deviation of these samples gives a measure of precision in the average.

## QUESTIONS AND EXERCISES

**A1.1.** Express each of the following quantities with a more appropriate SI prefix equivalent:

(*a*) 100,000,000 kg.

(*b*) 0.000000025 m.

(*c*) $16.9 \times 10^{-10}$ s.

**A1.2.** Plot the histogram and calculate the mean, standard deviation, median, mode, and geometric mean for the data obtained from Class Discussion Item A1.5.

**A1.3.** What are the total absolute and rms errors in the calculated maximum stress of a rectangular cantilever beam with an end load of 12,520 ± 10 N? The beam geometry, as measured by a metric ruler accurate to 0.5 mm, is 0.95 m long by 11.8 cm wide by 12.1 cm high. The maximum stress occurs at the wall on the surface of the beam and is given by

$$\sigma_{max} = \frac{Mc}{I}$$

where $M$ is the bending moment given by the product of the force and the beam length, $c$ is the distance from the neutral axis given by half the height, and $I$ is the area moment of inertia of the beam cross section given by

$$I = \frac{1}{12}wh^3$$

where $w$ and $h$ are the width and height of the beam cross section, respectively.

## BIBLIOGRAPHY

Beckwith, T., Marangoni, R., and Lienhard, J., *Mechanical Measurements,* Addison-Wesley, Reading, MA, 1993.

Chapra, S. and Canale, R., *Introduction to Computing for Engineers,* McGraw-Hill, New York, 1994.

Chatfield, C., *Statistics for Technology,* Penguin Books, Middlesex, England, 1970.

Doeblin, E., *Measurement Systems Applications and Design,* 4th Edition, McGraw-Hill, New York, 1990.

# Appendix 2

# PHYSICAL PRINCIPLES

OBJECTIVES: After you read, discuss, study, and apply ideas in this appendix, you will be able to:

- Determine methods for measuring nearly all physical quantities

## A2.1 PHYSICAL PRINCIPLES

Sensor and transducer design always involves the application of some law or principle of physics or chemistry that relates the quantity of interest to some measurable event. The list below summarizes many of the physical laws and principles that have potential application in sensor and transducer design. Some examples of applications are also provided. This list is extremely useful to a transducer designer who is searching for a method to measure a physical quantity. Practically every transducer applies one or more of these principles in its operation. The parameters related by the respective principles are highlighted.

- *Ampere's Law:* A ***current***-carrying conductor in a magnetic field experiences a ***force.***

  Based on this law, a galvanometer measures current by measuring the deflection of a pivoted coil in a permanent magnetic field.

- *Archimedes' Principle:* The buoyant ***force*** exerted on a submerged or floating object is equal to the weight of the fluid displaced. The ***volume*** displaced depends on the fluid ***density.***

  A ball submersion hydrometer uses this effect to measure the density of a fluid (e.g., automotive coolant).

- *Bernoulli's Equation:* Conservation of energy in a fluid predicts a relationship between ***pressure*** and ***velocity*** of the fluid.

  A pitot tube uses this effect to measure air speed of an aircraft.

- *Biot-Savart Law:* A conductor carrying a ***current*** is surrounded by a ***magnetic field.***

  A magnetic pickup sensor uses this effect as a nonintrusive method of measuring current in a conductor.

- *Biot's Law:* The rate of ***heat conduction*** through a medium is directly proportional to the ***temperature*** difference across the medium.

  This principle is basic to time constants associated with temperature transducers.

- *Blagdeno Law:* The ***freezing temperature*** of a liquid drops and the boiling temperature rises with ***concentration*** of impurities in the liquid.
- *Boyle's Law:* An ideal gas maintains a constant ***pressure-volume*** product with constant ***temperature.***
- *Bragg's Law:* The intensity of an x-ray beam diffracted by a ***crystal lattice*** is related to the crystal plane separation and the ***wavelength*** of the beam.

  An x-ray diffraction system uses this effect to measure the crystal lattice geometry of a crystalline specimen.

- *Brewster's Law:* The ***index of refraction*** of a material is related to the angle of ***polarized light*** reflection/transmission.

  A Brewster's window on a laser tube is used to extract some of the power in the form of a laser beam. Lasers are used extensively in measurement systems.

- *Butterfly Effect:* Chaotic nonlinear systems exhibit a sensitive dependence upon initial conditions.
- *Centrifugal Force:* A body ***moving*** along a curved path experiences an apparent inward ***force*** in line with the radius of ***curvature.***
- *Charles' Law:* An ideal gas maintains a constant ***pressure-temperature*** product with constant ***volume.***
- *Christiansen Effect:* Powders suspended in a liquid (i.e., a colloidal solution) result in altered fluid ***refraction*** properties.
- *Corbino Effect:* ***Current*** flow is induced in a conducting disk ***rotating*** in a ***magnetic field.***
- *Coriolis Effect:* A body ***moving*** relative to a ***rotating*** frame of reference (e.g., the earth) experiences a ***force*** relative to the frame.

  A coriolis flow meter uses this effect to measure mass flow rate in a u-tube in rotational vibration.

- *Coulomb's Law:* ***Electric charges*** exert a ***force*** between each other.
- *Curie-Weiss Law:* There is a ***transition temperature*** at which ferromagnetic materials exhibit ***paramagnetic*** behavior.
- *d'Alembert's Principle:* ***Acceleration*** of a ***mass*** is equivalent to an equal and opposite applied ***force.***

- *Debye Frequency Effect:* The ***conductance*** of an electrolyte increases (i.e., the ***resistance*** decreases) with ***frequency.***
- *Doppler Effect:* The ***frequency*** received from a wave source (e.g., sound or light) depends on the ***speed*** of the source.

  A laser Doppler velocimeter uses the frequency shift of laser light reflected off of particles suspended in a fluid to measure fluid velocity.

- *Edison Effect:* When metal is heated in a vacuum, it emits charged particles (i.e., ***thermionic emission***) at a rate dependent on ***temperature.***

  A vacuum tube amplifier is based on this effect where electrons are emitted and controlled to produce amplification of current.

- *Faraday's Law of Electrolysis:* The rate of ***ion deposition*** or depletion is proportional to the electrolytic ***current.***
- *Faraday's Law of Induction:* A coil resists a change in ***magnetic field*** linkage with an ***electromotive force.***

  The induced voltages in the secondary coils of a linear variable differential transformer (LVDT) are a result of this effect.

- *Gauss Effect:* The ***resistance*** of a conductor increases when ***magnetized.***
- *Gladstone-Dale Law:* The ***index of refraction*** of a substance is dependent upon ***density.***
- *Gyroscopic Effect:* A body rotating about one axis resists ***rotation*** about other axes.

  A navigation gyroscope uses this effect to track the position of a body with the aid of a gimbal-mounted flywheel that maintains constant orientation in space.

- *Hall Effect:* A ***voltage*** is generated perpendicular to ***current*** flow in a ***magnetic field.***

  A Hall Effect proximity sensor detects when a magnetic field changes due to the presence of a metallic object.

- *Hertz Effect:* ***Ultraviolet light*** affects the discharge of a ***spark*** across a gap.
- *Johnsen-Rahbek Effect:* ***Friction*** at interfaces between a conductor, semiconductor, or insulator increases with ***voltage*** across the interfaces.
- *Joule's Law:* ***Heat*** is produced by ***current*** flowing through a ***resistor.***

  The design of a hot-wire anemometer is based on this principle.

- *Kerr Effect:* Applying a ***voltage*** across a substance causes ***optical polarization***.

  Liquid crystal displays (LCDs) function as a result of this principle.

- *Kohlrausch's Law:* An ***electrolytic*** substance has a limiting conductance (minimum ***resistance***).
- *Lambert's Cosine Law:* The reflected ***luminance*** of a surface varies with the cosine of the ***angle of incidence.***

- *Lenz's Law:* A ***current***-carrying conductor moving in a ***magnetic field*** experiences an opposing ***force.***
- *Lorentz's Law:* A charged particle moving in an ***electric field*** and a ***magnetic field*** experiences forces due to these effects.
- *Magnus Effect:* When fluid ***flows*** over a rotating body, the body experiences a ***force*** in a direction perpendicular to the flow.
- *Meissner Effect:* A ***superconducting*** material within a ***magnetic field*** blocks this field and experiences no internal field.
- *Murphy's Law:* Whatever can go ***wrong*** will go ***wrong*** and at the ***wrong*** time in the ***wrong*** place.

  Your experiments in the laboratory will often demonstrate this law.

- *Nernst Effect:* ***Heat flow*** across ***magnetic field*** lines produces a ***voltage.***
- *Newton's Law:* ***Acceleration*** of an object is proportional to ***force*** acting on the object.
- *Ohm's Law:* ***Current*** through a ***resistor*** is proportional to the ***voltage*** drop across the resistor.
- *Parkinson's Law:* Human work expands to fill the time allotted for it.
- *Peltier Effect:* When ***current*** flows through the junction between two metals, ***heat*** is absorbed or liberated at the junction.

  Thermocouple measurements can be affected by this principle.

- *Photoconductive Effect:* When ***light*** strikes certain semiconductor materials, the ***resistance*** of the material decreases.

  A photodiode, which is used extensively in photodetector pairs, is based on this effect.

- *Photoelectric Effect:* When ***light*** strikes a metal cathode, electrons are emitted and attracted to an anode, resulting in ***current*** flow.

  The operation of a photomultiplier tube is based on this effect.

- *Photovoltaic Effect:* When ***light*** strikes a semiconductor in contact with a metal base, a ***voltage*** is produced.

  The operation of a solar cell is based on this effect.

- *Piezoelectric Effect:* ***Charge*** is displaced across a crystal when it is ***strained.***

  A piezoelectric accelerometer measures charge polarization across a piezoelectric crystal subject to deformations due to the inertia of a mass. A piezoelectric microphone's ability to convert sound pressure waves to a voltage signal is a result of this principle.

- *Piezoresistive Effect:* ***Resistance*** is proportional to an applied ***stress.***

  This effect is partially responsible for the behavior of a strain gage.

- *Pinch Effect:* The cross section of a ***liquid*** conductor reduces with ***current.***

- *Poisson Effect:* A material will ***deform*** in a direction perpendicular to an applied ***stress.***

  This effect is partially responsible for the behavior of a strain gage.

- *Pyroelectric Effect:* A crystal becomes ***polarized*** when its ***temperature*** changes.
- *Raleigh Criteria:* Relates the ***acceleration*** of a fluid to bubble formation.
- *Raoult's Effect:* ***Resistance*** of a conductor changes when its ***length*** is changed.

  This effect is partially responsible for the behavior of a strain gage.

- *Seebeck Effect:* Dissimilar metals in contact will result in a ***voltage*** difference across the junction that depends on ***temperature.***

  This is the primary effect that governs the behavior of a thermocouple.

- *Shape Memory Effect:* A heated metal will restore to its original shape after it has been deformed.
- *Snell's Law:* Reflected and refracted rays of ***light*** at an optical interface are related to the angle of incidence.
- *Stark Effect:* The ***spectral lines*** of an electromagnetic source split when the source is in a strong ***electric field.***
- *Stefan-Boltzmann Law:* The ***heat*** radiated from a black body is proportional to the fourth power of its ***temperature.***

  The design of a pyrometer is based on this principle.

- *Stokes' Law:* The ***wavelength*** of light emitted from a fluorescent material is always longer than that of the absorbed photons.
- *Tribo-Electric Effect*: Relative ***motion*** and ***friction*** between two dissimilar metals produces a ***voltage*** between the interface.
- *Wiedemann-Franz Law:* The ratio of ***thermal*** to ***electrical conductivity*** of a material is proportional to its absolute ***temperature.***
- *Wien Effect:* The ***conductance*** of an electrolyte increases (i.e., the ***resistance*** decreases) with applied ***voltage.***
- *Wien's Displacement Law:* As the ***temperature*** of an incandescent material increases, the spectrum of emitted ***light*** shifts toward blue.

# Appendix 3

# MECHANICS OF MATERIALS

**OBJECTIVES:** After you read, discuss, study, and apply ideas in this appendix, you will be able to:

- Understand the basic relationships between stress and strain
- Determine the principal stress values and directions for a general state of planar stress
- Construct Mohr's circle for a state of planar stress

## A3.1 STRESS/STRAIN RELATIONS

As shown in Figure A3.1, when a cylindrical rod is loaded axially, it will lengthen by an amount $\Delta L$ and deform radially by an amount $\Delta D$. The ***axial strain*** ($\varepsilon_{\text{axial}}$) is defined by the change in length per unit length:

$$\varepsilon_{\text{axial}} = \frac{\Delta L}{L} \tag{A3.1}$$

Note that strain is a dimensionless quantity. The ***axial stress*** ($\sigma_{\text{axial}}$) is related to axial strain through ***Hooke's Law,*** which states that for a uniaxially loaded linear elastic material the axial stress is directly proportional to the axial strain:

$$\sigma_{\text{axial}} = E\varepsilon_{\text{axial}} \tag{A3.2}$$

where $E$ is the constant of proportionality called the ***modulus of elasticity*** or ***Young's modulus.*** The axial stress in the rod is

$$\sigma_{\text{axial}} = F/A \tag{A3.3}$$

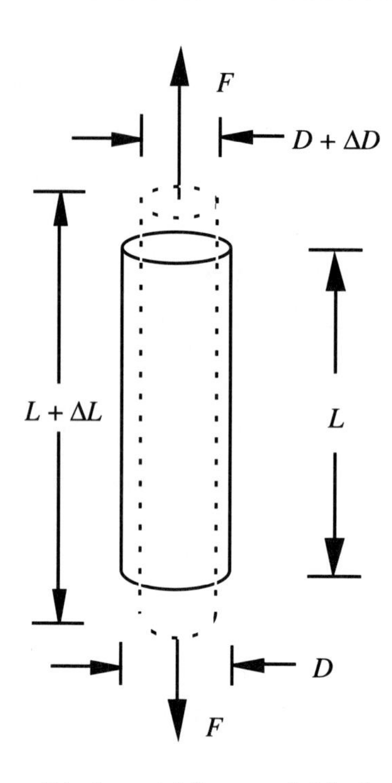


**FIGURE A3.1**
Axial and transverse deformation of a cylindrical bar.

where $F$ is the axial force and $A$ is the cross-sectional area of the rod. Therefore, the axial strain is related to the axial stress and load:

$$\varepsilon_{\text{axial}} = \frac{\sigma_{\text{axial}}}{E} = \frac{F/A}{E} \tag{A3.4}$$

The ***transverse strain*** is defined as the change in width divided by the original width:

$$\varepsilon_{\text{transverse}} = \frac{\Delta D}{D} \tag{A3.5}$$

The ratio of the transverse and axial strain is defined as ***Poisson's ratio*** ($\nu$):

$$\nu = -\frac{\varepsilon_{\text{transverse}}}{\varepsilon_{\text{axial}}} \tag{A3.6}$$

Note that for axial elongation ($\varepsilon_{\text{axial}} > 0$), $\varepsilon_{\text{transverse}}$ (from Equation A3.6), and therefore $\Delta D$ (from Equation A3.5) are negative, implying contraction in the transverse radial direction. Poisson's ratio for most metals is approximately 0.3, implying the transverse strain is –30% of the axial strain.

A general state of planar stress at a point, acting on a infinitesimal square element, is illustrated in Figure A3.2a. It includes two normal stress components ($\sigma_x$ and $\sigma_y$) and a shear stress component ($\tau_{xy}$) whose values depend on the orientation of the element. At any point, there is always an orientation of the element that results in the maximum normal stress magnitude and zero shear stress ($\tau_{xy} = 0$). The two orthogonal normal stress directions corresponding to this orientation are called the ***principal axes,*** and the normal stress magnitudes are referred to as the ***princi-***

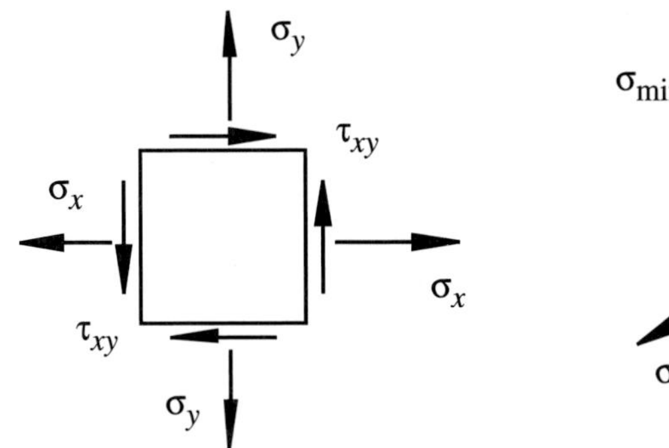


(*a*) general state of stress

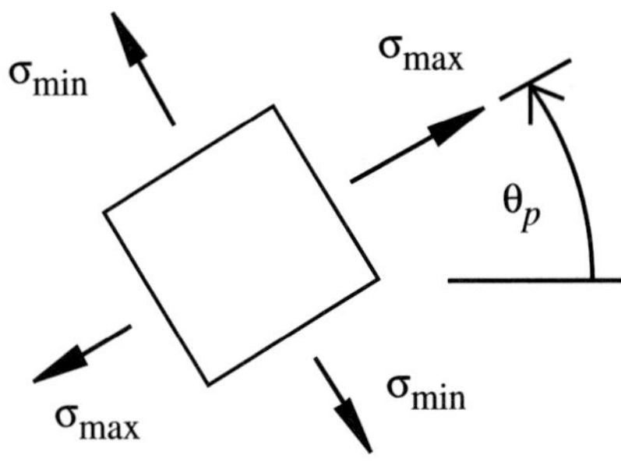


(*b*) principal stresses

**FIGURE A3.2**
General state of planar stress and principal stresses.

***pal stresses*** ($\sigma_{max}$ and $\sigma_{min}$). Figure A3.2b illustrates this orientation and its corresponding state of stress. The magnitude and direction of the principal stresses are related to the stresses in any other orientation by

$$\sigma_{max} = \left(\frac{\sigma_x + \sigma_y}{2}\right) + \sqrt{\left(\frac{\sigma_x - \sigma_y}{2}\right)^2 + \tau_{xy}^2} \tag{A3.7}$$

$$\sigma_{min} = \left(\frac{\sigma_x + \sigma_y}{2}\right) - \sqrt{\left(\frac{\sigma_x - \sigma_y}{2}\right)^2 + \tau_{xy}^2} \tag{A3.8}$$

$$\tan(2\theta_p) = \frac{2\tau_{xy}}{\sigma_x - \sigma_y} \tag{A3.9}$$

where $\theta_p$ is the angle between $\sigma_x$ and $\sigma_{max}$ measured counterclockwise.

The principal stresses are important quantities when determining if a material will yield or fail when loaded because they determine the maximum values of stress, which can be compared to the yield strength of the material. The maximum shear stress is also important when assessing failure and is given by

$$\tau_{max} = \sqrt{\left(\frac{\sigma_x - \sigma_y}{2}\right)^2 + \tau_{xy}^2} = \frac{\sigma_{max} - \sigma_{min}}{2} \tag{A3.10}$$

This relation can be used to rewrite Equations A3.7 and A3.8 as

$$\sigma_{max} = \sigma_{avg} + \tau_{max} \tag{A3.11}$$

$$\sigma_{min} = \sigma_{avg} - \tau_{max} \tag{A3.12}$$

where

$$\sigma_{avg} = \frac{\sigma_x + \sigma_y}{2} \tag{A3.13}$$

The orientation of the element that results in $\tau_{max}$ is given by

$$\tan(2\theta_s) = -\frac{\sigma_x - \sigma_y}{2\tau_{xy}} \tag{A3.14}$$

As with $\theta_p$, $\theta_s$ is also measured counterclockwise from the direction of $\sigma_x$. For the cylindrical bar in Figure A3.1, with an element oriented in the axial ($y$) direction, $\sigma_{max} = \sigma_y = F/A$; $\sigma_x = 0$; $\theta_p = 0$ since the element is aligned in the direction of the principal stress, $\theta_s = 45°$; and $\tau_{max} = \sigma_y/2 = F/2A$.

The state of stress and its relation to the magnitude and direction of the principal stresses are often illustrated with ***Mohr's circle,*** which displays the relationship between the shear stress and the normal stresses in different directions (see Figure A3.3). Remember, tensile normal stresses are positive, and compressive normal stresses are negative. For the example shown in Figure A3.3, corresponding to the element shown in Figure A3.2, both normal stresses are tensile. The sign of the shear stress is positive when it would cause the element to rotate clockwise about its center and negative when it would cause the element to rotate counterclockwise. For the element in Figure A3.2, $\tau_{xy}$ is negative on the $\sigma_x$ side of the element since it would cause the element to rotate counterclockwise, and $\tau_{xy}$ is positive on the $\sigma_y$ side for the opposite reason. Note that the angle between the original stress directions and the principal stresses ($\theta_p$) is measured in the same direction around the circle as with the actual element, but angles on the circle are twice the actual angles ($2\theta_p$). Since $\theta_p$ is measured counterclockwise from $\sigma_x$ to $\sigma_{max}$ in Figure A3.2, the angle between the $\sigma_x$ point and $\sigma_{max}$ is $2\theta_p$ counterclockwise in Figure A3.3. Also note that the orientation of the principal stresses and the orientation of the maximum shear stress are always 90° apart on Mohr's circle (45° apart on the actual element). This is why $\tan(2\theta_p)$ and $\tan(2\theta_s)$ are negative reciprocals of one another, as shown by comparing Equations A3.9 and A3.14.

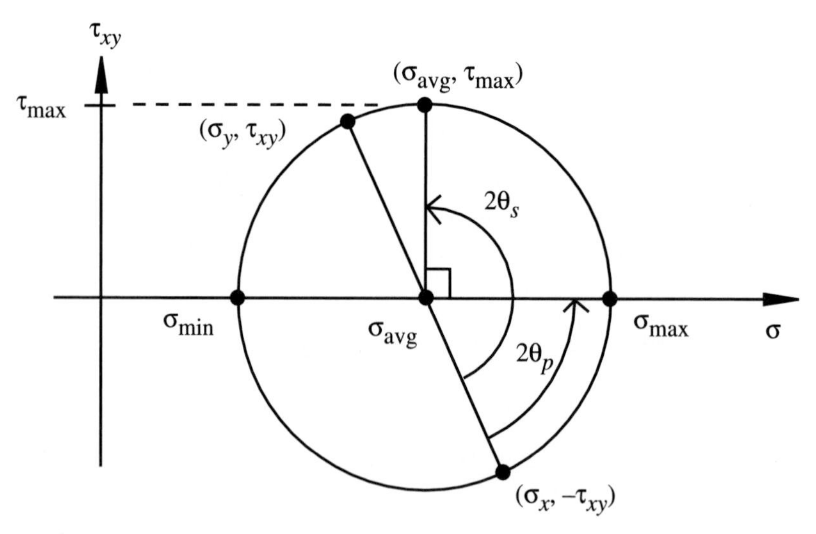


**FIGURE A3.3**
Mohr's circle of plane stresses.

▼▼▼ ***CLASS DISCUSSION ITEM A3.1. Fracture Plane Orientation in a Tensile Failure.*** When a metal bar fails under axial tension, the resulting fracture planes are oriented at 45° with respect to the bar's axis. Why?

## BIBLIOGRAPHY

Beer, F. and Johnston, E., *Mechanics of Materials,* McGraw-Hill, New York, 1981.

Dally, J. and Riley, W., *Experimental Stress Analysis,* 3rd Edition, McGraw-Hill, New York, 1991.

# INDEX

621 M685
690 427 K8am